AI for Climate Governance, Agriculture, and Earth Systems

Mehmet Ali Çelik
https://orcid.org/0000-0002-7729-6650
İgdir University, Turkey

IGI Global
Scientific Publishing
Publishing Tomorrow's Research Today

Vice President of Editorial	Melissa Wagner
Director of Acquisitions	Mikaela Felty
Director of Book Development	Jocelynn Hessler
Production Manager	Mike Brehm
Cover Design	Jose Rosado

Published in the United States of America by
IGI Global Scientific Publishing
701 East Chocolate Avenue
Hershey, PA, 17033, USA
Tel: 717-533-8845 | Fax: 717-533-7115
Website: https://www.igi-global.com E-mail: cust@igi-global.com

Library of Congress Cataloging-in-Publication Data

LCCN: (CIP Data Pending)
ISBN13: 9798260007785
Isbn13Softcover: 9798260007792
EISBN13: 9798260007808
British Cataloguing in Publication Data
A Cataloguing in Publication record for this book is available from the British Library.

Table of Contents

Detailed Table of Contents

Recent advances in the field of artificial intelligence (AI) have significantly altered the approach to the study of the Earth system; however, data-based models have difficulty in terms of physical consistency, interpretability, and generalization under non-stationary environmental conditions. Hybrid and physics-informed AI models have been developed to overcome these problems by using a combination of machine learning and physical laws and models. This chapter presents the major paradigms of hybrid modelling, including physics-informed neural networks, universal differential equations, differentiable modelling, and physics-encoded deep learning, along with their applications in climate studies, hydrology, atmospheric modelling, and environmental monitoring. In short, this chapter presents a synthesis of recent advances to demonstrate the potential of using hybrid and physics-informed AI models to bridge the gap between theory and data-based modelling.

Artificial intelligence (AI) and machine learning (ML) models have revolutionized Earth sciences, enabling unprecedented capabilities in climate prediction, remote sensing, and environmental monitoring. However, the reliability of AI-driven Earth science applications critically depends on rigorous uncertainty quantification, validation methodologies, and standardized benchmarking frameworks. This chapter comprehensively examines the state-of-the-art approaches for assessing AI model reliability in Earth sciences, covering probabilistic uncertainty estimation techniques,

AI can fulfill its potential to establish sustainable agriculture and global food security with favorable policies and increased usage.

Sk Muktadir Hossain, American International University-Bangladesh, Bangladesh
Sharif Eime Akhter, American International University-Bangladesh, Bangladesh
Ayswarjo Sarkar, American International University-Bangladesh, Bangladesh
Kazi Samia Mostofa, American International University-Bangladesh, Bangladesh
Md Sadiqur Rahman, American International University-Bangladesh, Bangladesh
Dipta Gomes, American International University-Bangladesh, Bangladesh
Mahfujur Rahman, American International University-Bangladesh, Bangladesh
Kazi Tanvir, American International University-Bangladesh, Bangladesh

This study introduces the Explainable Agro-Intelligence Machine Learning Framework (X-AMLF), which is a structured evaluation framework for yield risk classification in agriculture using aggregated country-year-crop data. A harmonised dataset was created by combining data on crop yield, annual average temperature, rainfall, pesticide use, crop identity, and country indicators to create 28,242 complete observations on the yield of 101 countries (1990–2016) for 10 major crops. Multiple classification models were evaluated, such as linear baselines and tree-based ensembles, and hyperparameters were optimised with Optuna to investigate variations in prediction and error behaviour. Under random splitting, ensemble models were able to get very high scores (e.g. Random forest F1 $\approx$ 0.976), but under a stricter evaluation, performance dropped. With country-holdout validation, the best-performing models had F1 $\approx$ 0.843, attenuation of performance when tested on unseen regions.

Agricultural systems are multifaceted adaptive systems that are constructed through dynamic processes of interaction between climate, soil, vegetation, and management activities, and which are progressively evidently, climate variability and extreme events. Conventional methods of monitoring have constraints of low spatial coverage, expensive nature and low sensitivity to the spatiotemporal heterogeneity. Current developments in satellite remote sensing have made available large amounts of multisensory data including optical, SAR, and thermal data, that when combined with the meteorological and edaphic data, can be used to fully monitor the agriculture. This chapter reviews the methods of artificial intelligence (AI) and machine learning, deep learning, and spatiotemporal models to monitor crops, predict yields and assess food security. The focus is put on the time-series analysis, multi-source data fusion, and early-warning applications, including the drought detection and yield-gap analysis.

Preface

INTRODUCTION

The 21st century has entered an era defined by unprecedented environmental challenges. Climate change is no longer a distant scientific concern but a lived global reality manifested through intensifying droughts, floods, wildfires, food insecurity, and ecosystem degradation. These challenges affect governance structures, economic systems, and social equity worldwide. Addressing such a multifaceted crisis requires not only traditional environmental policies but also transformative technological approaches capable of analyzing vast and complex datasets.

Artificial Intelligence (AI), together with big data analytics, remote sensing, and advanced computational models, has emerged as a powerful catalyst in understanding and managing Earth systems. These technologies enable governments, scientists, and policymakers to interpret environmental data more efficiently, forecast climate impacts, monitor agricultural productivity, and design evidence-based policies for sustainable development. As global environmental challenges become more interconnected, the integration of AI into climate governance and agricultural systems becomes not merely beneficial but essential.

This book, **"AI for Climate Governance, Agriculture, and Earth Systems,"** brings together interdisciplinary perspectives from researchers across multiple countries and academic disciplines. The chapters collectively explore how artificial intelligence can enhance environmental monitoring, improve agricultural resilience, and support data-driven governance. From theoretical frameworks to real-world applications, the contributions highlight the growing role of intelligent technologies in addressing climate-related risks while promoting sustainability and food security.

The volume is structured to guide readers through a progression of themes from climate governance and social justice perspectives to technical advancements in AI modeling and practical agricultural applications. Each chapter contributes to a comprehensive understanding of how AI-driven systems can transform environmental management and decision-making processes. Below is a chapter by chapter summary that may be used to supplement your reading.

CHAPTER OVERVIEW

Chapter 1: From Climate Crisis to Data-Driven Governance: Artificial Intelligence and Remote Sensing in Climate Policy Support

The opening chapter introduces the climate crisis as a complex global challenge that requires innovative governance solutions. It explores how artificial intelligence, big data, and remote sensing technologies can strengthen climate policy development and implementation. The chapter highlights the role of AI in supporting evidence-based decision-making, improving disaster risk management, and enabling governments to design more adaptive and forward-looking environmental policies.

Chapter 2: AI-Enhanced Bibliometric Mapping of Climate Change and Social Justice Research in Africa (1990–2023)

This chapter investigates the intersection of climate change and social justice within the African research context. Using bibliometric analysis, it maps scholarly developments from 1990 to 2023 and identifies major thematic clusters such as climate justice, gender equity, and sustainable development. The analysis demonstrates how research in this area has evolved into a rapidly expanding field with increasing global significance.

Chapter 3: Hybrid and Physics-Informed AI Models in Earth System Analysis

Chapter three explores the integration of machine learning with physical laws to overcome limitations of purely data-driven models. It introduces hybrid and physics-informed AI approaches, including physics-informed neural networks and differentiable modeling. These methods enhance interpretability and reliability in climate modeling, hydrology, and atmospheric studies.

Chapter 4: Uncertainty Analysis, Validation, and Benchmarking of AI Models in Earth Sciences

Reliable AI applications require rigorous validation and uncertainty assessment. This chapter focuses on methodological frameworks for evaluating AI models used in Earth science research. It examines probabilistic uncertainty techniques, validation strategies for geospatial data, and benchmarking practices that ensure robust and trustworthy environmental predictions.

Chapter 5: Deep Learning Architectures for Remote Sensing

This chapter provides a comprehensive overview of deep learning architectures used in remote sensing analysis. It examines Convolutional Neural Networks (CNNs), Recurrent Neural Networks (RNNs), Long Short-Term Memory (LSTM) networks, and Transformer models. These architectures are discussed in relation to their applications in land-use classification, crop monitoring, and environmental observation.

Chapter 6: Nanoenabled Artificial Intelligence for Sustainable and Digital Advancement in Precision Farming

Chapter six introduces the integration of nanotechnology with artificial intelligence to enhance precision agriculture. The chapter proposes innovative frameworks such as the Nano-AI Precision Farming Framework and the Nano-Agricultural Digital Twin. These concepts demonstrate how nanosensors, AI, and remote sensing can collectively support real-time agricultural intelligence and sustainable farming practices.

Chapter 7: AI Applications in Agricultural Monitoring, Crop Yield Estimation, and Food Security

This chapter highlights the transformative potential of AI in modern agriculture. It discusses how machine learning, IoT sensors, drones, and satellite data can be combined to monitor crops, predict yields, and detect environmental stress factors. The chapter also addresses challenges such as data accessibility and technological infrastructure in developing regions.

Chapter 8: A Framework for Robust and Interpretable Yield-Risk Classification Using Country–Year–Crop Data (X-AMLF)

Chapter eight introduces the Explainable Agro-Intelligence Machine Learning Framework (X-AMLF), a structured approach for classifying agricultural yield risks. Using a large multi-country dataset, the chapter evaluates multiple machine learning models and highlights the importance of explainability and cross-regional validation in agricultural forecasting.

Chapter 9: Advanced Artificial Intelligence and Multi-Source Remote Sensing for Agricultural Monitoring and Forecasting

The final chapter provides a comprehensive review of AI-driven monitoring systems for agriculture. It emphasizes the integration of satellite remote sensing, spatiotemporal modeling, and multi-source data fusion to improve crop monitoring and early warning systems. The chapter underscores the role of intelligent technologies in strengthening global food security under changing climate conditions.

CLOSING REMARKS

Together, the chapters in this book demonstrate that artificial intelligence is rapidly becoming a foundational tool in climate governance, Earth system science, and sustainable agriculture. By combining technological innovation with interdisciplinary research, AI-driven approaches can enhance our ability to understand complex environmental processes, support equitable policy decisions, and build resilient agricultural systems.

This volume aims to serve as a valuable resource for researchers, policymakers, environmental scientists, data scientists, and students interested in the intersection of artificial intelligence and environmental sustainability. The insights presented here highlight both the opportunities and responsibilities associated with applying advanced technologies to one of the most pressing challenges of our time: safeguarding the Earth's climate and ensuring sustainable development for future generations.

Chapter 1

From Climate Crisis to Data-Driven Governance: Artificial Intelligence and Remote Sensing in Climate Policy Support

Mehmet Ali Celik
http://orcid.org/0000-0002-7729-6650
Igdir University, Turkey

Adile Bilik
Independent Researcher, Turkey

Tuğba Yolcu
http://orcid.org/0000-0002-7131-7545
Independent Researcher, Turkey

ABSTRACT

The climate crisis is one of the most urgent and multifaceted problems of the 21st century, affecting ecosystems, economies, and societies by exacerbating extreme events and disasters such as droughts, floods, heat waves, and wildfires. The complexity of the climate crisis means that managing the problem requires strong public leadership, encompassing the capacity to internalize environmental externalities, set long-term goals, and implement mitigation and adaptation strategies through cross-sectoral coordination. In this context, artificial intelligence (AI), big data, and remote sensing technologies stand out as strategic tools that strengthen climate governance and support evidence-based policymaking.

DOI: 10.4018/979-8-2600-0778-5.ch001

1. INTRODUCTION

The climate crisis stands out as one of the most complex, multifaceted, and urgent global problems of the 21st century (Raftowicz, 2021; Babacan, 2025). The rapid increase in greenhouse gas concentrations in the atmosphere is leading to a significant rise in global average temperatures compared to pre-industrial levels (Wang et al., 2018; Nunes, 2023). This situation profoundly affects not only natural ecosystems but also economic, political, and social systems (Kumar et al., 2025; Alyemiş, 2025). Climate-related hazards such as more frequent heat waves, prolonged droughts, intense rainfall, floods, and wildfires have devastating consequences, particularly for vulnerable communities, and place serious pressures on food security, water resources, energy supply, and public health (Hadley et al., 2023; Sheehan, 2022).

Governments play a central role in managing this process (Meckling & Allan, 2020; Rashidi et al., 2025). Nationally developed climate policies, carbon neutrality targets, emissions trading systems, and renewable energy strategies are critical to keeping global warming below the 1.5°C threshold (Chen et al., 2022; Rabbi et al., 2022). However, the complexity of the climate system causes policymakers to face a high degree of uncertainty in their decision-making processes (Kumar et al., 2025; Dessai & Hulme, 2004). This is precisely where AI and data-driven technologies offer transformative potential in the fight against climate change (Favour et al., 2025; Cardellicchio et al., 2025). AI-powered systems are opening new horizons in the monitoring, modeling, and management of the climate crisis (Soomro et al., 2025; Kumari et al., 2026). Large-scale environmental data obtained through satellite observations, sensor networks, and big data infrastructures can be analyzed with machine learning (ML) algorithms to improve the accuracy of complex climate models (Cheng et al., 2025; Metheniti et al., 2025). For example, recent studies have demonstrated that deep learning (DL)-based models can more accurately predict temperature and precipitation changes, and estimate the probability of extreme weather events on a regional scale (Zhang et al., 2025; He & Rozos, 2025). In addition, AI-supported energy management systems improve the production forecasts of renewable energy sources (solar, wind, etc.), enabling more efficient management of the energy supply-demand balance (Khan, 2025; Bouquet et al., 2023). ML techniques integrated with remote sensing technologies analyze indicators such as forest fire risk, drought monitoring, vegetation change, and flood hazard with high accuracy, increasing the effectiveness of early warning systems (Biswas et al., 2025; Razavi-Termeh et al., 2025). Remote sensing is a key technology in understanding the dynamics of the climate system. High-resolution satellite data enable the monitoring of atmospheric changes, sea surface temperatures, glacier melting rates, and land-use transformations over the long term. When combined with AI algorithms, these data allow for the analysis of spatial and temporal patterns of climate change and provide strong analytical support to decision-makers in disaster management planning (Rodge & Rajan, 2026; Gu & Zeng, 2023).

However, AI technologies are not only important as a scientific modeling tool but also as policy and decision support mechanisms (Randeniya et al., 2025; Yılmaz & Alkan, 2026). Policymakers can analyze the potential socio-economic impacts of different climate policy scenarios (e.g., carbon taxes, energy subsidies, green infrastructure investments) in advance through AI-based simulations, strengthening evidence-based decision-making processes (Taheri Hosseinkhani, 2025; Islam, 2025). Furthermore, AI-powered data visualization and communication tools contribute to social adaptation and behavioral change processes by raising public awareness (Zamil Adib et al., 2025; Ayadi et al., 2025). At this point, Cortez (2023) emphasizes that the role of artificial intelligence and digital technologies in climate policies should be assessed not only in terms of technical efficiency but also with regard to democratic governance. According to the author, while AI-based tools have the potential to accelerate the design and implementation of climate policies, they also pose risks, such as the technocratization of decision-making processes and the weakening of democratic oversight. Cortez argues that the "speed versus voice" dilemma frequently encountered in climate governance can be partially addressed through carefully designed, AI-supported digital tools.

In this context, AI-supported policy tools facilitate public engagement by enabling citizens and stakeholders to engage directly with different climate policy scenarios, thereby encouraging bottom-up participation and strengthening participatory climate governance. In doing so, artificial intelligence goes beyond being merely an analytical tool for decision-makers and becomes a governance component that enhances the social legitimacy of climate policies while supporting public awareness and behavioral alignment with policy goals.

In conclusion, artificial intelligence, big data, and remote sensing technologies are strategic tools that enable not only an understanding of the climate crisis but also its effective management (Patro et al., 2025; Vidhya et al., 2026). However, for these technologies to be used effectively, ethical, transparent, and inclusive data governance principles must be adopted, and international cooperation strengthened (Trived & Juyal, 2026; Skotnicky et al., 2025). Thus, technological innovations can become a cornerstone of a sustainable and equitable transformation in humanity's collective struggle against the climate crisis (Zolfaghari, 2026; Sharma et al., 2026).

Figure 1. Conceptual framework of climate governance supported by artificial intelligence and remote sensing

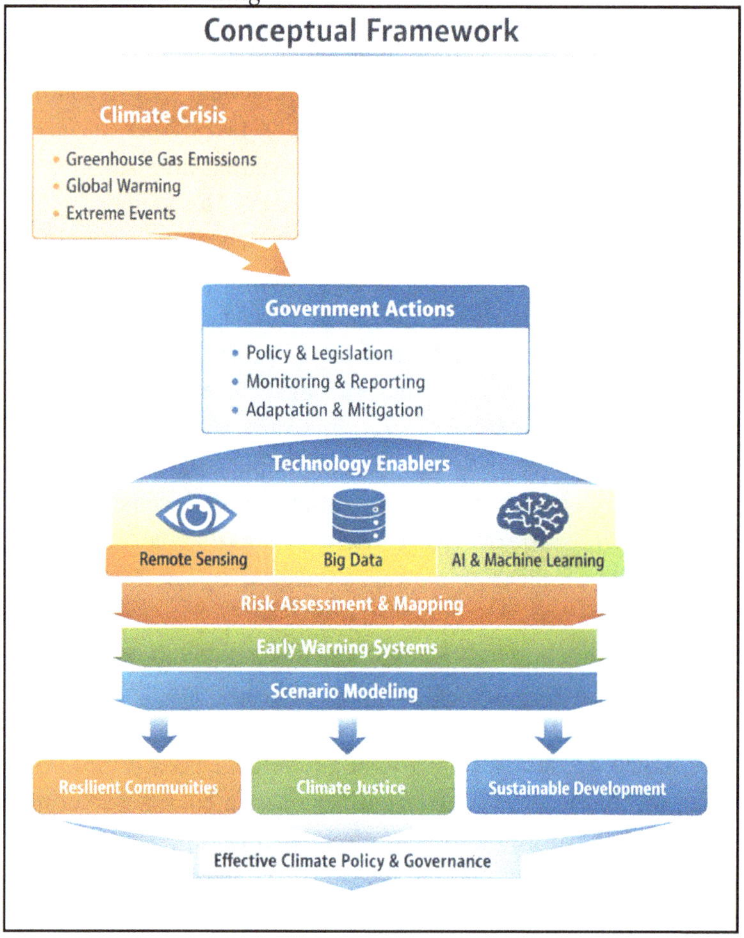

(This figure was generated with the support of AI-based image generation tools).

In this context, artificial intelligence, big data analytics, and remote sensing technologies are increasingly shaping data-driven climate governance. Rather than positioning AI as an autonomous decision-maker, this study examines how AI-based decision support systems and satellite-based Earth Observation data can strengthen policy design, monitoring, and evaluation processes within institutional and democratic governance frameworks. Bringing together public policy analysis, climate governance debates, and digital transformation perspectives, this chapter offers an integrative conceptual framework that links technological innovation to state capacity, accountability, and policy legitimacy. In doing so, it contributes to the broader discussion on how governments can manage the transition to data-driven climate governance by balancing efficiency, transparency, and public acceptance.

2. THE KEY ROLE OF GOVERNMENTS IN THE CLIMATE CRISIS

The climate crisis is a complex collective action problem that cannot be solved by market mechanisms or individual behavioral changes alone. Therefore, while the private sector, technological innovations, and civil society play important complementary roles, governments are the key actors with the capacity to determine the institutional, regulatory, and normative framework for climate action. Governments have unique powers to internalize environmental externalities, set long-term goals, and redistribute the costs and benefits of climate transformation across society. Governments are the most critical actors in combating the climate crisis, acting as bridges between social, economic, and environmental systems (Castañeda Rodriguez et al., 2026; Lahn et al., 2025). States hold a decisive position in managing the crisis, both because they have the authority to make policies and regulations at the national level and because they represent their countries in international climate diplomacy (Madani et al., 2025; Orkuma, 2025). The fundamental roles of governments in this process can be examined under three main headings: policy and legislation development, monitoring, evaluation and reporting, and adaptation and mitigation strategies (Balan et al., 2025; Benedetto et al., 2025). Table 1 summarizes the most widely used governmental policy instruments for climate action, ranging from regulatory and market-based mechanisms to data-driven monitoring systems supported by AI and remote sensing.

Table 1. Key Government Policy Instruments for Climate Action.

Instrument	Core Objective	Typical Scale	Representative Example
Carbon tax	Internalize carbon cost and discourage fossil-fuel dependence	National	Sweden, Canada
Emissions trading system (ETS)	Reduce emissions under a cap via market-based allocation	Regional/National	EU ETS, China ETS
Renewable subsidies / incentives	Accelerate renewable deployment and investment	National	Germany EEG
Energy efficiency standards	Reduce consumption in buildings, appliances, industry	National/Municipal	EU EPBD
National adaptation plans	Build resilience to climate risks across sectors	National/Local	UK NAP

continued on following page

Table 1. Continued

Instrument	Core Objective	Typical Scale	Representative Example
Disaster risk reduction policies	Strengthen preparedness and risk management for extremes	National	Sendai Framework
Green finance instruments	Mobilize capital for low-carbon transition	National/Market	Green bonds
Carbon border adjustment (CBAM-type)	Limit carbon leakage via trade-linked instruments	Regional	EU CBAM
Climate-smart agriculture support	Increase food-system resilience and lower emissions	National/Regional	CAP eco-schemes
Monitoring–Reporting–Verification (MRV)	Ensure transparency and compliance in mitigation policies	National/ International	UNFCCC MRV
AI/RS-enabled monitoring systems	Provide near-real-time risk mapping and evidence for decisions	National/ International	Copernicus, NASA EOS

2.1. Policy and Legislation Development

A primary responsibility of governments is to integrate climate change into the national policy agenda and to create legal frameworks for reducing greenhouse gas emissions (Maranhão & Marchi 2025; Ilyas et al., 2026). In this context, governments define national carbon reduction targets, renewable energy strategies, and climate action plans (Zhao et al., 2026; Oraby & Orabi, 2025).

Internationally, global agreements such as the Paris Agreement (2015), the United Nations Framework Convention on Climate Change (UNFCCC), and the Sustainable Development Goals (SDG 13: Climate Action) shape countries' nationally determined contributions (NDCs). These declarations include countries' commitments to greenhouse gas reduction and adaptation for a specific period (Janetschek et al., 2019; Khan et al., 2021). Furthermore, many countries are guiding the private sector toward low-carbon investments through carbon pricing mechanisms, ETS, and green finance arrangements (Campiglio, 2016; Narassimhan et al., 2018). The European Union's Green Deal can be cited as a transformative policy example in this regard. Such policies do not only aim at environmental protection; It also aims to increase economic competitiveness and promote sustainable growth (Maričić & Pantić, 2025; Ioannou et al., 2025).

At this very point, AI-based decision support systems stand out as tools that enhance governments' policy capacity in the design and evaluation processes of climate policies. AI-supported modeling and simulations enable the analysis of the economic, social, and environmental impacts of different policy scenarios in advance, facilitating more comprehensive and evidence-based policymaking. Thus, artificial intelligence becomes a governance component that not only enables the implementation of climate legislation and policy tools but also facilitates their strategic rethinking.

2.2. Monitoring, Evaluation, and Reporting

Strong monitoring and evaluation mechanisms are essential to ensure the effectiveness of climate policies (Juel et al., 2026; Hsu et al., 2018). Governments monitor atmospheric changes, water resources, forest cover, and soil moisture through meteorological and hydrological observation networks (Trenberth et al., 2015; Running et al., 2004). This data is collected and regularly evaluated by national meteorological institutions, environmental agencies, and scientific research institutes (Boers et al., 2019; Morice et al., 2021).

National greenhouse gas inventories and climate impact assessment reports are key documents submitted to the United Nations (UN). These reports play a crucial role in measuring countries' emission trends, energy efficiency performance, and the effectiveness of climate policies (URL-1, 2024; Grassi et al., 2018; Huang et al., 2023). They also enable the early detection of climate-related disasters (e.g., floods, droughts, wildfires) through early warning systems and risk mapping tools.

In recent years, the integration of artificial intelligence and remote sensing technologies into these processes has significantly increased monitoring capacity (Pandey et al., 2026; Singh & Jeganathan 2025). Thanks to satellite imagery and ML analyses, governments can monitor land-use changes, water scarcity trends, and the status of carbon sinks more quickly and accurately (Bojer et al., 2025; Uyar & Uyar, 2025). This technological capacity policymaking and ensures the efficient use of public resources.

These developments provide a significant increase in capacity for monitoring and evaluating climate policies, while also raising the issue of how and within what institutional framework these technologies are utilized. Artificial intelligence and remote sensing-based monitoring systems enable decision-makers to evaluate policy implementations more quickly and in greater detail; however, data quality, algorithmic transparency, and institutional coordination deficiencies can limit the impact of these technologies on policy outcomes. Therefore, the effectiveness of monitoring and evaluation mechanisms depends not only on technological capacity but also on the extent to which this capacity can be integrated into institutional learning, policy feedback processes, and democratic accountability. In this context, for AI-supported monitoring systems to have a lasting and meaningful impact on climate governance, technical innovations must be supported by strong institutional structures and transparent decision-making processes.

2.3. Adaptation and Mitigation Strategies

There are two complementary aspects to responding to the climate crisis: mitigation and adaptation (Pavlova et al., 2026; Yu et al., 2026).

- Mitigation targets the causes of climate change; that is, it aims to reduce greenhouse gas emissions (Amini et al., 2026; Rizwan & Saleem, 2025).
- Adaptation, on the other hand, aims to increase the resilience of societies, economies, and ecosystems against the inevitable impacts of change (Fauchald et al., 2026; Nair et al., 2026).

Governments must implement these two processes simultaneously, especially in sectors such as agriculture, water management, urban planning, energy, transportation, and health (Sharifi 2021; Hurlimann et al., 2021).

- In agriculture: Development of drought-resistant seeds, increasing irrigation efficiency, and soil management policies that act as carbon sinks are implemented (Lobell et al., 2011; Lesk et al., 2016).
- In the energy sector: Investment in renewable energy (solar, wind, hydro, biomass), energy efficiency incentives, and the reduction of fossil fuel subsidies support a low-carbon transition (Steckel et al., 2016; Creutzig et al., 2018).
- In urban planning: Green infrastructure projects, climate-resilient transportation systems, and waste management policies reduce the carbon footprint of cities (Grimm et al., 2008; Anguelovski et al., 2017).

Furthermore, the social dimension should not be overlooked by governments (Schlosberg & Collins 2014; Klinsky 2017). From a climate justice perspective, protecting low-income or vulnerable communities from climate risks and ensuring a just transition is fundamental to long-term sustainability (Robinson & Shine 2018; Newell & Mulvaney 2013). Therefore, many countries aim to increase societal resilience through initiatives such as National Adaptation Plans (NAPs) and Green Employment Programmes (Neil Adger et al., 2005; Dzebo & Stripple, 2015).

In conclusion, the role of governments in combating the climate crisis is not only regulatory but also coordinative, directive, and innovative. Effective climate governance must be based on strong institutional capacity, science-based decision-making, international cooperation, and societal participation. Such a framework makes it possible to both manage climate risks and accelerate the transition to low-carbon development. The success of mitigation and adaptation policies addressing the climate crisis depends on these two approaches being addressed in a holistic and coordinated manner, rather than in isolation. Ensuring consistency among sectoral policies, effectively managing the tensions between short-term economic goals and long-term climate goals, and fairly distributing the social costs of climate policies are fundamental governance challenges facing governments. In this context, effective climate governance is only possible through the joint development of strong institutional capacity, science-based decision-making, social participation, and political legitimacy, rather than solely through technical solutions and regulatory tools. Such an approach facilitates the management of climate risks while also contributing to accelerating the transition to low-carbon development. Within this institutional and political framework, artificial intelligence technologies stand out as important tools reshaping the capacity of states to design and implement climate policies.

3. THE EMERGING ROLE OF ARTIFICIAL INTELLIGENCE IN CLIMATE POLICIES

AI stands out as a transformative technological tool in managing the climate crisis, strengthening decision-support mechanisms and increasing scientific prediction capacity (Rolnick et al., 2022; Reichstein et al., 2019). Governments are increasingly utilizing AI-based systems to reduce uncertainties, improve risk estimation, and optimize resource use when shaping climate policies (Galaz et al., 2021; Vinuesa et al., 2020). Big data analytics, ML, and DL techniques can reveal complex climate patterns where traditional statistical modeling is limited; enabling policymakers to make more dynamic, adaptable, and science-based decisions. The rising role of AI in climate policies can be assessed along three main axes: climate modeling and prediction systems, early warning and disaster management, and policy analysis and scenario assessment (Reichstein, 2017; Irrgang et al., 2021).

3.1. Climate Modeling and Prediction Systems

Climate systems are highly complex, multivariate dynamic structures where the interactions between atmospheric circulation, ocean currents, glaciers, soil moisture, and the biosphere are extremely intricate (Stevens & Bony, 2013; Held, 2005). Traditional climate models (e.g., general circulation models – GCMs) play a significant role in explaining this complexity, but they have limitations in terms of high spatial resolution and local-level prediction. AI-based models come into play here, increasing the accuracy of climate projections. ML algorithms can learn patterns from past climate data, predicting future temperature, precipitation, wind, and drought patterns with higher resolution (Brenowitz, 2018; Rasp et al., 2018).

- Deep neural networks can model nonlinear relationships between climate variables (Ham et al., 2019; Chattopadhyay et al., 2020).
- Reinforcement learning methods enable the development of adaptive planning strategies in energy systems based on climate conditions (Vázquez-Canteli & Nagy, 2019; Nik & Javanroodi, 2025).
- Satellite-based remote sensing data is processed by AI to analyze temporal trends of indicators such as glacier melting rate, vegetation change, and sea surface temperatures (Zhong et al., 2019; Yao et al., 2012).

These technologies help governments update their national climate strategies, produce local risk maps, and determine regional policy priorities (Gorelick et al., 2017; Cramer et al., 2018).

3.2. Early Warning and Disaster Management

The frequency and intensity of extreme weather events, a direct consequence of climate change, have increased significantly in recent years (Diffenbaugh et al., 2019; Coumou & Rahmstorf, 2012). This situation necessitates a restructuring of disaster management capacities, particularly for governments. Artificial intelligence significantly enhances the effectiveness of early warning systems in this area (Walker et al., 2018; Clark et al., 2025).

- ML algorithms can detect early signs of events such as floods, heat waves, wildfires, tornadoes, and storms by analyzing historical meteorological data and satellite imagery (Abraham et al., 2023; Verma et al., 2023).
- DL-based image recognition systems identify fire outbreaks or sudden temperature increases by detecting anomalies from remote sensing data (Mandla et al., 2025; Meimand et al., 2025).
- These systems enable not only disaster prediction but also real-time impact analysis (e.g., potential economic damage, population impact, infrastructure risk) (Cicilio et al., 2021; Mukherjee & Hastak, 2017).

Thus, governments can make early evacuation decisions in at-risk areas, effectively direct emergency resources, and create response plans that minimize human casualties. For example, in countries with high climate risk such as India, the Philippines, and Japan, AI-based disaster prediction platforms are integrated with public institutions and used at the operational level (Baltazar et al., 2024; Upadhyaya et al., 2025).

3.3. Policy Analysis and Scenario Evaluation

The success of climate policies depends not only on technical expertise but also on a holistic assessment of economic, environmental, and social impacts. In this context, artificial intelligence offers governments a powerful tool for evidence-based policy analysis (Ukoba et al., 2025; Rohde et al., 2021). AI-powered simulation systems and scenario models allow for a comparative assessment of the short- and long-term impacts of different policy options (e.g., carbon tax, elimination of fossil fuel subsidies, increased renewable energy investment) (Wang et al., 2024; Alestra et al., 2020).

- Data-driven optimization algorithms analyze the balance between carbon reduction and economic growth (Zhang et al., 2014; Wang et al., 2021).
- Natural language processing (NLP) techniques can reveal trends and policy gaps by performing text mining on international climate agreements, legal texts, and policy reports (Stede & Patz, 2021; Mallick et al., 2024).
- Multi-criteria decision support systems (MCDA) are used to identify the most sustainable policy combinations among different scenarios (Davidsdottir et al., 2024; Karakosta & Papathanasiou, 2024).

In this way, governments can optimize climate policies not only according to emission targets but also along the axes of energy security, social justice, and economic transformation (Leong, 2025; Olawade et al., 2024). AI is also important for transparency and accountability (Figure 2). Algorithms that monitor policy performance can identify inconsistencies in emission reduction reporting and provide data-driven feedback to decision-makers (Alder et al., 2024; Mohsen, 2025).

AI technologies are fundamentally transforming governments' capacity to respond to the climate crisis at scientific, technical, and managerial levels, moving climate governance toward a data-driven, proactive, and systematic structure (Ullah et al., 2025). By analyzing multidimensional datasets, AI enables a better understanding of the causes, dynamics, and impacts of climate change; it also provides decision-makers with a powerful decision-support infrastructure in areas such as risk estimation, resource planning, energy management, and disaster response. However, the adoption of certain fundamental principles is essential for the effective and ethical use of these technologies (Reichstein et al., 2025; Tasha, 2025). Firstly, data privacy and security are critical in big data systems related to climate; governments must ensure that data is managed accurately, securely, and ethically. Furthermore, algorithmic transparency and accountability are indispensable for democratic legitimacy; ensuring that how AI systems make decisions is clear, understandable, and auditable increases public trust (McCormack et al., 2025; Feng & Chandra,

2025). Finally, international cooperation is inevitable because the climate crisis is a global problem; technology transfer, data sharing, and global harmonization of ethical standards will strengthen the climate adaptation capacities of developing countries in particular. (Adebola et al., 2024; Sharman, 2022). For AI to be used effectively, it is also crucial to improve the technical capacity of public institutions, train expert human resources, and establish strategic collaborations with universities. When all these elements come together, artificial intelligence becomes not just a technological innovation, but a strategic transformative tool at the heart of sustainable development, just transition, and evidence-based climate governance (Ibrahim et al., 2025; Priya et al., 2025).

Figure 2. Conceptual overview of the emerging role of artificial intelligence (AI) in climate policies, illustrating three key application axes: (i) climate modeling and prediction systems, (ii) early warning and disaster management, and (iii) policy analysis and scenario evaluation (This figure was generated with the support of AI-based image generation tools).

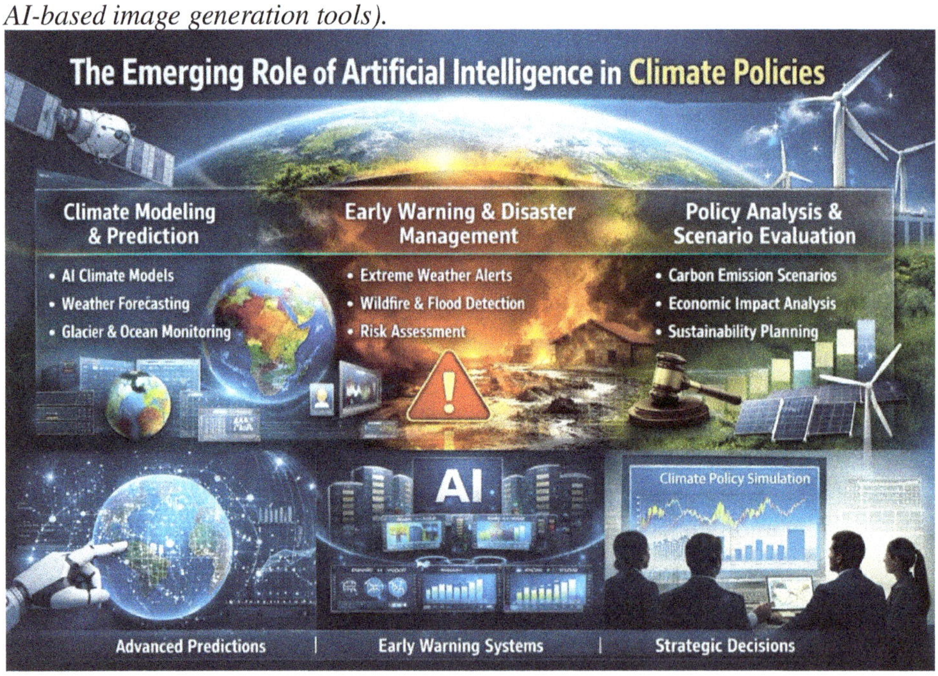

4. CURRENT APPROACHES AND INTERNATIONAL EXAMPLES

Today, many countries are transitioning to data-driven governance by placing artificial intelligence (AI) at the heart of their climate policies (Mehryar et al., 2024; Machen & Pearce, 2025). The European Union (EU), in particular, stands out as a pioneer in this field; through initiatives such as the Copernicus Earth Observation Program, DestinE (Digital Twin Earth), and AI4Climate, it combines remote sensing data and AI-based analyses to produce high-resolution climate risk maps, air quality prediction models, and ecosystem monitoring systems. (d'Andrimont et al., 2024; Hoffmann et al., 2023). These systems enable EU countries to predict disaster risks such as floods, droughts, and wildfires more accurately and to shape regional adaptation strategies accordingly. In the United States, the National Ocean and Atmospheric Administration (NOAA) and the Department of Energy (DOE) are integrating AI-powered climate simulations into energy efficiency, renewable energy production, and carbon management processes (Zscheischler et al., 2018; Nair et al., 2025). Private sector initiatives, particularly Google DeepMind and IBM Green Horizon, provide technical support to public policies with smart energy grids and carbon emission prediction models. China widely uses artificial intelligence in both industrial emission monitoring and city-level carbon management policies (Ejiyi et al., 2025; Chen et al., 2022). Through projects like "AI for Earth," national-scale data on air pollution, energy consumption, and agricultural production are analyzed in real time, accelerating progress toward the country's carbon neutrality goals. Similarly, countries like Canada and Japan are integrating AI into climate adaptation policies, developing early warning systems in various areas from wildfire prediction to water resource management (Xie et al., 2022; Xia et al., 2025). In Africa, Kenya and South Africa are supporting small-scale farmers in building resilience to drought and yield risks through AI-based agricultural climate advisory systems. All these examples demonstrate that artificial intelligence is becoming not only a tool for monitoring the climate crisis but also a governance paradigm that empowers policy decision-making with data. This allows governments to more accurately model complex climate systems and take effective measures, building a transparent, measurable, and adaptable climate policy framework (Chavula et al., 2024; Maheshwari, 2025).

5. DISCUSSION

This section discusses the role of artificial intelligence (AI) and remote sensing technologies in climate governance and policy support processes, in light of current approaches in the literature and the conceptual framework addressed in this study.

The findings and examples clearly demonstrate that data-driven and technological solutions are increasingly becoming central to managing the climate crisis (Overpeck et al., 2011; Stephens et al., 2020). This situation indicates that the production and use of information in climate governance has undergone a fundamental transformation, revealing that decision-making processes have evolved from intuitive and reactive approaches to analytical and predictive models.

One of the key points discussed is the evolution of climate governance from its traditional, reactive, and often sectorally fragmented structure toward a data-driven and holistic model (Hughes et al., 2020; Drakopulos et al., 2022). Remote sensing systems allow for the monitoring of atmospheric, terrestrial, and oceanic processes with high spatial and temporal resolution; artificial intelligence algorithms can then derive meaningful patterns from these large and complex datasets. This makes it possible for climate policies to be based not only on past trends but also on predictive scenarios (Anderson et al., 2017; Huntingford et al., 2019). While this transformation strengthens the role of information in the policy-making process, it also raises critical questions about how data and model results are interpreted and translated into political decisions.

One of the most significant challenges in climate policy is uncertainty. The nonlinear nature of the climate system makes risk assessment processes difficult for policymakers. The examples discussed show that while AI-based models do not completely eliminate these uncertainties, they make decision-making processes more rational through probability-based risk analyses and scenario comparisons. In particular, multiple scenario analyses and simulations offer the possibility of evaluating the short-term and long-term impacts of policy options together (Stainforth et al., 2019; Hall et al., 2012). This situation reveals that making uncertainty manageable rather than eliminating it is a more realistic goal in climate governance.

The international examples discussed in this study demonstrate that the integration of artificial intelligence and remote sensing has a transformative effect on disaster management (Kerle et al., 2019; Zhu et al., 2017). A critical point in this discussion is that these technologies not only increase technical capacity but also transform institutional reflexes and response speed. However, the effectiveness of these systems depends not only on technological infrastructure but also on institutional coordination, data sharing, and local governance capacity (Mergel et al., 2019; Gupta et al., 2010).

Although artificial intelligence offers a powerful decision-making support tool in climate policy, there are significant limitations that need to be discussed (Jasanoff, 2010; Mökander & Floridi, 2021). AI systems are only as powerful as their inputs; incomplete, erroneous, or biased data can lead to problematic policy recommendations. Furthermore, the lack of transparency in algorithmic decision-making processes poses risks to democratic legitimacy and societal acceptance. Therefore, artificial intelligence should be positioned not as the ultimate decision-maker, but as a scientific and technical support tool (Rudin, 2019).

Another important dimension of the discussion is the differences in technological capacity between countries. While developed countries are rapidly implementing AI and remote sensing-based climate policies, developing countries may lag behind due to a lack of data infrastructure, expert human resources, and funding. This creates an inequality that contradicts the global nature of the climate crisis (Ziervogel et al., 2016; Hansen et al., 2019). Therefore, technology transfer, open data policies, and international cooperation mechanisms are critical in reducing this inequality (Cash et al., 2003; Dechezleprêtre et al., 2009).

Finally, the success of artificial intelligence and remote sensing technologies in climate governance depends on the strength of science-policy-society interaction (Lemos et al., 2012; Sarewitz & Pielke, 2007). AI-based analyses need to be reflected in policy documents, presented to the public in an open and understandable manner, and supported by public awareness. In this context, data visualization tools and open communication strategies are important components that ensure technological solutions find a societal response (Kouziokas & Perakis 2017; Meadow et al., 2015). As long as this interaction cannot be achieved, even the most advanced technological solutions will remain insufficient in creating policy impact.

When evaluated within this framework, artificial intelligence and remote sensing technologies emerge not only as tools that increase technical capacity in climate governance but also as structural elements that transform the nature of decision-making processes. However, for this transformation to be effective and legitimate, technological innovations must be integrated with strong institutional structures, transparent governance mechanisms, and inclusive policy processes. Otherwise, data-driven solutions risk being reduced to technocratic practices disconnected from democratic oversight. Therefore, an effective governance model for the climate crisis should position artificial intelligence not as a central decision-maker, but as a complementary policy tool that strengthens scientific knowledge production, manages uncertainties, and is supported by social participation. This approach will enable both more effective management of climate risks and accelerate a low-carbon and equitable transition process.

6. CONCLUSION

This study emphasizes that artificial intelligence should be positioned not as an independent decision-maker in climate governance, but as a complementary tool supporting policy-making processes. The climate crisis is not only an environmental challenge, but also an economic, social, and governance challenge. In this context, the role of governments is no longer limited to producing policies and legislation; it also includes the responsibility of strategically integrating innovative technologies. AI stands out as one of the most powerful tools in the hands of governments in this transformation process. Thanks to AI components such as big data, ML, and remote sensing, the complex dynamics of climate systems can be analyzed more accurately, risks can be detected early, and policies can be built on more predictable foundations. Data-driven and scientific decision-making mechanisms facilitate governments in developing proactive, rapid, and effective responses to climate change. In the future, the central role of AI in climate governance will strengthen not only environmental sustainability but also economic resilience, energy security, and social justice. However, for these technologies to be used effectively, strong steps need to be taken in areas such as ethically based data management, algorithmic transparency, international information sharing, and institutional capacity building. The integration of AI into climate policies will accelerate the digital transformation of public institutions, provide decision-makers with real-time analytical support, and enhance democratic legitimacy by strengthening citizen participation. In this context, the climate governance of the future will be built on a holistic model that aims not only to reduce emissions but also to build informed, adaptable, and resilient societies. Ultimately, a future where AI technologies are properly designed, ethically managed, and inclusively implemented will play a decisive role in increasing global resilience to the climate crisis.

Rather than reiterating the normative principles discussed throughout the chapter, the key contribution of this study lies in demonstrating how data-driven tools are reshaping the relationship between technological capacity and governance practices. By integrating AI-based analytics with climate policy processes, governments are moving toward more anticipatory and adaptive decision-making models. However, this transformation also requires rethinking institutional roles, redefining expertise within policymaking, and balancing technological efficiency with democratic legitimacy.

7. RECOMMENDATIONS

In line with the AI, remote sensing, and climate governance framework discussed in this section, the following are key recommendations for increasing the effectiveness, transparency, and sustainability of climate policies. Governments should integrate AI-powered decision support systems into climate governance frameworks at the institutional level. Establishing AI and climate analytics units within critical ministries such as environment, energy, agriculture, and disaster management will strengthen the link between scientific data and policy-making processes. High-quality, reliable, and interoperable data infrastructures are essential for effective climate governance. Investing in open data platforms that integrate remote sensing data, ground observation stations, socio-economic indicators, and climate model outputs is recommended. Common data standards will facilitate cooperation at national and international levels. Ensuring that artificial intelligence systems used in climate policies are ethical, transparent, and accountable is critical to democratic legitimacy. Adopting ethical AI guidelines covering data privacy, algorithmic bias, and accountability will enhance public trust, particularly in high-risk decision-making processes. Given the frequency and severity of climate-related disasters, AI-powered early warning systems need to be widely implemented, especially in vulnerable areas. Integrating ML and satellite data will enable earlier and more accurate predictions of floods, droughts, heat waves, and wildfires. The effective use of artificial intelligence in climate governance depends on qualified human resources. Therefore, it is recommended to support interdisciplinary education and capacity-building programs that combine climate science, data science, remote sensing, and public policy. Strategic collaborations with universities and research institutions will strengthen this process. Since the climate crisis is a global problem, international cooperation in AI and remote sensing-based solutions is inevitable. Developed countries and international organizations should support the climate adaptation and mitigation capacities of developing countries through technology transfer, data sharing, and joint research programs. AI should be used not only in policy design but also in the continuous monitoring and evaluation of policy performance. AI-based monitoring systems can analyze the effectiveness of climate policies in real time, contributing to the creation of an adaptive governance structure that can adjust to changing climate and socio-economic conditions. Implementing these recommendations will enable governments to move away from reactive approaches to the climate crisis and toward a proactive, data-driven, and resilient governance model. When used within an ethical and inclusive framework, artificial intelligence and remote sensing technologies will be key tools in achieving sustainable development and climate resilience goals.

In this context, innovative approaches to the design and timing of policy processes are needed to enhance the effectiveness of climate policies based on artificial intelligence and remote sensing. In particular, it is recommended that AI models used in climate policies be tested and updated through periodic ex-post evaluations. Such an approach will strengthen institutional learning in climate governance by revealing which policy tools are effective under which conditions. Furthermore, the systematic evaluation of rapid interventions following crises and disasters through AI-supported analysis can contribute to the development of long-term structural policy adjustments rather than temporary solutions. Future research may further explore how AI-mediated policy processes influence political accountability, public trust, and cross-sectoral coordination, particularly in contexts where digital capacity remains uneven.

REFERENCES

Abraham, K., Abdelwahab, M., & Abo-Zahhad, M. (2023). Classification and detection of natural disasters using machine learning and deep learning techniques: A review. *Earth Science Informatics, 17*(2), 869–891. DOI: 10.1007/s12145-023-01205-2

Alder, N., Ebert, K., Herbrich, R., & Hacker, P. (2024). AI, Climate, and Transparency: Operationalizing and Improving the AI Act. DOI: 10.2139/ssrn.4941352

Alestra, C., Cette, G., Chouard, V., & Lecat, R. (2020). Long-term Growth Impact of Climate Change and Policies: The Advanced Climate Change Long-term (ACCL) Scenario Building Model. SSRN Electronic Journal. DOI: 10.2139/ssrn.3569499

Alyemiş, H. (2025). *The Impacts of Climate Change on Social Risks: An Evaluation within the Framework of Green Social Policy.* TAM Akademi Journal., DOI: 10.58239/tamde.2025.02.001.x

Amini Toosi, H., Famiglietti, J., & Motta, M. (2026). *Climate Change and Clean Energy Solution Nexus. Global Nexus Handbook.* Portico., DOI: 10.1002/9781394248087.ch21

Anderson, K., Ryan, B., Sonntag, W., Kavvada, A., & Friedl, L. (2017). Earth observation in service of the 2030 Agenda for Sustainable Development. *Geo-Spatial Information Science, 20*(2), 77–96. DOI: 10.1080/10095020.2017.1333230

Anguelovski, I., Connolly, J. J. T., Masip, L., & Pearsall, H. (2017). Assessing green gentrification in historically disenfranchised neighborhoods: A longitudinal and spatial analysis of Barcelona. *Urban Geography, 39*(3), 458–491. DOI: 10.1080/02723638.2017.1349987

Ayadi, R., Forouheshfar, Y., & Moghadas, O. (2025). Enhancing system resilience to climate change through artificial intelligence: A systematic literature review. *Frontiers in Climate, 7*, 1585331. Advance online publication. DOI: 10.3389/fclim.2025.1585331

Babacan, H. (2025). From Crisis to Policy Action: Tackling Climate Change Through SDG 13. *The International Journal of Community and Social Development, 7*(2), 205–221. DOI: 10.1177/25166026251351370

Balan, S., Vorotynskyy, V., Patalakha, V., Rybak, I., & Tarasiuk, V. (2025). The Role of State Information Policy in Shaping Democratic Transformation Amid Hybrid Threats. DOI: 10.1590/1678-98732433e015

Baltazar, R., Florencio, B., Vicente, A., & Belizario, P. (2024). The Role of Artificial Intelligence in Disaster Prediction, Mitigation, and Response in the Philippines: Challenges and Opportunities. *International Journal of Artificial Intelligence*, *11*(1), 37–51. DOI: 10.36079/lamintang.ijai-01101.675

Benedetto, M., Sousa, A., Reis, J. L., Pratas, J., & Pimenta, P. (2025). Unleashing Open Government Data: Implications for Urban Sustainability and Public Management.

Biswas, U., Mahato, S., & Joshi, P. K. (2025). Spatial prediction of forest fires in India: A machine learning approach for improved risk assessment and early warning systems. *Environmental Science and Pollution Research International*, *32*(8), 4856–4878. DOI: 10.1007/s11356-025-35982-8 PMID: 39891812

Boers, N., Goswami, B., Rheinwalt, A., Bookhagen, B., Hoskins, B., & Kurths, J. (2019). Complex networks reveal global pattern of extreme-rainfall teleconnections. *Nature*, *566*(7744), 373–377. DOI: 10.1038/s41586-018-0872-x PMID: 30700912

Bojer, A. K., Abshare, M. W., Mesfin, F., & Al-Quraishi, A. M. F. (2025). Assessing climate and land use impacts on surface water yield using remote sensing and machine learning. *Scientific Reports*, *15*(1), 18477. Advance online publication. DOI: 10.1038/s41598-025-03493-8 PMID: 40425713

Bouquet, P., Jackson, I., Nick, M., & Kaboli, A. (2023). AI-based forecasting for optimised solar energy management and smart grid efficiency. *International Journal of Production Research*, *62*(13), 4623–4644. DOI: 10.1080/00207543.2023.2269565

Brenowitz, N. D., & Bretherton, C. S. (2018). Prognostic Validation of a Neural Network Unified Physics Parameterization. Geophysical Research Letters, 45(12), 6289–6298. Portico. DOI: 10.1029/2018GL078510

Campiglio, E. (2016). Beyond carbon pricing: The role of banking and monetary policy in financing the transition to a low-carbon economy. *Ecological Economics*, *121*, 220–230. DOI: 10.1016/j.ecolecon.2015.03.020

Cardellicchio, A., Renò, V., Guadagno, C. R., Cellini, F., & Amitrano, C. (2025). Editorial: Agricultural innovation in the age of climate change: a 4.0 approach. *Frontiers in Plant Science*, *16*, 1715677. Advance online publication. DOI: 10.3389/fpls.2025.1715677 PMID: 41158684

Cash, D. W., Clark, W. C., Alcock, F., Dickson, N. M., Eckley, N., Guston, D. H., Jäger, J., & Mitchell, R. B. (2003). Knowledge systems for sustainable development. *Proceedings of the National Academy of Sciences of the United States of America*, *100*(14), 8086–8091. DOI: 10.1073/pnas.1231332100 PMID: 12777623

Castañeda Rodriguez, L., Galvez-Nieto, A., Aguilar Chunga, Y. A., Ccalla Chusho, J. A., & Salinas Romero, M. E. (2026). Transformative Urban Resilience and Collaborative Participation in Public Spaces: A Systematic Review of Theoretical and Methodological Insights. *Urban Science (Basel, Switzerland), 10*(1), 51. DOI: 10.3390/urbansci10010051

Chattopadhyay, A., Nabizadeh, E., & Hassanzadeh, P. (2020). Analog Forecasting of Extreme-Causing Weather Patterns Using Deep Learning. Journal of Advances in Modeling Earth Systems, 12(2). Portico. DOI: 10.1029/2019MS001958

Chavula, P., Kayusi, F., & Juma, L. (2024). Leveraging Artificial Intelligence for Enhancing Wheat Yield Resilience Amidst Climate Change in Sub-Saharan Africa. *LatIA, 2*, 88. DOI: 10.62486/latia202588

Chen, L., Msigwa, G., Yang, M., Osman, A. I., Fawzy, S., Rooney, D. W., & Yap, P.-S. (2022). Strategies to achieve a carbon neutral society: A review. *Environmental Chemistry Letters, 20*(4), 2277–2310. DOI: 10.1007/s10311-022-01435-8 PMID: 35431715

Chen, P., Gao, J., Ji, Z., Liang, H., & Peng, Y. (2022). Do Artificial Intelligence Applications Affect Carbon Emission Performance?—Evidence from Panel Data Analysis of Chinese Cities. *Energies, 15*(15), 5730. DOI: 10.3390/en15155730

Cheng, Y., Shen, L., Zhang, J., He, H., Gu, X., Wang, S., & Ma, T. (2025). A Multi-Stage Deep Learning Framework for Multi-Source Cloud Top Height Retrieval from FY-4A/AGRI Data. *Atmosphere (Basel), 16*(11), 1288. DOI: 10.3390/atmos16111288

Cicilio, P., Glennon, D., Mate, A., Barnes, A., Chalishazar, V., Cotilla-Sanchez, E., Vaagensmith, B., Gentle, J., Rieger, C., Wies, R., & Kapourchali, M. H. (2021). Resilience in an Evolving Electrical Grid. *Energies, 14*(3), 694. DOI: 10.3390/en14030694

Clark, M., Fröhner, C., Jørgensen, A., Pienkowski, T., Yekela, S., Isacs, A., Crowe, O., Andrews, J., Smaldino, P., Gallizioli, I., Arena, G., & Mills, M. (2025). Eliminating climate change mitigation trade-offs in African rangelands. DOI: 10.21203/rs.3.rs-7750461/v1

Cortez, F. (2023). Artificial Intelligence, Climate Change and Innovative Democratic Governance. *European Journal of Risk Regulation, 14*(3), 484–503. DOI: 10.1017/err.2023.60

Coumou, D., & Rahmstorf, S. (2012). A decade of weather extremes. *Nature Climate Change, 2*(7), 491–496. DOI: 10.1038/nclimate1452

Cramer, W., Guiot, J., Fader, M., Garrabou, J., Gattuso, J.-P., Iglesias, A., Lange, M. A., Lionello, P., Llasat, M. C., Paz, S., Peñuelas, J., Snoussi, M., Toreti, A., Tsimplis, M. N., & Xoplaki, E. (2018). Climate change and interconnected risks to sustainable development in the Mediterranean. *Nature Climate Change*, *8*(11), 972–980. DOI: 10.1038/s41558-018-0299-2

Creutzig, F., Roy, J., Lamb, W. F., Azevedo, I. M. L., Bruine de Bruin, W., Dalkmann, H., Edelenbosch, O. Y., Geels, F. W., Grubler, A., Hepburn, C., Hertwich, E. G., Khosla, R., Mattauch, L., Minx, J. C., Ramakrishnan, A., Rao, N. D., Steinberger, J. K., Tavoni, M., Ürge-Vorsatz, D., & Weber, E. U. (2018). Towards demand-side solutions for mitigating climate change. *Nature Climate Change*, *8*(4), 260–263. DOI: 10.1038/s41558-018-0121-1

d'Andrimont, R., Yordanov, M., Sedano, F., Verhegghen, A., Strobl, P., Zachariadis, S., Camilleri, F., Palmieri, A., Eiselt, B., Rubio Iglesias, J. M., & van der Velde, M. (2024). Advances in LUCAS Copernicus 2022: Enhancing Earth observations with comprehensive in situ data on EU land cover and use. *Earth System Science Data*, *16*(12), 5723–5735. DOI: 10.5194/essd-16-5723-2024

Davidsdottir, B., Ásgeirsson, E. I., Fazeli, R., Gunnarsdottir, I., Leaver, J., Shafiei, E., & Stefánsson, H. (2024). Integrated Energy Systems Modeling with Multi-Criteria Decision Analysis and Stakeholder Engagement for Identifying a Sustainable Energy Transition. *Energies*, *17*(17), 4266. DOI: 10.3390/en17174266

Dechezleprêtre, A., Glachant, M., & Ménière, Y. (2009). Technology transfer by CDM projects: A comparison of Brazil, China, India and Mexico. *Energy Policy*, *37*(2), 703–711. DOI: 10.1016/j.enpol.2008.10.007

Dessai, S., & Hulme, M. (2004). Does climate adaptation policy need probabilities? *Climate Policy*, *4*(2), 107–128. DOI: 10.1080/14693062.2004.9685515

Diffenbaugh, N. S., & Burke, M. (2019). Global warming has increased global economic inequality. *Proceedings of the National Academy of Sciences of the United States of America*, *116*(20), 9808–9813. DOI: 10.1073/pnas.1816020116 PMID: 31010922

Drakopulos, L., Havice, E., & Campbell, L. (2022). Architecture, agency and ocean data science initiatives: Data-driven transformation of oceans governance. *Earth System Governance*, *12*, 100140. DOI: 10.1016/j.esg.2022.100140

Dzebo, A., & Stripple, J. (2015). Transnational adaptation governance: An emerging fourth era of adaptation. *Global Environmental Change*, *35*, 423–435. DOI: 10.1016/j.gloenvcha.2015.10.006

Ejiyi, C. J., Cai, D., Thomas, D., Obiora, S., Osei-Mensah, E., Acen, C., Eze, F. O., Sam, F., Zhang, Q., & Bamisile, O. O. (2025). Comprehensive review of artificial intelligence applications in renewable energy systems: Current implementations and emerging trends. *Journal of Big Data*, *12*(1), 169. Advance online publication. DOI: 10.1186/s40537-025-01178-7

Fauchald, P., Christensen, T. R., & Christensen, T. (2026). Climate change impacts on Arctic ecosystems and associated feedbacks. *Frontiers in Environmental Science*, *13*, 1747632. Advance online publication. DOI: 10.3389/fenvs.2025.1747632

Favour, N.Favour N. EzeAdepeju Nafisat SanusiIsrael Jonathan IheomaChijioke Cyriacus EkechiMicheal Adeolu OlatunbosunFavour Chizurum UkasoanyaMuhdawwal Aremu Eleshin. (2025). Artificial Intelligence in Climate Change Mitigation and Adaptation: A Review of Emerging Technologies and Real-World Applications. *Global Journal of Engineering and Technology Advances*, *24*(2), 235–250. DOI: 10.30574/gjeta.2025.24.2.0247

Feng, N., & Chandra, Y. (2025). Accountability in Government Use of AI: Citizen Concerns and Preferences. *Public Administration*, padm.70030. Advance online publication. DOI: 10.1111/padm.70030

Galaz, V., Centeno, M. A., Callahan, P. W., Causevic, A., Patterson, T., Brass, I., Baum, S., Farber, D., Fischer, J., Garcia, D., McPhearson, T., Jimenez, D., King, B., Larcey, P., & Levy, K. (2021). Artificial intelligence, systemic risks, and sustainability. *Technology in Society*, *67*, 101741. DOI: 10.1016/j.techsoc.2021.101741

Gorelick, N., Hancher, M., Dixon, M., Ilyushchenko, S., Thau, D., & Moore, R. (2017). Google Earth Engine: Planetary-scale geospatial analysis for everyone. *Remote Sensing of Environment*, *202*, 18–27. DOI: 10.1016/j.rse.2017.06.031

Grassi, G., House, J., Kurz, W. A., Cescatti, A., Houghton, R. A., Peters, G. P., Sanz, M. J., Viñas, R. A., Alkama, R., Arneth, A., Bondeau, A., Dentener, F., Fader, M., Federici, S., Friedlingstein, P., Jain, A. K., Kato, E., Koven, C. D., Lee, D., & Zaehle, S. (2018). Reconciling global-model estimates and country reporting of anthropogenic forest CO2 sinks. *Nature Climate Change*, *8*(10), 914–920. DOI: 10.1038/s41558-018-0283-x

Grimm, N. B., Faeth, S. H., Golubiewski, N. E., Redman, C. L., Wu, J., Bai, X., & Briggs, J. M. (2008). Global Change and the Ecology of Cities. *Science*, *319*(5864), 756–760. DOI: 10.1126/science.1150195 PMID: 18258902

Gu, Z., & Zeng, M. (2023). The Use of Artificial Intelligence and Satellite Remote Sensing in Land Cover Change Detection: Review and Perspectives. *Sustainability (Basel)*, *16*(1), 274. DOI: 10.3390/su16010274

Gupta, J., Termeer, C., Klostermann, J., Meijerink, S., van den Brink, M., Jong, P., Nooteboom, S., & Bergsma, E. (2010). The Adaptive Capacity Wheel: A method to assess the inherent characteristics of institutions to enable the adaptive capacity of society. Environmental Science &amp. *Environmental Science & Policy*, *13*(6), 459–471. DOI: 10.1016/j.envsci.2010.05.006

Hadley, K., Wheat, S., Rogers, H. H., Balakumar, A., Gonzales-Pacheco, D., Davis, S. S., Linstadt, H., Cushing, T., Ziska, L. H., Piper, C., & Sorensen, C. (2023). Mechanisms underlying food insecurity in the aftermath of climate-related shocks: A systematic review. *The Lancet. Planetary Health*, *7*(3), e242–e250. DOI: 10.1016/S2542-5196(23)00003-7 PMID: 36774944

Hall, J. W., Lempert, R. J., Keller, K., Hackbarth, A., Mijere, C., & McInerney, D. J. (2012). Robust Climate Policies Under Uncertainty: A Comparison of Robust Decision Making and Info-Gap Methods. Risk Analysis, 32(10), 1657–1672. Portico. DOI: 10.1111/j.1539-6924.2012.01802.x

Ham, Y.-G., Kim, J.-H., & Luo, J.-J. (2019). Deep learning for multi-year ENSO forecasts. *Nature*, *573*(7775), 568–572. DOI: 10.1038/s41586-019-1559-7 PMID: 31534218

He, M., & Rozos, E. (2025). Editorial: Harnessing artificial intelligence to address climate-induced challenges in water resources management. *Frontiers in Water*, *7*, 1637341. Advance online publication. DOI: 10.3389/frwa.2025.1637341

Held, I. M. (2005). The Gap between Simulation and Understanding in Climate Modeling. *Bulletin of the American Meteorological Society*, *86*(11), 1609–1614. DOI: 10.1175/BAMS-86-11-1609

Hoffmann, J., Bauer, P., Sandu, I., Wedi, N., Geenen, T., & Thiemert, D. (2023). Destination Earth – A digital twin in support of climate services. *Climate Services*, *30*, 100394. DOI: 10.1016/j.cliser.2023.100394

Hsu, A., Höhne, N., Kuramochi, T., Roelfsema, M., Weinfurter, A., Xie, Y., Lütkehermöller, K., Chan, S., Corfee-Morlot, J., Drost, P., Faria, P., Gardiner, A., Gordon, D. J., Hale, T., Hultman, N. E., Moorhead, J., Reuvers, S., Setzer, J., Singh, N., & Widerberg, O. (2018). A research roadmap for quantifying non-state and subnational climate mitigation action. *Nature Climate Change*, *9*(1), 11–17. DOI: 10.1038/s41558-018-0338-z

Huang, Z., Wang, J., Bing, L., Qiu, Y., Guo, R., Yu, Y., Ma, M., Niu, L., Tong, D., Andrew, R. M., Friedlingstein, P., Canadell, J. G., Xi, F., & Liu, Z. (2023). Global carbon uptake of cement carbonation accounts 1930–2021. *Earth System Science Data*, *15*(11), 4947–4958. DOI: 10.5194/essd-15-4947-2023

Huntingford, C., Jeffers, E. S., Bonsall, M. B., Christensen, H. M., Lees, T., & Yang, H. (2019). Machine learning and artificial intelligence to aid climate change research and preparedness. *Environmental Research Letters*, *14*(12), 124007. DOI: 10.1088/1748-9326/ab4e55

Hurlimann, A., Moosavi, S., & Browne, G. R. (2021). Urban planning policy must do more to integrate climate change adaptation and mitigation actions. *Land Use Policy*, *101*, 105188. DOI: 10.1016/j.landusepol.2020.105188

Ibrahim, R. E., Motshegwa, T., Benouar, D., Khodja, M., Ead, H., Zhang, L., Crawley, F., Mchunu, N. P., Hachigonta, S., Hodson, S., Mcbride, V., & Yohannes, T. (2025). Sustainable North-South Africa Collaboration for Disaster and Crisis Management: A Strategic Capacity Development Framework using Open Science, Artificial Intelligence and Geoinformatics. *Data Science Journal*, *24*, 11. Advance online publication. DOI: 10.5334/dsj-2025-011

Ilyas, M., Liaqat, U., Eqan, M., Boškailo, E., Safdar, M., Ajanovic, T., Yaqoob, H., Chovancová, J., & Onifade, S. T. (2026). Policy frameworks and regulations for net zero. Net ZERO: Foundations, Technologies and Strategic Innovation, 189–247. DOI: 10.1016/B978-0-443-36426-6.00010-1

Ioannou, K., Karasmanaki, E., & Tsantopoulos, G. (2025). Environmental Policy as a Tool for Sustainable Development. *Sustainability (Basel)*, *17*(8), 3674. DOI: 10.3390/su17083674

Irrgang, C., Boers, N., Sonnewald, M., Barnes, E. A., Kadow, C., Staneva, J., & Saynisch-Wagner, J. (2021). Towards neural Earth system modelling by integrating artificial intelligence in Earth system science. *Nature Machine Intelligence*, *3*(8), 667–674. DOI: 10.1038/s42256-021-00374-3

Islam, F. A. S. (2025). Artificial Intelligence-driven Optimization of Nature-based Carbon Sequestration: A Scalable Architecture for Urban Climate Resilience. *International Journal of Environment and Climate Change*, *15*(7), 252–277. DOI: 10.9734/ijecc/2025/v15i74928

Janetschek, H., Brandi, C., Dzebo, A., & Hackmann, B. (2019). The 2030 Agenda and the Paris Agreement: Voluntary contributions towards thematic policy coherence. *Climate Policy*, *20*(4), 430–442. DOI: 10.1080/14693062.2019.1677549

Jasanoff, S. (2010). A New Climate for Society. Theory, Culture &amp. *Theory, Culture & Society*, *27*(2–3), 233–253. DOI: 10.1177/0263276409361497

Juel, R., Whitmee, S., Green, R., Davies, M., Milner, J., & Turcu, C. (2026). Integrated indicators for tracking climate mitigation and adaptation: A protocol for a rapid review with a focus on health co-benefits and trade-offs. *Wellcome Open Research*, *11*, 30. DOI: 10.12688/wellcomeopenres.25619.1

Karakosta, C., & Papathanasiou, J. (2024). Climate-Driven Sustainable Energy Investments: Key Decision Factors for a Low-Carbon Transition Using a Multi-Criteria Approach. *Energies*, *17*(21), 5515. DOI: 10.3390/en17215515

Kerle, N., Nex, F., Gerke, M., Duarte, D., & Vetrivel, A. (2019). UAV-Based Structural Damage Mapping: A Review. *ISPRS International Journal of Geo-Information*, *9*(1), 14. DOI: 10.3390/ijgi9010014

Khan, A. (2025). AI-Optimized Renewable Energy Management: Enhancing Efficiency and Grid Stability Using Machine Learning. *International Journal of Research & Technology*, *13*(1), 24–32.

Khan, M., Mfitumukiza, D., & Huq, S. (2021). Capacity building for implementation of Nationally Determined Contributions under the Paris Agreement. Making Climate Action More Effective, 95–106. .DOI: 10.4324/9781003162407-8

Klinsky, S. (2017). An initial scoping of transitional justice for global climate governance. *Climate Policy*, *18*(6), 752–765. DOI: 10.1080/14693062.2017.1377594

Kouziokas, G. N., & Perakis, K. (2017). Decision Support System Based On Artificial Intelligence, Gis And Remote Sensing For Sustainable Public And Judicial Management. *European Journal of Sustainable Development*, *6*(3). Advance online publication. DOI: 10.14207/ejsd.2017.v6n3p397

Kumar, D., Shekhar, S., & Tewary, T. (2025). Editorial: AI and data analytics for climate data management. *Frontiers in Environmental Science*, *13*, 1679608. Advance online publication. DOI: 10.3389/fenvs.2025.1679608

Kumar, M. S., Palanichamy, N. V., Shivakumar, K. M., Chandrakumar, M., Kalpana, M., & Murugananthi, D. (2025). Impact of climate change on global economy: A comprehensive review. *Regional Sustainability*, *6*(6), 100274. DOI: 10.1016/j.regsus.2025.100274

Kumari, P., Lavania, P., Dobriyal, M., Srivastav, M., & Kumar, P. (2026). Artificial intelligence and forest management. Advanced Geospatial and Ground Based Techniques in Forest Monitoring, 255–272. .DOI: 10.1016/B978-0-443-18949-4.00002-6

Lahn, B., Sognnaes, I., & Hermansen, E. A. T. (2025). Analysing the use of emissions scenarios in practice. *Npj Climate Action*, *4*(1), 117. Advance online publication. DOI: 10.1038/s44168-025-00320-8 PMID: 41450564

Lemos, M. C., Kirchhoff, C. J., & Ramprasad, V. (2012). Narrowing the climate information usability gap. *Nature Climate Change*, *2*(11), 789–794. DOI: 10.1038/nclimate1614

Leong, W. Y. (2025). Artificial Intelligence-Driven Analytics for Monitoring and Mitigating Climate Change Impacts. *IEEE ICEIB*, *7*, 7. Advance online publication. DOI: 10.3390/engproc2025108007

Lesk, C., Rowhani, P., & Ramankutty, N. (2016). Influence of extreme weather disasters on global crop production. *Nature*, *529*(7584), 84–87. DOI: 10.1038/nature16467 PMID: 26738594

Lobell, D. B., Schlenker, W., & Costa-Roberts, J. (2011). Climate Trends and Global Crop Production Since 1980. *Science*, *333*(6042), 616–620. DOI: 10.1126/science.1204531 PMID: 21551030

Machen, R., & Pearce, W. (2025). Anticipating the Challenges of <scp>AI</scp> in Climate Governance: An Urgent Dilemma for Democracies. WIREs Climate Change, 16(2). Portico. DOI: 10.1002/wcc.70002

Madani, M., Gray, S., Dyussenov, M., & Tadjine, F. (2025). Editorial: Climate change, natural resources, and human security in governance and society: vulnerabilities and adaptation strategies. *Frontiers in Political Science*, *7*, 1621475. Advance online publication. DOI: 10.3389/fpos.2025.1621475

Maheshwari, G. (2025). Governance of Nature-Based Solutions in Indian Climate Policy Framework. Handbook of Nature-Based Solutions to Mitigation and Adaptation to Climate Change, 2353–2368. DOI: 10.1007/978-3-031-34967-6_73

Mallick, T., Murphy, J., Bergerson, J. D., Verner, D. R., Hutchison, J. K., & Levy, L. A. (2024). Analyzing regional impacts of climate change using natural language processing techniques. arXiv preprint arXiv:2401.06817

Mandla, V. R., Zhu, X., Tiwari, A., & Shoab, M. (2025). Editorial: Emerging trends and advancements of geoinformatics applications in earth and environmental systems. *Frontiers in Earth Science (Lausanne)*, *13*, 1641510. Advance online publication. DOI: 10.3389/feart.2025.1641510

Maranhão, R. de A., & Marchi, C. M. D. F. (2025). Urban Solid Waste Management in Salvador and Climate Mitigation: Theoretical Foundations, Legal Framework, Actions and Discourses of Public Authorities. *Revista de Gestão Social e Ambiental*, *19*(7), e012975. DOI: 10.24857/rgsa.v19n7-101

Maričić, T., & Pantić, M. (2025). Just Green Transition of Coal-Mining Cities. The Palgrave Handbook of Just Green Transitions in the Western Balkans and Beyond, 183–202. DOI: 10.1007/978-3-031-95075-9_11

McCormack, L., Bendechache, M., Lewis, D., & Huyskes, D. (2025). Trust and transparency in AI: Industry voices on data, ethics, and compliance. AI &amp. *AI & Society*. Advance online publication. DOI: 10.1007/s00146-025-02654-7

Meadow, A. M., Ferguson, D. B., Guido, Z., Horangic, A., Owen, G., & Wall, T. (2015). Moving toward the Deliberate Coproduction of Climate Science Knowledge. *Weather, Climate, and Society*, *7*(2), 179–191. DOI: 10.1175/WCAS-D-14-00050.1

Meckling, J., & Allan, B. B. (2020). The evolution of ideas in global climate policy. *Nature Climate Change*, *10*(5), 434–438. DOI: 10.1038/s41558-020-0739-7

Mehryar, S., Yazdanpanah, V., & Tong, J. (2024). AI and climate resilience governance. *iScience*, *27*(6), 109812. DOI: 10.1016/j.isci.2024.109812 PMID: 38784017

Meimand, H. M., Chen, J., Kneeshaw, D., Bakhtyari, M., & Peng, C. (2025). Burned Area Detection in the Eastern Canadian Boreal Forest Using a Multi-Layer Perceptron and MODIS-Derived Features. *Remote Sensing (Basel)*, *17*(13), 2162. DOI: 10.3390/rs17132162

Mergel, I., Edelmann, N., & Haug, N. (2019). Defining digital transformation: Results from expert interviews. *Government Information Quarterly*, *36*(4), 101385. DOI: 10.1016/j.giq.2019.06.002

Metheniti, V., Parasyris, A., Pereira, R. S., & Kazanjian, G. (2025). Recent Advancements and Challenges in Artificial Intelligence for Digital Twins of the Ocean. *Climate (Basel)*, *14*(1), 3. DOI: 10.3390/cli14010003

Mökander, J., & Floridi, L. (2021). Ethics-Based Auditing to Develop Trustworthy AI. *Minds and Machines*, *31*(2), 323–327. DOI: 10.1007/s11023-021-09557-8

Morice, C. P., Kennedy, J. J., Rayner, N. A., Winn, J. P., Hogan, E., Killick, R. E., Dunn, R. J. H., Osborn, T. J., Jones, P. D., & Simpson, I. R. (2021). An Updated Assessment of Near-Surface Temperature Change From 1850: The HadCRUT5 Data Set. Journal of Geophysical Research: Atmospheres, 126(3). Portico. DOI: 10.1029/2019JD032361

Mukherjee, S., & Hastak, M. (2017). A Novel Methodological Approach to Estimate the Impact of Natural Hazard-Induced Disasters on Country/Region-Level Economic Growth. *International Journal of Disaster Risk Science*, *9*(1), 74–85. DOI: 10.1007/s13753-017-0156-3

Nair, A., Smitha Chandran, S., & Varangalil, N. (2025). Modelling and Forecasting of Climate Change Effects Using Artificial Intelligence Techniques. Climate Resilience: Impact of Quantum Computing and Artificial Intelligence on Urban Planning, 127–143. DOI: 10.1007/978-3-032-06791-3_7

Nair, M. N., Mani, S. K., Polinga, I. M., & Baiju, K. R. (2026). Eco-Social Resilience and Nature-Based Solutions: Local Strategies for Disaster Risk Reduction Across the Globe. Eco-Social Work for Disaster Risk Reduction, 71–87. DOI: 10.1007/978-981-95-4885-9_4

Narassimhan, E., Gallagher, K. S., Koester, S., & Alejo, J. R. (2018). Carbon pricing in practice: A review of existing emissions trading systems. *Climate Policy*, *18*(8), 967–991. DOI: 10.1080/14693062.2018.1467827

Neil Adger, W., Arnell, N. W., & Tompkins, E. L. (2005). Successful adaptation to climate change across scales. *Global Environmental Change*, *15*(2), 77–86. DOI: 10.1016/j.gloenvcha.2004.12.005

Newell, P., & Mulvaney, D. (2013). The political economy of the 'just transition.' The Geographical Journal, 179(2), 132–140. Portico. DOI: 10.1111/geoj.12008

Nik, V. M., & Javanroodi, K. (2025). Adaptive reinforcement learning for energy management – A progressive approach to boost climate resilience and energy flexibility. *Advances in Applied Energy*, *17*, 100213. DOI: 10.1016/j.adapen.2025.100213

Nunes, L. J. R. (2023). The Rising Threat of Atmospheric CO2: A Review on the Causes, Impacts, and Mitigation Strategies. *Environments (Basel, Switzerland)*, *10*(4), 66. DOI: 10.3390/environments10040066

Olawade, D. B., Wada, O. Z., David-Olawade, A. C., Fapohunda, O., Ige, A. O., & Ling, J. (2024). Artificial intelligence potential for net zero sustainability: Current evidence and prospects. *Next Sustainability*, *4*, 100041. DOI: 10.1016/j.nxsust.2024.100041

Oraby, G. A. E. M., & Orabi, A. A. E. M. (2025). Government Policies in Egypt and Indonesia to Mitigate Carbon Emissions: A Comparative Study. *Journal of Geographical Sciences and Education*, *3*(4), 343–356. DOI: 10.69606/geography.v3i4.371

Orkuma, M. (2025). The Geopolitics of Climate Change: Global Power Relations and the Pursuit of Climate Justice. .DOI: 10.14293/PR2199.002193.v1

Overpeck, J. T., Meehl, G. A., Bony, S., & Easterling, D. R. (2011). Climate Data Challenges in the 21st Century. *Science*, *331*(6018), 700–702. DOI: 10.1126/science.1197869 PMID: 21311006

Pandey, M., Mishra, V. N., Kumari, M., Janizadeh, S., & Costache, R. (2026). Editorial: Advanced geospatial data analytics for environmental sustainability: current practices and future prospects. *Frontiers in Remote Sensing*, *6*, 1761905. Advance online publication. DOI: 10.3389/frsen.2025.1761905

Patro, P., M, S., N, S., Das, S., V P, A. K., & R, D. (2025). Big Data-Based Remote-Sensed Intelligent Visual Analytics for Environmental and Earth Monitoring Systems. Journal of Applied Science and Technology Trends, 35–50. .DOI: 10.38094/jastt605589

Pavlova, O., Liashenko, O., Pavlov, K., Demianiuk, O., Vitkovskyi, Y., Jakóbik, K., Piwowarczyk, Z., & Karpinska, N. (2026). Utility-Based Evaluation of National Climate Policies: A Multi-Criteria Framework for Global Assessment. DOI: 10.20944/preprints202601.1360.v1

Priya, N. K., Khatri, A., Karki, P., Samota, S. D., Vishwakarma, S. K., Sukdeve, E. K., Tripathi, S., & Pathak, A. K. (2025). The Important Role of Extension Services in Strengthening the Capacity of Farmers' Resilience to Climate Change in India. *Journal of Experimental Agriculture International*, *47*(3), 204–223. DOI: 10.9734/jeai/2025/v47i33329

Rabbi, M. F., Popp, J., Máté, D., & Kovács, S. (2022). Energy Security and Energy Transition to Achieve Carbon Neutrality. *Energies*, *15*(21), 8126. DOI: 10.3390/en15218126

Raftowicz, M. (2021). The Climate Crisis as a Product of Globalization. SHS Web of Conferences, 92, 06029. DOI: 10.1051/shsconf/20219206029

Randeniya, J. N., Haigh, R., & Amaratunga, D. (2025). Responsible intelligence: Ethical AI governance for climate prediction in the Australian context. *AI and Ethics*, *6*(1), 81. Advance online publication. DOI: 10.1007/s43681-025-00909-z

Rashidi Mohammadi, A., Petrović, E. K., & Perkins, N. (2025). Mapping forestry practices against sustainable development goal targets. *Frontiers in Sustainability*, *6*, 1534449. Advance online publication. DOI: 10.3389/frsus.2025.1534449

Rasp, S., Pritchard, M. S., & Gentine, P. (2018). Deep learning to represent subgrid processes in climate models. *Proceedings of the National Academy of Sciences of the United States of America*, *115*(39), 9684–9689. DOI: 10.1073/pnas.1810286115 PMID: 30190437

Razavi-Termeh, S. V., Sadeghi-Niaraki, A., Ali, F., Pradhan, B., & Choi, S.-M. (2025). Optimizing ensemble learning for satellite-based multi-hazard monitoring and susceptibility assessment of landslides, land subsidence, floods, and wildfires. *Scientific Reports*, *15*(1), 30968. Advance online publication. DOI: 10.1038/s41598-025-15381-2 PMID: 40846725

Reichstein, M. (2017). *How Can Artificial Intelligence Enhance Our Understanding of the Earth System?* Latest Thinking., DOI: 10.21036/LTPUB10819

Reichstein, M., Benson, V., Blunk, J., Camps-Valls, G., Creutzig, F., Fearnley, C. J., Han, B., Kornhuber, K., Rahaman, N., Schölkopf, B., Tárraga, J. M., Vinuesa, R., Dall, K., Denzler, J., Frank, D., Martini, G., Nganga, N., Maddix, D. C., & Weldemariam, K. (2025). Early warning of complex climate risk with integrated artificial intelligence. *Nature Communications*, *16*(1), 2564. Advance online publication. DOI: 10.1038/s41467-025-57640-w PMID: 40089483

Reichstein, M., Camps-Valls, G., Stevens, B., Jung, M., Denzler, J., Carvalhais, N., & Prabhat, . (2019). Deep learning and process understanding for data-driven Earth system science. *Nature*, *566*(7743), 195–204. DOI: 10.1038/s41586-019-0912-1 PMID: 30760912

Rizwan, A., & Saleem, Y. (2025). Stakeholders Role to Combat Climate Change: Initiatives Adopted and Resource Integration in Response to Sustainable Development Goal 13 in Pakistan. Disasters, Climate Change and Sustainability in South Asia, 231–253. DOI: 10.1007/978-3-032-04750-2_11

Robinson, M., & Shine, T. (2018). Achieving a climate justice pathway to 1.5 °C. *Nature Climate Change*, *8*(7), 564–569. DOI: 10.1038/s41558-018-0189-7

Rodge, R. R., & Rajan, R. (2026). Frontier technologies and remote sensing applications for climate change mitigation and adaptation. Geospatial Techniques in Climate Change Monitoring and Environmental Mapping, 1–16. .DOI: 10.1016/B978-0-443-36396-2.00015-9

Rohde, F., Gossen, M., Wagner, J., & Santarius, T. (2021). Sustainability challenges of Artificial Intelligence and Policy Implications. Ökologisches Wirtschaften - Fachzeitschrift, 36(O1), 36–40. DOI: 10.14512/OEWO360136

Rolnick, D., Donti, P. L., Kaack, L. H., Kochanski, K., Lacoste, A., Sankaran, K., Ross, A. S., Milojevic-Dupont, N., Jaques, N., Waldman-Brown, A., Luccioni, A. S., Maharaj, T., Sherwin, E. D., Mukkavilli, S. K., Kording, K. P., Gomes, C. P., Ng, A. Y., Hassabis, D., Platt, J. C., & Bengio, Y. (2022). Tackling Climate Change with Machine Learning. *ACM Computing Surveys*, *55*(2), 1–96. DOI: 10.1145/3485128

Rudin, C. (2019). Stop explaining black box machine learning models for high stakes decisions and use interpretable models instead. *Nature Machine Intelligence*, *1*(5), 206–215. DOI: 10.1038/s42256-019-0048-x PMID: 35603010

Running, S. W., Nemani, R. R., Heinsch, F. A., Zhao, M., Reeves, M., & Hashimoto, H. (2004). A Continuous Satellite-Derived Measure of Global Terrestrial Primary Production. *Bioscience*, *54*(6), 547. DOI: 10.1641/0006-3568(2004)054[0547:ACSMOG]2.0.CO;2

Sarewitz, D., & Pielke, R. A.Jr. (2007). The neglected heart of science policy: Reconciling supply of and demand for science. Environmental Science &amp. *Environmental Science & Policy*, *10*(1), 5–16. DOI: 10.1016/j.envsci.2006.10.001

Schlosberg, D., & Collins, L. B. (2014). From environmental to climate justice: climate change and the discourse of environmental justice. WIREs Climate Change, 5(3), 359–374. Portico. .DOI: 10.1002/wcc.275

Sharifi, A. (2021). Co-benefits and synergies between urban climate change mitigation and adaptation measures: A literature review. *The Science of the Total Environment*, *750*, 141642. DOI: 10.1016/j.scitotenv.2020.141642 PMID: 32858298

Sharma, R., Bansal, R., Munna, A. S., & Mrabet, J. (2026). Reimagining Development: Environmental Sustainability and the Global Green Transition. Aligning ESG Goals with the UNs Sustainable Development Goals Planning, 249–263. DOI: 10.1007/978-3-032-01715-4_13

Sharman, N. (2022). Inter-State climate technology transfer under the UNFCCC: A benefit-sharing approach. Review of European, Comparative & International Environmental Law, 31(3), 435–446. Portico. DOI: 10.1111/reel.12454

Sheehan, M. C. (2022). 2021 Climate and Health Review – Uncharted Territory: Extreme Weather Events and Morbidity. *International Journal of Health Services*, *52*(2), 189–200. DOI: 10.1177/00207314221082452 PMID: 35229682

Singh, S. S., & Jeganathan, C. (2025). Examining the Role of Geospatial Technologies and AI in Disaster Risk Management and Response, Including Hazard Mapping, Vulnerability Assessment, and Real-Time Monitoring. Advanced Geospatial Intelligence and AI for Environmental Resilience and Sustainable Development, 335–366. DOI: 10.1007/978-3-032-03714-5_14

Skotnicky, P., Puccio, A., & Das, S. (2025). The AI Compass: Navigating Ethical Dilemmas in Tech-Driven Sustainability. Generative AI for a Net-Zero Economy, 111–128. DOI: 10.1007/978-981-96-8015-3_7

Soomro, J. A., Suhrab, M., & Abbasi, B. A. (2025). Role of Emerging Technologies in Promoting Sustainable Environment and Public Health. Environment and Public Health, 485–507. .DOI: 10.1007/978-3-031-99770-9_230

Stainforth, D. A., Allen, M. R., Tredger, E. R., & Smith, L. A. (2007). Confidence, uncertainty and decision-support relevance in climate predictions. Philosophical Transactions of the Royal Society A: Mathematical, Physical and Engineering Sciences, 365(1857), 2145–2161. DOI: 10.1098/rsta.2007.2074

Steckel, J. C., Jakob, M., Flachsland, C., Kornek, U., Lessmann, K., & Edenhofer, O. (2016). From climate finance toward sustainable development finance. WIREs Climate Change, 8(1). Portico. DOI: 10.1002/wcc.437

Stede, M., & Patz, R. (2021). The Climate Change Debate and Natural Language Processing. *Proceedings of the 1st Workshop on NLP for Positive Impact.* DOI: 10.18653/v1/2021.nlp4posimpact-1.2

Stephens, G. L., Slingo, J. M., Rignot, E., Reager, J. T., Hakuba, M. Z., Durack, P. J., Worden, J., & Rocca, R. (2020). Earth's water reservoirs in a changing climate. *Proceedings - Royal Society. Mathematical, Physical and Engineering Sciences, 476*(2236), 20190458. Advance online publication. DOI: 10.1098/rspa.2019.0458 PMID: 32398926

Stevens, B., & Bony, S. (2013). What Are Climate Models Missing? *Science, 340*(6136), 1053–1054. DOI: 10.1126/science.1237554 PMID: 23723223

Taheri Hosseinkhani, N. (2025). Artificial Intelligence and Large Language Models in Energy Systems and Climate Strategies: Economic Pathways to Cost-Effective Emissions Reduction and Sustainable Growth. SSRN Electronic Journal. DOI: 10.2139/ssrn.5385513

Tasha, S. (2025). A Review of Artificial Intelligence Applications in Climate Change Mitigation. *International Journal of Environment and Climate Change, 15*(5), 443–449. DOI: 10.9734/ijecc/2025/v15i54864

Trenberth, K. E., Fasullo, J. T., & Shepherd, T. G. (2015). Attribution of climate extreme events. *Nature Climate Change, 5*(8), 725–730. DOI: 10.1038/nclimate2657

Trivedi, A., & Juyal, S. A. (2026). Smart Cities and Urban Circularity: Advancing Sustainable Transformation through Digital Twins, AI, and Capacity Building. Green Horizons: Advancing Climate Solutions Through Green Finance, Circular Economy, and ESG, 339–356. DOI: 10.1007/978-3-032-03484-7_18

Ukoba, K., Onisuru, O. R., Jen, T.-C., Madyira, D. M., & Olatunji, K. O. (2025). Predictive modeling of climate change impacts using Artificial Intelligence: A review for equitable governance and sustainable outcome. *Environmental Science and Pollution Research International, 32*(17), 10705–10724. DOI: 10.1007/s11356-025-36356-w PMID: 40183866

Ullah, I., Ali, E., Osama, S., Murtaza, A., & Ullah, F. (2025). <b>Harnessing AI for Climate Change Mitigation: Strategies Challenges and Opportunities</b>. *Journal of Political Stability Archive, 3*(3), 963–977. . DOI: 10.63468/jpsa.3.3.65

Upadhyaya, A. N., Debnath, P., Rani, P., Selvakumar, P., Manjunath, T. C., & Bhattacharya, S. (2025). AI-Based Disaster Prediction and Early Warning Systems. AI and Emerging Technologies for Emergency Response and Smart Cities, 55–74. DOI: 10.4018/979-8-3693-9770-1.ch003

URL-1. UN National Inventory Reports. https://unfccc.int/process-and-meetings/transparency-and-reporting/reporting-and-review/reporting-and-review-under-the-paris-agreement/national-inventory-reports last access date: 24.01.2026

Uyar, N., & Uyar, A. (2025). Assessing Climate Change Impacts on Cropland and Greenhouse Gas Emissions Using Remote Sensing and Machine Learning. *Atmosphere (Basel), 16*(4), 418. DOI: 10.3390/atmos16040418

Vázquez-Canteli, J. R., & Nagy, Z. (2019). Reinforcement learning for demand response: A review of algorithms and modeling techniques. *Applied Energy, 235*, 1072–1089. DOI: 10.1016/j.apenergy.2018.11.002

Verma, S., Srivastava, K., Tiwari, A., & Verma, S. (2023). Deep learning techniques in extreme weather events: A review. arXiv preprint arXiv:2308.10995.

Vidhya, S., Sushith, M., Angelo, K. M., & Geetha, R. (2026). Climate Change Prediction Model on Smart Forestry 5.0 Using Novel Enhanced Streaming Gradient Boosted Trees-Based Optimisation Framework. Smart Forestry 5.0 for Powered AI Sustainable Environment and Climate, 75–102. .DOI: 10.1007/978-3-031-99086-1_5

Vinuesa, R., Azizpour, H., Leite, I., Balaam, M., Dignum, V., Domisch, S., Felländer, A., Langhans, S. D., Tegmark, M., & Fuso Nerini, F. (2020). The role of artificial intelligence in achieving the Sustainable Development Goals. *Nature Communications, 11*(1), 233. Advance online publication. DOI: 10.1038/s41467-019-14108-y PMID: 31932590

Walker, D. W., Vergopolan, N., Cavalcante, L., Smith, K. H., Agoungbome, S. M. D., Almagro, A., Apurv, T., Dahal, N. M., Hoffmann, D., Singh, V., & Xiang, Z. (2024). Flash Drought Typologies and Societal Impacts: A Worldwide Review of Occurrence, Nomenclature, and Experiences of Local Populations. *Weather, Climate, and Society*, *16*(1), 3–28. DOI: 10.1175/WCAS-D-23-0015.1

Wang, M., Fang, X., & Zhang, K. (2024). Exploring a pathway to optimise the carbon tax policy in terms of the economy, the environment and health: A scenario-based system dynamics approach. *Heliyon*, *10*(10), e31093. DOI: 10.1016/j.heliyon.2024.e31093 PMID: 38803911

Wang, M., Yu, H., Yang, Y., Lin, X., Guo, H., Li, C., Zhou, Y., & Jing, R. (2021). Unlocking emerging impacts of carbon tax on integrated energy systems through supply and demand co-optimization. *Applied Energy*, *302*, 117579. DOI: 10.1016/j.apenergy.2021.117579

Wang, X., Jiang, D., & Lang, X. (2018). Climate Change of 4°C Global Warming above Pre-industrial Levels. *Advances in Atmospheric Sciences*, *35*(7), 757–770. DOI: 10.1007/s00376-018-7160-4

Xia, S., Yang, Y., & Poon, J. P. H. (2025). Ensuring a carbon-neutral future for artificial intelligence. *The Innovation Energy*, *2*(1), 100071. DOI: 10.59717/j.xinn-energy.2024.100071

Xie, L., Huang, T., Zheng, X., Liu, Y., Wang, M., Vittal, V., Kumar, P. R., Shakkottai, S., & Cui, Y. (2022). Energy system digitization in the era of AI: A three-layered approach toward carbon neutrality. *Patterns (New York, N.Y.)*, *3*(12), 100640. DOI: 10.1016/j.patter.2022.100640 PMID: 36569552

Yao, T., Thompson, L., Yang, W., Yu, W., Gao, Y., Guo, X., Yang, X., Duan, K., Zhao, H., Xu, B., Pu, J., Lu, A., Xiang, Y., Kattel, D. B., & Joswiak, D. (2012). Different glacier status with atmospheric circulations in Tibetan Plateau and surroundings. *Nature Climate Change*, *2*(9), 663–667. DOI: 10.1038/nclimate1580

Yılmaz, O., & Alkan, M. (2026). The Potential of Machine Learning and Multi-Criteria Decision-Making for Data-Driven Land Policies in Türkiye. The International Archives of the Photogrammetry, Remote Sensing and Spatial Information Sciences, XLVIII-4/W17-2025, 339–344. DOI: 10.5194/isprs-archives-XLVIII-4-W17-2025-339-2026

Yu, Y., Gao, S., Wang, J., Cheng, Q., Deng, H., & Shen, Y. (2026). Unraveling climate–land cover interactions: How SSP–RCP scenarios drive ecosystem service trade-offs in contrasting Yangtze and Yellow River Basins, China. *Landscape Ecology*, *41*(2), 34. Advance online publication. DOI: 10.1007/s10980-026-02295-1

Zamil Adib, A. A., Chan, G., & Orji, R. (2025). Combining Data Visualization and Interactive Narrative: A Persuasive Approach to Raise Climate Change Awareness. Human Behavior and Emerging Technologies, 2025(1). Portico. .DOI: 10.1155/hbe2/7275480

Zhang, H., Wang, S., & Ding, J. (2025). Driving mechanism and nonlinear threshold identification of vegetation in China: Based on causal inference and machine learning. *Journal of Arid Land, 17*(10), 1341–1360. DOI: 10.1007/s40333-025-0110-y

Zhang, S., Bauer, N., Luderer, G., & Kriegler, E. (2014). Role of technologies in energy-related CO2 mitigation in China within a climate-protection world: A scenarios analysis using REMIND. *Applied Energy, 115*, 445–455. DOI: 10.1016/j.apenergy.2013.10.039

Zhao, S., Qing, L., & Wang, C. (2026). Editorial: Moving towards sustainable development: exploring the impact of land-use policies on land green utilization efficiency. *Frontiers in Environmental Science, 14*, 1767889. Advance online publication. DOI: 10.3389/fenvs.2026.1767889

Zhong, L., Hu, L., & Zhou, H. (2019). Deep learning based multi-temporal crop classification. *Remote Sensing of Environment, 221*, 430–443. DOI: 10.1016/j.rse.2018.11.032

Zhu, X. X., Tuia, D., Mou, L., Xia, G.-S., Zhang, L., Xu, F., & Fraundorfer, F. (2017). Deep Learning in Remote Sensing: A Comprehensive Review and List of Resources. *IEEE Geoscience and Remote Sensing Magazine, 5*(4), 8–36. DOI: 10.1109/MGRS.2017.2762307

Ziervogel, G., Cowen, A., & Ziniades, J. (2016). Moving from Adaptive to Transformative Capacity: Building Foundations for Inclusive, Thriving, and Regenerative Urban Settlements. *Sustainability (Basel), 8*(9), 955. DOI: 10.3390/su8090955

Zolfaghari, A. (2026). Role of Technology and Innovation in Sustainable Entrepreneurship. Sustainable Entrepreneurship, 121–140. .DOI: 10.1007/978-3-032-04909-4_7

Zscheischler, J., Westra, S., van den Hurk, B. J. J. M., Seneviratne, S. I., Ward, P. J., Pitman, A., AghaKouchak, A., Bresch, D. N., Leonard, M., Wahl, T., & Zhang, X. (2018). Future climate risk from compound events. *Nature Climate Change, 8*(6), 469–477. DOI: 10.1038/s41558-018-0156-3

Chapter 2

AI-Enhanced Bibliometric Mapping of Climate Change and Social Justice Research in Africa (1990-2023)

Fatma Şencan
http://orcid.org/0009-0007-3223-6463
Medipol Üniversitesi, Turkey

ABSTRACT

This study performs a bibliometric analysis to systematically map the scholarly production at the intersection of climate change and social justice research focused on the African context from 1990 to 2023. Utilizing a rigorous, data-driven approach, the analysis is based on 145 documents retrieved from the Scopus database. Both performance and structural analyses were executed, including scientific output mapping, productivity analysis, and a one-way Analysis of Variance (ANOVA) test for statistical impact assessment. The findings reveal that this research domain has evolved from a minimal concern in the 1990s to a rapidly accelerating and vital research frontier, marked by a significant surge in publications after 2010. The core intellectual structure is predominantly organized around justice-based concepts, with terms like "climate justice," "environmental justice," "gender," "energy justice," and "sustainable development" serving as the dominant thematic hubs.

DOI: 10.4018/979-8-2600-0778-5.ch002

1 INTRODUCTION

Climate change represents one of the most critical and complex challenges of the 21st century (Cohen and Waddell, 2009; Iqbal and Ghauri, 2011; Çelik et al., 2026a), transcending environmental degradation to become a fundamental issue of socio-economic development and justice (Oreskes, 2004). While the physical impacts of climate change such as rising sea levels, extreme weather events, and resource scarcity are globally recognized, the resultant burdens are distributed highly unevenly. This disparity forms the core of the climate justice debate, which posits that those least responsible for historical emissions often bear the greatest consequences (Schlosberg, 2007).

The continent of Africa is consistently highlighted as being particularly vulnerable to the effects of climate change due to its reliance on rain-fed agriculture, high levels of poverty, limited institutional capacity, and low adaptive capabilities (Adisa et al., 2020; Intergovernmental Panel on Climate Change, 2023; Amsalu et al., 2026). Consequently, issues of social justice, equity, and human rights are intrinsically linked to climate action across African nations. Climate-induced disasters disproportionately affect marginalized groups, including women, indigenous communities, and the poor, exacerbating existing inequalities and threatening peace and stability (Adger, 2006). Despite this critical vulnerability and the urgent need for context-specific solutions, the visibility and thematic integration of social justice within climate change research emanating from Africa in the global academic discourse remains a subject requiring systematic scrutiny (Leiserowitz, 2005).

Existing academic literature offers numerous qualitative reviews and case studies on African climate policy and social vulnerability (Thornton et al., 2004; Tucker et al., 2014; Dumenu and Obeng, 2016). However, there is a distinct lack of large-scale, quantitative assessment that comprehensively maps the intellectual landscape, thematic evolution, and collaborative structures of this interdisciplinary research domain over an extended period. Understanding the growth trajectory, core research clusters, influential authors, and key knowledge gaps in the "Climate Change - Social Justice - Africa" nexus is essential for future research prioritization and evidence-based policymaking (Wang et al., 2018).

This need necessitates a rigorous, data-driven approach. Bibliometric analysis is a powerful methodology that provides statistical insights into the research output, impact, and knowledge structure of a field, allowing for an objective evaluation of trends and patterns in scholarly communication (van Eck and Waltman, 2009).

In response to this gap, this study performs a bibliometric analysis to systematically map the scholarly production at the intersection of climate change and social justice research focused on Africa.

Specifically, the primary objectives of this study are:

- To analyze the evolutionary trajectory and overall research output on the nexus between climate change and social justice in Africa from 1990 to 2023.
- To identify the most influential contributors (authors, institutions, countries) and the knowledge collaboration networks driving this research.
- To determine the key thematic clusters and intellectual structure of the field through co-citation and co-word analysis.
- To highlight the emerging research frontiers and persistent gaps in the current body of literature to guide future research agendas.

1.1 Why is it Important to Focus on Climate Change and Social Justice in Africa?

Focusing on the intersection between climate change and social justice in the African context is central to current global issues, both scientifically and ethically. The relevance of this area of research stems from the depth of the climate crisis's impact on the continent and the way these impacts intersect with existing socioeconomic inequalities (Adebisi et al., 2025; Bailey and Okoduwa, 2025).

The African continent stands out as one of the most vulnerable regions to the impacts of climate change (Nwokolo et al., 2023; Omotoso and Omotayo, 2023). This vulnerability is fueled by structural factors such as the continent's high dependence on rain-fed agriculture, high levels of poverty, and limited institutional and adaptive capacity (Hope, 2009).

Double Jeopardy: Climate change-related risks such as drought, flooding, food insecurity, and livelihood precariousness intersect with the deep-rooted socioeconomic inequalities already present on the continent (Adaawen et al., 2019). This intersection transforms the climate crisis from a mere environmental problem into a fundamental development and social justice challenge (Williams et al., 2023; Apraku et al., 2025).

Global Justice Principle: A focus on social justice underpins the climate justice debate (Schlosberg and Collins, 2014). This argument argues that those least responsible for historical emissions (most African countries and communities) bear the greatest burden of the consequences. This creates an ethical imperative centered on the principles of global responsibility and reparations.

The impacts of climate change do not affect everyone equally (Addaney, 2025; Chari and Munoriyarwa, 2025). Social justice research (Mugambiwa, 2025) is vital to understanding who suffers the most and why:

Marginalized Groups: Climate-related disasters disproportionately impact marginalized groups, including women, indigenous communities, and the poor.

A Justice Framework: Concepts such as climate justice and environmental justice, which underpin the research, confirm that academic discourse has successfully shifted from merely diagnosing environmental impacts to exploring the ethical and distributional dimensions of the crisis. This perspective requires that climate action also address existing inequalities based on gender, class, and ethnicity.

A social justice lens is a fundamental requirement in climate policy design.

Developing Equitable Solutions: Research has shifted in recent years from diagnosing climate vulnerabilities to exploring sustainable and equitable transformation pathways, focusing on themes such as energy justice, sustainable development, and land-use change.

Policy Translation: Adaptation and mitigation strategies designed without a social justice focus can increase the risk of maladaptation. Therefore, national governments need to formally incorporate climate justice principles into their adaptation and mitigation strategies.

Data-Driven Impact: High-impact (citation) studies are often interdisciplinary and integrative studies that successfully integrate areas such as health, governance, rights, and resilience with climate change, demonstrating that such justice-centered research is a fundamental tool for informing effective and equitable climate action in Africa.

In summary, focusing on climate change and social justice in Africa not only reflects current scientific trends but also provides the critical, ethical and socio-political framework necessary for decision-makers to design just, effective and inclusive responses to the multi-layered crises facing the region.

2 DATA AND METHODS

This analysis is based on **145 documents** retrieved from the Scopus database, focusing on the intersection of climate change and social justice research in the African context between 1990 and 2023.

The bibliometric dataset used in this study was retrieved from the Scopus database, one of the most comprehensive indexing platforms for peer-reviewed scientific literature. Scopus was selected due to its extensive coverage of multidisciplinary research and its suitability for bibliometric analyses.

The literature search was conducted on 4 December 2025 using the following query applied to the Title, Abstract, and Keywords fields:

TITLE-ABS-KEY (“climate change” AND “social justice” AND Africa)

This search strategy was designed to capture scholarly publications addressing the intersection between climate change, social justice, and the African context. Only documents indexed in Scopus at the time of the search were included in the dataset. The retrieved records were subsequently exported in compatible formats for bibliometric analysis, including metadata such as authors, publication year, source title, affiliations, keywords, and citation information. These data formed the basis for the performance analysis, thematic mapping, and statistical evaluations conducted in the study.

2.1 Data Source and Collection Strategy

The dataset for this bibliometric analysis was sourced from the Scopus database, one of the most comprehensive repositories of peer-reviewed literature (Erfanmanesh, 2017; Celik, 2020; Çelik and Çelik, 2024). The data collection strategy was focused on identifying the intersection of "Climate Change" and "Social Justice" research within the "Africa" context to ensure high relevance to the research questions. The bibliometric fields extracted for analysis included Authors, Publication Year, Source Title (Journal/Book), Cited by Count, Author Keywords, and Index Keywords. The initial search was performed using a combination of key terms in the document title, abstract, and keywords. The analysis was strictly confined to the time period of 1990 to 2023 to map the historical evolution of the field (Table 1).

Table 1. Bibliometric Search and Dataset Characteristics

Criteria	Value
Database	Scopus
Analysis Period	1990 – 2023
Document Types Included	Article, Review, Book, Book Chapter, Editorial
Total Records Retrieved	197
Final Records Analyzed (1990-2023)	145

2.2 Bibliometric Analysis Methods

To comprehensively map the structure, impact, and evolution of the literature, both performance and structural analyses were performed, complemented by statistical testing.

This study employs a combination of bibliometric mapping techniques and artificial intelligence–assisted analytical tools to examine the intellectual structure of research on climate change and social justice in Africa. After retrieving the bibliographic records from the Scopus database, the dataset was cleaned and organized by removing duplicate entries and standardizing author names, keywords, and institutional affiliations.

Bibliometric network analyses were conducted using VOSviewer. This software was used to generate visualization maps of keyword co-occurrence, co-authorship networks, and thematic clusters, allowing the identification of major research themes, collaboration patterns, and emerging topics within the field. Network visualization techniques such as clustering and link strength analysis were applied to reveal the intellectual structure and thematic evolution of the literature. VOSviewer and artificial intelligence–based tools have been widely used in numerous bibliometric studies to analyze large volumes of scientific literature, identify collaboration networks, and detect emerging research themes (Yu et al., 2020; Çelik et al., 2026b). Their ability to process complex bibliographic datasets and generate visual knowledge maps makes them particularly effective for exploring the intellectual structure and evolution of research fields. Consequently, the integration of bibliometric visualization software such as VOSviewer with artificial intelligence–assisted analytical approaches has become increasingly common in recent bibliometric research.

In addition, artificial intelligence–assisted tools were used to support data pre-processing, text normalization, and thematic interpretation of large bibliographic datasets. These tools facilitated efficient handling of metadata, identification of semantic relationships between keywords, and improved classification of thematic clusters. The integration of bibliometric mapping and AI-supported analytical approaches enabled a comprehensive and systematic exploration of research trends related to climate change and social justice in Africa.

2.2.1 Performance and Impact Analysis

1. Annual Scientific Production: The chronological evolution of research was mapped by counting publications per year. The period was subsequently segmented into two sub-periods (1990-2010: Early Period and 2011-2023: Acceleration Period) to analyze thematic shifts over time.
2. Productivity Analysis: The Top 10 Most Productive Sources (journals or books) and Top 10 Most Productive Authors were identified by counting the frequency of their appearances in the filtered dataset.
3. Impact Analysis: The academic influence of the literature was assessed by identifying the Top 10 Most Cited Documents based on the total number of citations received up to the data collection date.

2.2.2 Structural and Thematic Analysis

1. Thematic Hub Analysis: The principal research themes were identified using frequency analysis of both Author Keywords and Index Keywords. The Index

Keywords were used to proxy the overall subject categories of the published works.

2. Thematic Evolution: To capture the changing focus of the field, a comparative analysis of the most frequent Index Keywords was performed across the defined two periods (1990-2010 vs. 2011-2023).
3. Keyword Co-occurrence Network Data: A co-occurrence analysis was conducted on the Index Keywords to generate a dataset of keyword pairs and their simultaneous appearance frequency. This dataset provides the necessary input for visualizing the thematic structure and clustering of the research field using a network map.

2.2.3. Statistical Analysis (ANOVA)

A One-way Analysis of Variance (ANOVA) test was employed to statistically determine if the mean citation counts differed significantly across various document types with sufficient sample sizes (Jiang et al., 2021; Yeh et al., 2022).

- Dependent Variable: Cited by Count (measure of impact)
- Independent Variable (Factor): Document Type (e.g., Article, Review, Book)
- Significance Level: alpha = 0.05

The ANOVA test examined the null hypothesis: that the mean citation counts across all valid document types are equal.

2.3. Software and Visualization

All data manipulation, cleaning, filtering, and statistical calculations, including the ANOVA test and co-occurrence matrix generation, were executed using the Python programming language, relying heavily on the Pandas library. All visual charts and plots (Annual Production, Sources, Keyword Hubs, and the ANOVA Box Plot) were generated using the Altair visualization library. The raw co-occurrence data for the network analysis was exported in CSV format to be compatible with specialized network visualization software (e.g., VOSviewer or Gephi).

The initial phase of the bibliometric study, encompassing data acquisition and preprocessing, was meticulously designed to ensure the validity and relevance of the final dataset. The reliance on the Scopus database is a key methodological choice, guaranteeing comprehensive coverage of peer-reviewed scientific literature and reliable citation metrics.

The search strategy itself was highly targeted, focusing on the explicit intersection of "Climate Change," "Social Justice," and "Africa." This precision ensured that the resulting dataset is directly relevant to the core research question: mapping the nexus between these three critical components. The extracted fields including Authors, Year, Cited by Count, and Keywords are standard components necessary for executing comprehensive performance, impact, and structural analyses.

The time frame, set between 1990 and 2023, is strategically chosen to capture the field's entire history, from its nascent stages in the late 20th century to its contemporary acceleration. The rigorous filtering process reduced the initial retrieval of 197 records to a core set of 145 documents for final analysis, excluding certain document types and records outside the chronological scope. This step is crucial for maintaining the quality and comparability of the analyzed units (Table 2).

Finally, the subsequent segmentation of the data into the Early Period (1990–2010) and the Acceleration Period (2011–2023) is a critical design feature. This division allows the study to move beyond simple output counting and perform a comparative thematic evolution analysis, providing deep insight into when and how the core conceptual and geographical focus of climate and social justice research shifted on the African continent.

Table 2. Data Acquisition and Preprocessing Steps

Step	Process	Tools / Output
1.1. Data Collection & Scoping	Perform targeted search in Scopus for "Climate Change," "Social Justice," and "Africa" (1990–2023).	Scopus Database (197 Records)
1.2. Filtering & Cleaning	Filter records by analysis period (1990–2023) and document type (Article, Review, Book, etc.). Execute data cleaning and keyword normalization.	Python (Pandas) 145 Final Records
1.3. Period Segmentation	Divide the 145 records into two sub-periods for evolutionary analysis.	**Period 1:** 1990–2010 (Early Period) **Period 2:** 2011–2023 (Acceleration Period)

Workflow chart given in **Fig. 1** systematically outlines the methodological process for the bibliometric study titled "Mapping the Nexus: A Bibliometric Analysis of Climate Change and Social Justice Research in Africa (1990–2023)." The study is divided into three distinct and sequential phases: Data Acquisition & Preprocessing (Phase 1), Analytical Methods (Phase 2), and Segmentation for Evolutionary Analysis (Phase 3), leading finally to the Results & Discussion.

Phase 1: Data Acquisition & Preprocessing

The process begins with Data Collection & Scoping (1.1), where a highly targeted search using the intersecting keywords "Climate Change," "Social Justice," and "Africa" is executed in the Scopus Database for the timeframe 1990–2023. This initial retrieval yields 197 records. This step is crucial as it defines the universe of the analysis.

Next is Filtering & Cleaning (1.2). The records are refined by including specific document types (Article, Review, Book, etc.) and undergoing data cleaning using Python (Pandas), resulting in the final, high-quality dataset of 145 records. The loss of 52 records indicates a rigorous selection process to maintain relevance and consistency.

Phase 2: Analytical Methods

The final 145 documents are subjected to a comprehensive set of analyses to address the research objectives:

1. Performance & Impact Analysis (2.1): This pillar focuses on who is producing the research and what its academic influence is. It involves calculating annual scientific output, identifying top productive authors and sources, and assessing impact through citation counts.
2. Structural & Thematic Analysis (2.2): This pillar maps the intellectual structure of the field using keywords. Key methods include thematic hub analysis, and the generation of a Keyword Co-occurrence Network Data matrix, which is the necessary input for visualization software like VOSviewer or Gephi.
3. Statistical Analysis (ANOVA) (2.3): A specialized step involving a One-way Analysis of Variance (ANOVA) test. This is used to statistically test for significant differences in mean citation counts (impact) across various document types, adding a layer of statistical rigor to the impact findings.

Phase 3: Segmentation for Evolutionary Analysis

A key feature of the methodology is the strategic division of the 145 records into two periods: 1990–2010 (Early Period) and 2011–2023 (Acceleration Period). This segmentation is not a separate process but a design feature integrated into the analysis. It enables the comparison of thematic trends and productivity between the field's nascent stages and its more recent, accelerated development phase, providing deep insight into thematic evolution.

The entire process is heavily reliant on the Python programming language (Pandas, Altair) for data manipulation, statistical testing, and generating visual charts. Specialized software like VOSviewer/Gephi is used for the advanced visualization of the co-occurrence network, ensuring robust and professional output (van Eck and Waltman, 2017)

The provided workflow diagram clearly illustrates a robust, three-phased methodology for a bibliometric analysis focusing on the intersection of climate change and social justice research in Africa between 1990 and 2023.

The methodology begins with a highly targeted data acquisition process from the Scopus database, which initially yielded 197 records. This number was then rigorously filtered and cleaned using Python's Pandas library to establish a final, reliable dataset of 145 documents, ensuring the research's high relevance.

The study employs a dual analytical approach: Performance Analysis assesses the field's productivity and academic impact through metrics like annual output and citation counts, while Structural and Thematic Analysis maps the intellectual core using keyword co-occurrence networks. This structure allows the study to answer both “who” is publishing and “what” the core research topics are.

A critical element of the design is the use of a One-way ANOVA test for statistical rigor, specifically checking if the impact (citation count) differs significantly based on the document type. Furthermore, the dataset is strategically divided into two periods an Early Period (1990–2010) and an Acceleration Period (2011–2023) to conduct a meaningful evolutionary analysis, revealing how the field's focus has shifted over time. The entire process relies on Python for processing and specialized tools (like VOSviewer) for visualization, confirming a high standard of data handling and analytical output, a methodology widely adopted in numerous research studies (José de Oliveira et al., 2019; van Eck and Waltman, 2021; Arruda et al., 2022; Bukar et al., 2023).

Figure 1. Workflow chart of study.

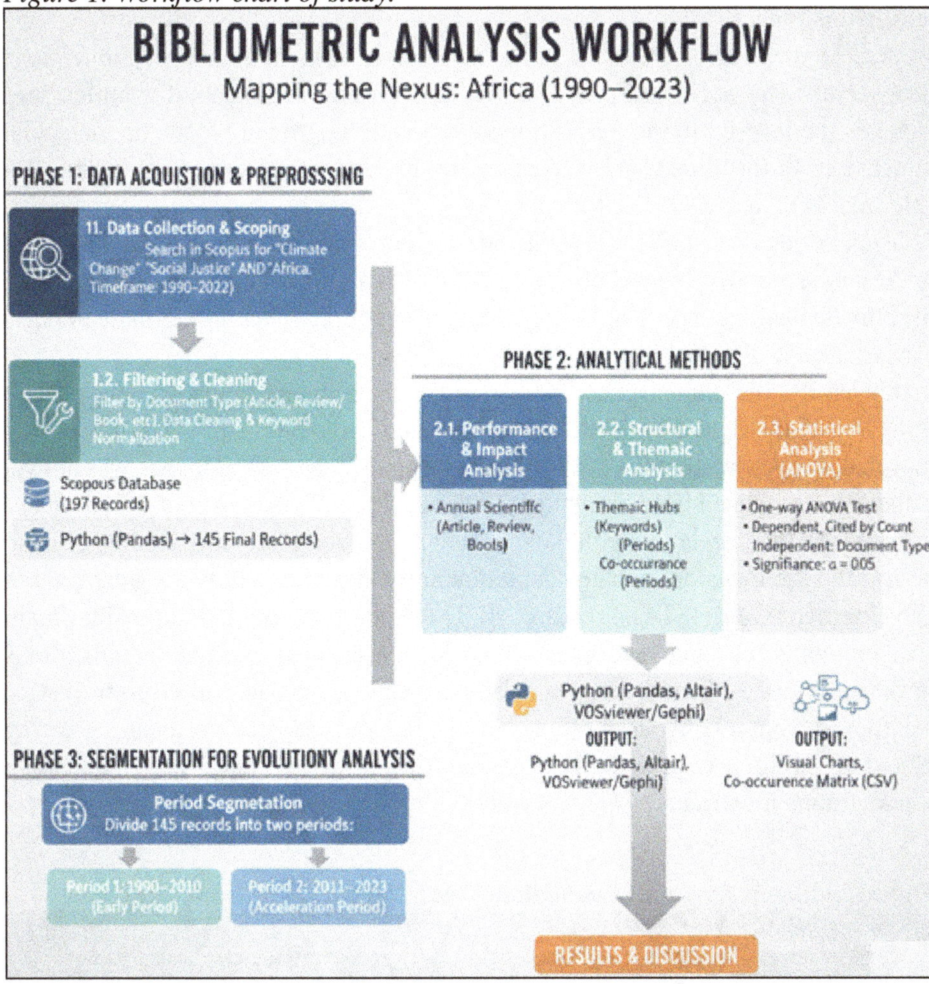

3 RESULTS

Figure 2 illustrates the evolution of research output on the climate change and social justice nexus in Africa over time. The trend in the chart clearly indicates a significant surge in scientific interest and production in this field over the past two decades.

Early Period (1990s - Mid-2000s): The number of publications remained very low or non-existent. This suggests that the intersection of climate change and social justice was not a primary academic focus, or these topics were studied in isolation during this time.

Acceleration Period (Post-2010): A marked increase in the number of publications is observed. This acceleration is likely driven by global climate policy milestones (such as the lead-up to the Paris Agreement) and the growing focus on the social dimensions of the climate crisis, particularly in the context of the Sustainable Development Goals (SDGs).

Peak Period (Recent Years): Publication counts reach their peak towards the end of the analysis period (especially post-2020). This aligns with the mainstreaming of the climate justice concept and increasing global concern over how climate impacts exacerbate social inequalities in Africa.

Generally, the annual publication trend reveals a clear and dynamic growth trajectory in the field. For most of the 1990s and early 2000s, research output was minimal, suggesting that the nexus of climate change and social justice was not yet a distinct, formalized field of inquiry. The scientific activity began to gain momentum after **2010**, correlating with the increasing global urgency surrounding climate policy and the focus on climate adaptation and mitigation within the **Sustainable Development Goals (SDGs)** framework. The field experienced its most significant acceleration in the final years of the study, with a substantial surge in publications in the early 2020s. This latest peak confirms that the concept of **climate justice** has moved from a fringe topic to a central academic concern, reflecting growing global recognition of how climate change disproportionately affects vulnerable social groups in Africa.

Figure 2. Annual Scientific Production (1990-2023)

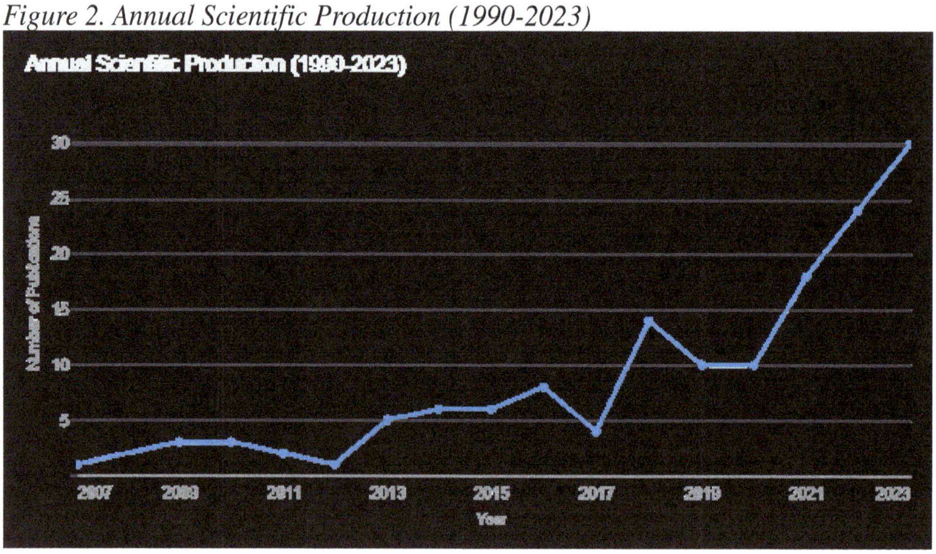

Bradford's Law describes how scientific literature on a given topic is distributed across journals. When journals are ranked by productivity (articles published), they form three concentric "zones," each containing roughly one-third of the total articles but requiring increasingly more journals to accumulate the same output.

In the **Fig. 3**:

- The first red dashed line represents the boundary of Zone 1 (Bradford Core) a very small number of highly productive journals that publish the largest share of the articles.
- The second line marks the end of Zone 2, consisting of a larger group of moderately productive journals.
- The third line indicates Zone 3, containing a large number of journals that each publish only a small number of papers.

The steep initial slope shows that the core journals contribute disproportionately to the total number of papers. As the curve progresses, the slope becomes more gradual, illustrating that additional journals contribute fewer articles individually, consistent with Bradford's theoretical productivity pattern.

This result confirms that the dataset follows a classical Bradford distribution, highlighting:

- A small elite group of core journals dominating the publication landscape.
- A broad long-tail distribution of journals with lower productivity.
- A typical "scatter phenomenon" where publications become increasingly dispersed across many outlets

Figure 3. Bradford law distribution of journals.

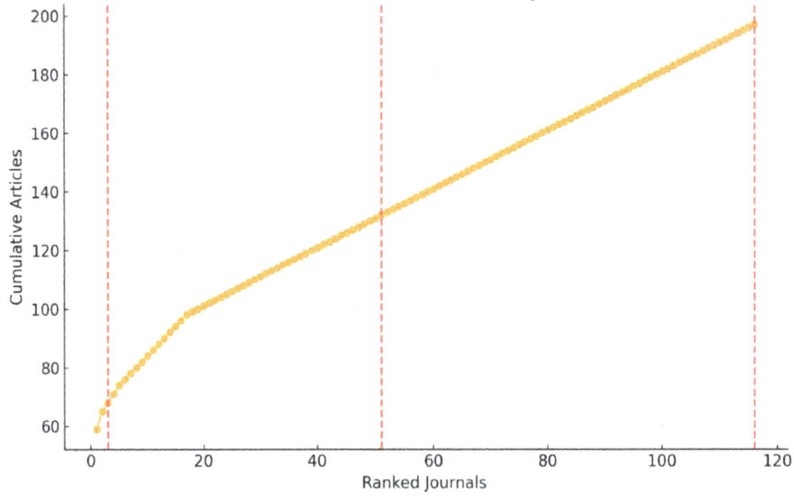

Fig. 4 identifies the most frequent and influential journals or book titles where researchers published their work The source distribution highlights the highly interdisciplinary nature of the research:

Sustainability Focus: Journals like Sustainability (Switzerland) and Journal of Cleaner Production lead the list. This indicates that a large portion of the research is framed within sustainable development and environmental management paradigms.

Energy and Social Sciences: The presence of titles like Energy Research and Social Science and Agenda (likely a social sciences/policy journal) underscores the importance of studies focusing on the social consequences of climate policies and energy transitions.

Health and Human Rights: The inclusion of high-impact journals such as The Lancet and the South African Journal on Human Rights demonstrates that critical work addresses the social justice dimension of climate change impacts on public health and legal/human rights frameworks.

High-Impact Visibility: The presence of Nature indicates that some research in this specific African nexus achieves high visibility within the top-tier global scientific community.

Figure 4. Top 10 Most Productive Sources (1990-2023).

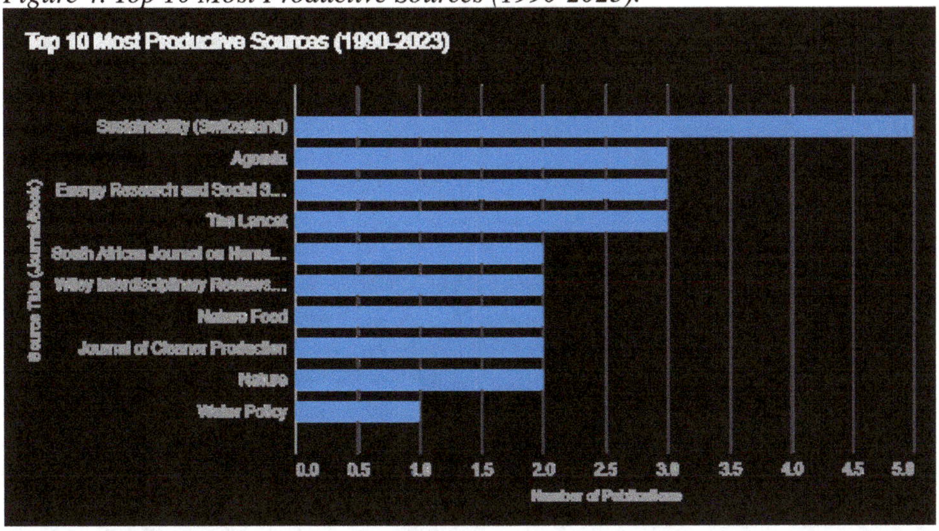

The analysis of the most cited documents clearly reveals the foundational papers and key thematic shifts that have shaped the field of climate change and social justice research (**Fig. 5**).

The list is dominated by a few highly influential papers that established the research agenda. The most cited document, "Managing the health effects of climate change..." (2009), from The Lancet and UCL, carries a disproportionately high citation count. This landmark review was crucial in positioning the topic as a major public health and social justice issue globally, thereby setting the stage for subsequent research, including work focusing on Africa. Furthermore, highly cited theoretical works like "Environmental justice: Concepts, evidence and politics" (2012) and "Dynamic sustainabilities..." (2010) underscore that the field's high impact is deeply rooted in robust theoretical and conceptual frameworks that successfully bridge environment, technology, and social equity.

Research explicitly focused on the continent, such as "Why is socially-just climate change adaptation in sub-Saharan Africa so challenging?..." (2015) and "Climate change adaptation and inequality in Africa..." (2021), highlights that empirical, region-specific challenges related to adaptation, energy, and resource security are areas of sustained influence. This impact is further confirmed by the high citation counts of papers related to specific thematic hubs: Energy Justice (e.g., "Contextualizing the Covid-19 pandemic..." (2020) and "Community acceptance of large-scale solar energy..." (2016)) and Gender and Water (e.g., "Women and water management in times of climate change..." (2013)). These documents underscore the critical need for intersectional research focused on the social equity of energy transitions and resource access for vulnerable groups.

In essence, the most impactful documents are those that either successfully established the foundational theoretical connection between climate and justice or provided critical, empirically grounded analysis on the implementation challenges of just transitions in the African context.

Figure 5. Top 10 most cited documents

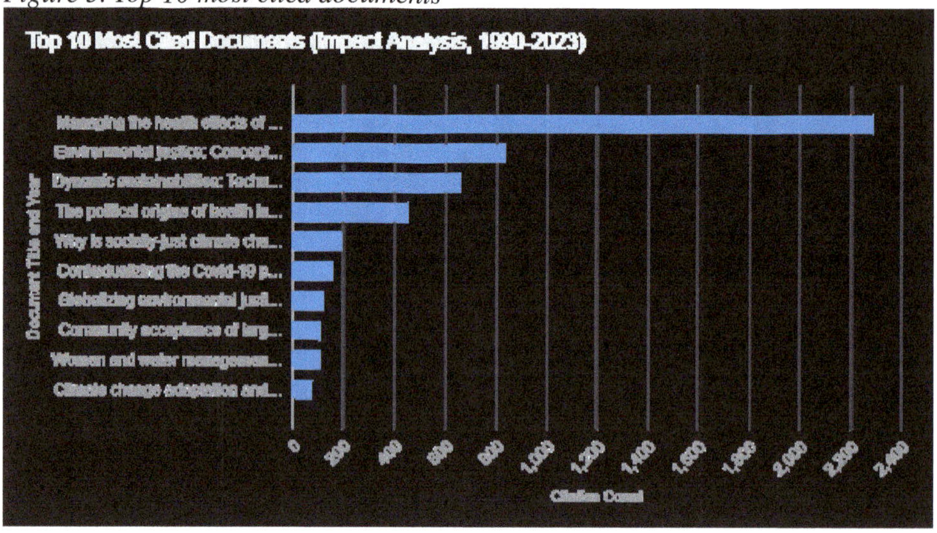

The most dominant thematic cluster is centered on the intersection of "climate change" and "social justice" and "environmental justice". This high-frequency grouping demonstrates that the core research community shares common references and focuses on the ethical and equity dimensions of the climate crisis. The frequent appearance of "human" and "humans" further emphasizes that this cluster is predominantly concerned with the anthropogenic and social impacts of these issues (**Fig. 6**).

A strong Geographic Cluster is evident through the prominence of "south africa" and "africa". South Africa emerges as the most widely referenced and researched case study, confirming it as a key nexus for academic collaboration and comparison within the continent. This cluster is supported by the presence of "sub-saharan africa", highlighting a broader regional focus. Finally, a separate Sectoral and Policy Cluster is established by keywords such as "agriculture" and "sustainable development". This cluster indicates significant co-referenced research focusing on practical applications, particularly concerning food security, livelihoods, and broader sustainable policy frameworks affected by climate change.

In summary, the field structurally organizes itself around three major clusters: theoretical justice frameworks, geographical case studies (led by South Africa), and sectoral issues like agriculture and sustainability.

The thematic hub analysis of author keywords provides a clear map of the research field's core concerns. While "climate change" is the most frequent term, it is immediately followed by a robust cluster of justice concepts: "environmental justice," "climate justice," and "social justice." This pattern unequivocally confirms that the primary academic focus extends beyond climate impacts to the crucial ethical and distributional aspects of the crisis, establishing these justice terms as the theoretical backbone of the field. Geographically, "south africa" emerges as the second most frequent keyword, positioning the country as the dominant case study and academic locus within the continent, likely owing to its deep structural inequalities and energy-intensive economy. The simultaneous presence of "africa" as a major keyword confirms a balanced approach that incorporates both specific national case studies and broader, continental-scale analyses. Furthermore, key intersecting themes include "sustainable development" and, crucially, "energy justice" and "gender." The high frequency of "energy justice" demonstrates that policy research is intensely focused on the equity issues surrounding Africa's necessary energy transition. Similarly, the focus on "gender" highlights a strong research interest in understanding and mitigating how climate vulnerabilities are disproportionately shaped by gender-based inequalities. In summary, the bibliometric landscape reveals a rapidly maturing, interdisciplinary field rooted in the core principles of social and environmental equity, with concentrated efforts on South Africa, energy systems, and gender dynamics (**Fig. 6**).

Figure 6. Top 10 most frequent keywords (upper) and top 10 most frequent author keywords (Thematic Hub Analysis) (bottom).

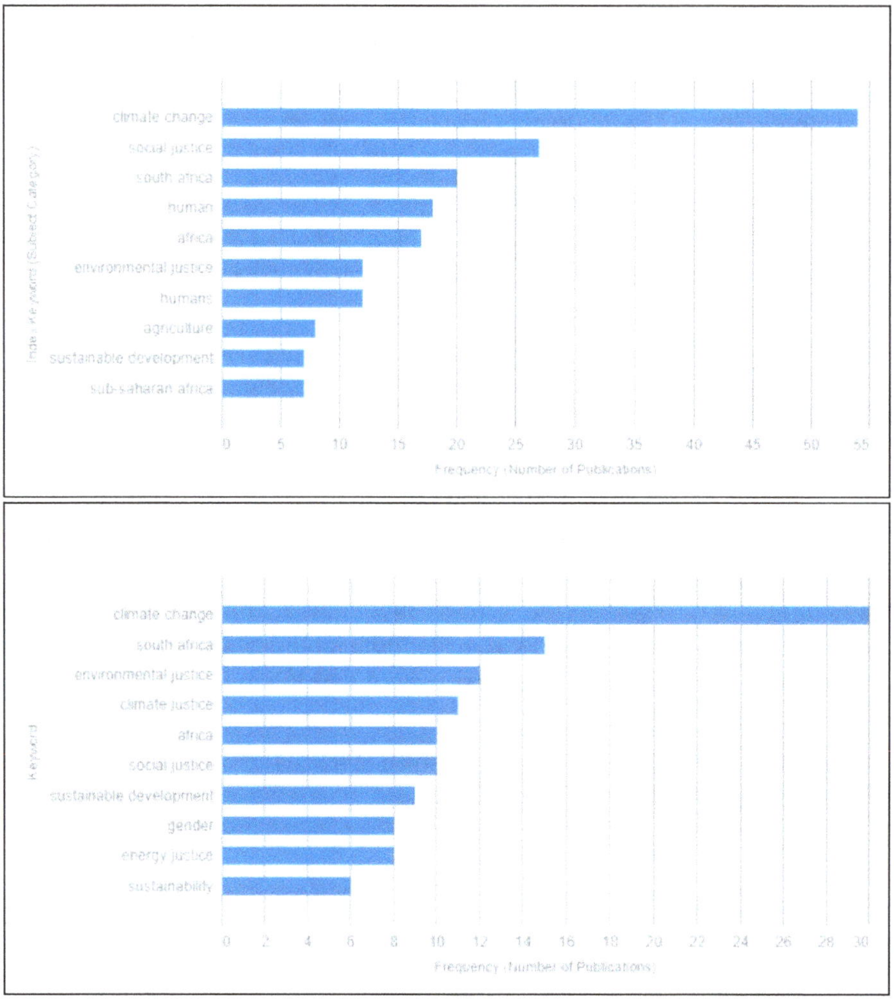

The four-panel keyword distribution reveals a clear and meaningful evolution in thematic priorities across the literature over the last two decades (**Fig. 7**). Although each period exhibits its own dominant concepts, the overall trend shows a strong and steady convergence toward climate-related, social, and environmental justice discourses.

In the 2005–2009 period, the vocabulary is heavily shaped by themes such as *health*, *climate*, *change*, and *emissions*. This indicates that early scholarly focus was primarily oriented toward the direct environmental and public-health implications of climate change. Development-related terms appear but remain secondary, reflecting an emerging not yet fully dominant concern for socio-economic dimensions.

During 2010–2014, the panel demonstrates a noticeable shift. While *climate*, *health*, and *change* remain prominent, keywords like *justice*, *political*, *policy*, and *Africa* rise significantly. This transition suggests that climate change began to be framed more explicitly as a political and ethical issue, with growing attention to inequality, governance, and regional vulnerability in the Global South. The appearance of terms such as *environmental* and *international* underscores increasing global coordination and policy engagement in climate discourse.

Between 2015 and 2019, the emphasis on *climate* becomes even more dominant, far surpassing other high-frequency terms. Themes such as *development*, *social*, *carbon*, and *environmental* reflect a more integrated narrative combining climate science with socio-economic transformation and decarbonization strategies. The appearance of *research*, *world*, and *economic* suggests a period of rapid expansion in interdisciplinary climate studies, linking environmental change with global development pathways.

From 2020 to 2024, the trend intensifies. *Climate* and *change* reach their highest frequencies across all periods, indicating that the climate crisis has become the central anchor of contemporary academic production. New terms *energy*, *sustainable*, *transition*, *land* highlight the rapid rise of energy transitions, low-carbon pathways, and land-use change as defining elements of recent research. *Social*, *Africa*, *justice*, and *water* remain highly relevant, demonstrating persistent concern for regional vulnerability, resource inequality, and environmental justice frameworks.

Overall, the combined 5-year keyword panel illustrates a clear thematic transformation: from early climate–health–emissions concerns toward a far more integrated research landscape that links climate change with social justice, political structures, resource management, sustainable development, and energy transition. The progression reflects both the maturation of climate science and the expansion of its socio-political implications, positioning climate change as the dominant cross-cutting issue of contemporary global scholarship.

Figure 7. Comparative Thematic Evolution via Frequent Keyword Distribution (5-Year Segments)

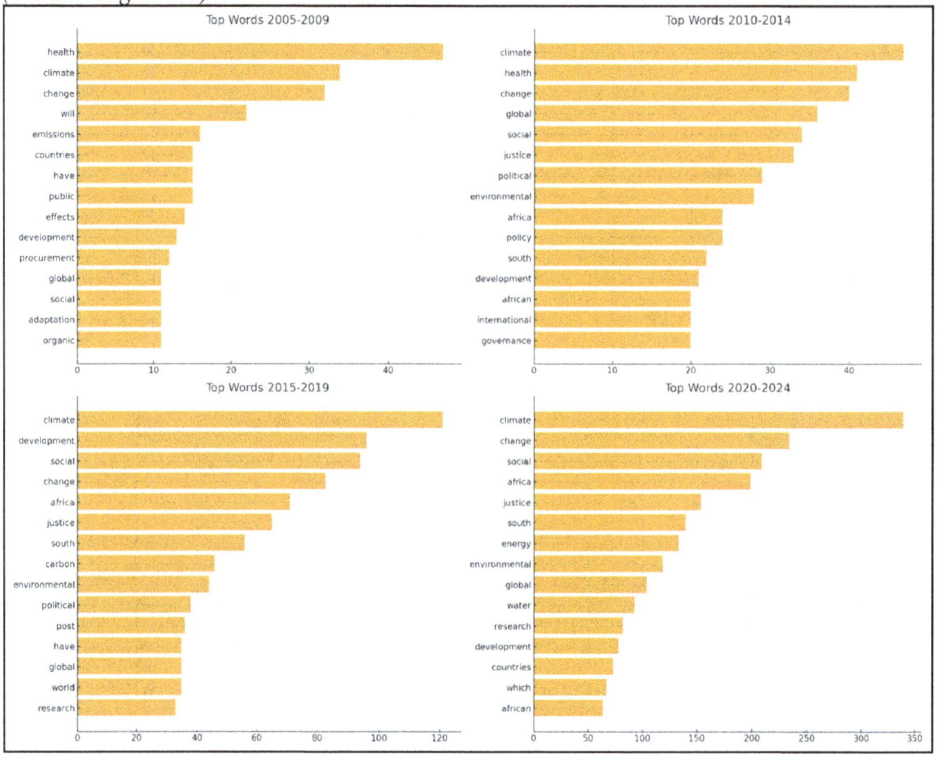

The heatmap of the top 20 core authors shows that the co-authorship structures of the most central researchers in the dataset are extremely dense. The yellow color of the matrix indicates that almost all of these researchers have collaborated with each other at least once. This suggests that scientific production occurs within a highly integrated and collaborative research network, rather than scattered individual contributions.

The most striking finding in the heatmap is that high-degree centrality nodes like "Maslin, M." and "Costello, A." maintain strong connections to numerous authors. These researchers, both in terms of their volume of collaboration and their position within the network, assume a role of integrator and coordinator. This role corresponds to a structure typically seen in large-scale project consortia and interdisciplinary research teams.

The presence of large yellow blocks outside the diagonal of the matrix suggests that the community exhibits more of a homogeneous core group rather than a core-periphery structure. This structure demonstrates that the research group is structured around a common thematic area and that the majority of the production is produced by researchers who interact with each other.

On the other hand, the very limited purple color in the heatmap indicates that there are no low-connected subgroups or isolated clusters among the Top 20 researchers. This dense network structure reflects the network's high degree of cohesion, information flow, and scientific coordination capacity. Such a structure can increase the rate of collective publication production and facilitate the rapid circulation of new research ideas or datasets within the team.

In conclusion, the Top 20 heatmap demonstrates a highly integrated, interactive, and collaborative scientific community in this research area. This structure indicates a strong network organization in terms of both publication performance and interdisciplinary integration. Furthermore, this close interaction among core researchers confirms the existence of a strategic collaborative backbone that shapes the field's development direction.

Figure 8. Heatmap

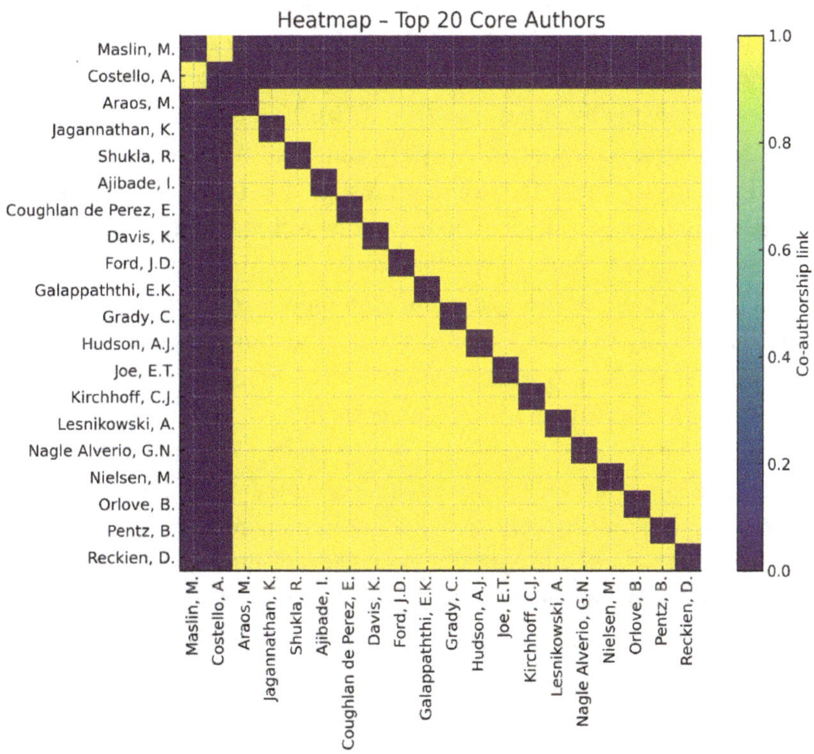

3.1 Anova Test

The three-panel ANOVA visualization provides a comparative overview of how citation counts vary across publication years, journals, and author-affiliated countries (**Fig. 9**). Together, these distributions reveal meaningful structural patterns in the dataset, shedding light on temporal effects, outlet-specific influence, and geographical citation dynamics.

The first panel clearly demonstrates substantial temporal variation in citation counts. Older publications exhibit noticeably higher medians and a greater spread, reflecting the cumulative nature of citation accrual. Certain years show distinct high-impact outliers, indicating that a limited number of papers from those years gained exceptional visibility. Conversely, the most recent years display expectedly lower citation values, as these publications have had limited time to accumulate citations. The statistically significant ANOVA result ($p < 0.01$) confirms that publication year exerts a meaningful effect on citation performance within the dataset.

The second panel highlights differences in citation distributions among the journals with at least three indexed publications. Although *The Lancet* shows a dramatically higher citation range compared with *Sustainability (Switzerland)*, *Agenda*, and *Energy Research and Social Science*, the ANOVA test does not identify these differences as statistically significant ($p > 0.10$). This is primarily due to the small sample sizes within each journal category and large within-group variance. Still, the visual pattern indicates that journal prestige and impact factor likely play an influential though statistically inconclusive role in citation outcomes in this dataset.

The country-based panel shows that while some nations (such as the United Kingdom and the United States) contain high-citation outliers, the overall citation distributions broadly overlap. Countries with large numbers of publications, such as South Africa, exhibit moderate citation levels with limited variability. The non-significant ANOVA result ($p > 0.30$) suggests that, within this dataset, geographical affiliation alone does not systematically predict citation performance. This implies that other factors such as research topic, collaboration networks, or journal selection—may be more influential in shaping citation impact than country of origin.

The combined results indicate that publication year is the strongest differentiator of citation performance, mainly driven by time-dependent accumulation. In contrast, journal and country effects, while visually suggestive, are not statistically significant given the current sample structure. These findings reinforce the importance of accounting for temporal exposure when evaluating citation-based research performance and highlight the need for larger and more balanced samples when comparing journals or geographical contexts.

Figure 9. Box Plot Visualization of Citation Distribution by Year, Journal, and Country (ANOVA Context)

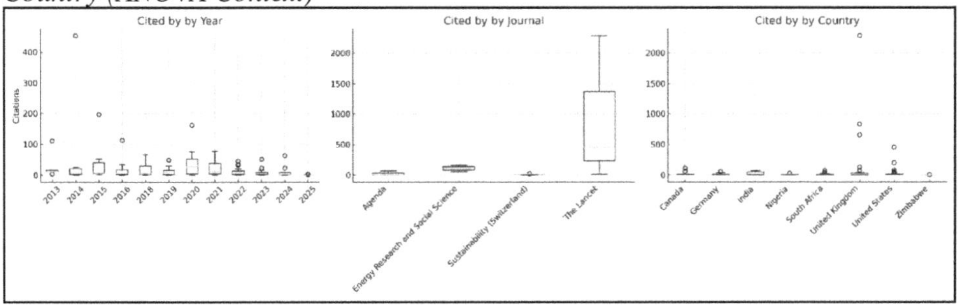

4 DISCUSSIONS

The results of this bibliometric analysis clearly demonstrate that the nexus between climate change and social justice in Africa has evolved into an increasingly central research domain over the last three decades. The notable surge in publications after 2010 reflects both a global recognition of climate change as a multidimensional crisis and a growing awareness of its disproportionate impacts on already vulnerable African populations. This trend is strongly aligned with the IPCC's assessment that Africa remains one of the regions most exposed to climate-related risks, including droughts, floods, food insecurity, and livelihood instability. At the same time, these risks intersect with deep-rooted socio-economic inequalities, which amplifies the importance of justice-oriented research frameworks.

The thematic structure revealed through keyword co-occurrence networks and author keyword analysis confirms that the field is predominantly organized around concepts such as *climate justice*, *environmental justice*, *gender*, *energy justice*, and *sustainable development*. This indicates that climate change research in Africa is no longer limited to environmental impacts but increasingly incorporates ethical, distributive, and human-rights-based dimensions. South Africa's prominence as a core geographical hub highlights how contexts of high inequality, rapid urbanization, and carbon-intensive energy systems provide fertile ground for examining climate–justice linkages (Vercillo et al., 2021; Obi and Ojo, 2025).

A significant observation is the persistent rise of themes related to energy transitions, land-use change, and water justice, especially after 2020. These emerging themes suggest that scholars are shifting their focus from diagnosing climate vulnerabilities to exploring pathways for sustainable and equitable transformation. In parallel, the increasing presence of gender-related keywords reveals an expanding research interest in intersectional vulnerabilities, where women, indigenous groups, and impoverished communities face disproportionately higher climate risks (Bond, 2013; Tantoh et al., 2022).

The citation-based results reinforce the centrality of justice-oriented scholarship: the most influential papers in the dataset are conceptual or interdisciplinary works that connect climate change with health, governance, rights, and resilience. These findings suggest that impactful research in this domain is grounded in cross-sectoral, integrative frameworks rather than single-discipline approaches. Furthermore, the ANOVA results highlight that citation performance is significantly shaped by publication year emphasizing the cumulative nature of research influence whereas geographical affiliation or journal identity does not independently predict citation success. This underscores that impactful research on climate and justice is widely recognized regardless of where it originates, provided that it addresses globally relevant questions.

Taken together, these findings reveal a rapidly maturing research field that is increasingly capable of informing equitable climate action in Africa. However, pronounced gaps remain. Certain subregions of Africa particularly Central and Northern Africa are underrepresented. There is also limited engagement with justice issues related to climate-induced displacement, conflicts over water, and cross-border environmental governance. Additionally, despite the rise of energy justice research, the literature still lacks robust analyses of how decarbonization pathways may create new inequalities if not implemented carefully.

4.1 Policy Implications and Recommendations

Based on the observed bibliometric patterns and thematic gaps, several actionable policy recommendations emerge:

National governments should formally embed *climate justice* principles into adaptation and mitigation strategies. Policies must consider the unequal distribution of climate impacts across gender, class, ethnicity, and geography, emphasizing fairness in access to climate finance, early-warning systems, and adaptation technologies (Schlosberg et al., 2017; Prakash et al., 2025).

The dominance of theoretical and conceptual publications suggests a need for stronger translation of academic knowledge into policy action. Governments, regional organizations, and research institutions should create sustained platforms where scholars, policymakers, and community stakeholders co-produce climate solutions (Turnhout et al., 2016; Klenk et al., 2017).

As Africa accelerates its renewable energy deployment, ensuring affordability, community participation, and equitable land-use decisions is crucial. Policies should require social impact assessments for major energy projects and guarantee that local populations benefit from energy transitions (Scoville-Simonds et al., 2020; Nu ă et al., 2025).

Given the growing research emphasis on gendered impacts, climate adaptation programs should explicitly target the vulnerabilities of women and youth. This includes improving access to credit, land rights, sustainable farming resources, and climate-resilient livelihoods (Perez et al., 2015; Abegunde et al., 2019).

The uneven geographic distribution in the literature underscores the need for capacity-building efforts, research funding, and regional climate observatories in understudied African regions, especially Central Africa and the Sahel (Dinku et al., 2014; Obi and Ojo, 2025).

Many African nations lack long-term, high-resolution climate and socio-economic datasets. Improving climate monitoring networks, supporting open-access repositories, and promoting data literacy will enhance both research quality and policy design (Dinku et al., 2014; Carr et al., 2019).

In conclusion, this bibliometric analysis highlights that the intersection of climate change and social justice has become a vital research frontier in Africa. As the continent navigates intensifying climate pressures, justice-centered research provides essential insights to ensure that climate action is not only effective but also equitable. The findings underscore the urgency of designing policies that safeguard vulnerable communities, promote fair transitions, and build resilient socio-ecological systems. Strengthening collaboration between academia, policymakers, and communities will be fundamental for moving the continent toward a more just and sustainable climate future.

5 CONCLUSIONS

This bibliometric analysis systematically mapped the scholarly production at the intersection of climate change and social justice in the African context from 1990 to 2023. The findings unequivocally demonstrate that this nexus has evolved from a peripheral concern in the 1990s and early 2000s to a vital, rapidly maturing research frontier in contemporary global scholarship.

The key conclusions drawn from the performance and structural analyses are:

- Accelerated Growth and Centrality: The field experienced a significant surge in publications after 2010, confirming the global recognition of climate change as a multidimensional crisis with profound socio-economic and ethical implications for vulnerable African populations.
- Justice-Focused Intellectual Structure: The core research landscape is structurally organized around justice-based concepts, with "climate justice," "en-

vironmental justice," "gender," "energy justice," and "sustainable development" serving as the dominant thematic hubs. This confirms that academic discourse has successfully transitioned from merely diagnosing environmental impacts to exploring ethical and distributive dimensions.

- Geographical and Sectoral Focus: South Africa emerged as the single most prominent national case study, reflecting its complex context of deep inequality and carbon dependency. Key sectoral areas of intense focus include energy transitions, land-use change, and water justice.
- Impact Driven by Integration: The most impactful (most cited) documents are theoretical and interdisciplinary works that successfully connect climate change with health, governance, rights, and resilience. This underscores that influential research relies on cross-sectoral, integrative frameworks.
- Methodological Gaps Remain: Despite the overall maturity, the literature exhibits pronounced geographical gaps, particularly the underrepresentation of Central and Northern Africa. There is also limited engagement with critical emerging issues like climate-induced displacement and cross-border environmental governance.

Based on the observed bibliometric patterns and identified thematic gaps, several actionable policy implications and recommendations emerge:

- Embed Justice in Policy: National governments must formally embed climate justice principles into all adaptation and mitigation strategies.
- Address Unequal Distribution: Policies must explicitly consider the unequal distribution of climate impacts across gender, class, ethnicity, and geography, ensuring fairness in access to climate finance, early-warning systems, and adaptation technologies.
- Co-production Platforms: Given the dominance of theoretical publications, governments, regional organizations, and research institutions should create sustained platforms where scholars, policymakers, and community stakeholders can co-produce climate solutions.
- Enhance Data Quality: African nations should invest in improving climate monitoring networks and supporting open-access repositories to enhance both research quality and evidence-based policy design.
- Equitable Energy Deployment: As Africa accelerates its renewable energy deployment, policies must ensure affordability, community participation, and equitable land-use decisions.
- Mandatory Social Impact Assessment: Policies should require social impact assessments for major energy projects to guarantee that local populations directly benefit from energy transitions and are not further marginalized.

- Targeted Gender Programs: Given the growing research emphasis on gendered impacts, climate adaptation programs must explicitly target the vulnerabilities of women and youth.
- Livelihood Support: This includes improving access to credit, land rights, sustainable farming resources, and climate-resilient livelihoods for vulnerable groups.
- Capacity Building: The uneven geographic distribution in the literature—with underrepresentation in Central and Northern Africa—underscores the need for targeted capacity-building efforts, research funding, and regional climate observatories in understudied African regions, especially the Sahel.

In conclusion, as the continent navigates intensifying climate pressures, justice-centered research provides essential insights to ensure that climate action is not only effective but also equitable. The findings underscore the urgency of designing policies that safeguard vulnerable communities, promote fair transitions, and build resilient socio-ecological systems. Strengthening collaboration between academia, policymakers, and communities will be fundamental for moving the continent toward a more just and sustainable climate future.

STATEMENTS AND DECLARATIONS

Acknowledgement: The author gratefully acknowledges the Scopus Database for providing datasets.

Funding: The authors received no financial support from public, commercial, or not-for-profit funding agencies to conduct this research.

Conflicts of interest: The authors declare that they have no known competing financial interests or personal relationships that could have appeared to influence the work reported in this manuscript.

Data availability: The datasets generated and/or analyzed during the current study are available from the corresponding author upon reasonable request.

Authors' contributions: F.Ş. M.A.Ç. and M.B.A.: Study concept and design, data collection. F.Ş. M.A.Ç. and M.B.A: Analyses, interpretation of results, manuscript preparation, and critical revision. F.Ş., İ.G. and D.D.: Reviewed and approved the final version of the manuscript.

Ethics Approval: Not applicable.

Consent to Participate: Not applicable.

Consent to Publish: Not applicable.

REFERENCES

Abegunde, V. O., Sibanda, M., & Obi, A. (2019). The Dynamics of Climate Change Adaptation in Sub-Saharan Africa: A Review of Climate-Smart Agriculture among Small-Scale Farmers. *Climate (Basel)*, *7*(11), 132. DOI: 10.3390/cli7110132

Adaawen, S., Rademacher-Schulz, C., Schraven, B., & Segadlo, N. (2019). Drought, migration, and conflict in sub-Saharan Africa: what are the links and policy options? Drought Challenges, 15–31. https://doi.org/DOI: 10.1016/B978-0-12-814820-4.00002-X

Addaney, M. (2025). Addressing the impacts of climate change on human rights through adaptation-focused litigation: Cases before the African Union's human rights bodies. *Journal of Human Rights*, *24*(3), 353–369. DOI: 10.1080/14754835.2025.2507588

Adebisi, Y. A., Akande-Sholabi, W., Jimoh, N. D., Lali, H., Adebowale, K. A., Ouachhou, A., Egwu, K. C., Nachibwede, L. V., Shomuyiwa, D. O., Gilbert, A., Olpengs, D., Sesay, N., & Oladunjoye, I. O. (2025). Leveraging a people-centred approach to combat antimicrobial resistance in Africa. *Tropical Medicine and Health*, *53*(1), 177. Advance online publication. DOI: 10.1186/s41182-025-00863-w PMID: 41316445

Adger, W. N. (2006). Vulnerability. *Global Environmental Change*, *16*(3), 268–281. DOI: 10.1016/j.gloenvcha.2006.02.006

Adisa, O. M., Masinde, M., Botai, J. O., & Botai, C. M. (2020). Bibliometric analysis of methods and tools for drought monitoring and prediction in Africa. *Sustainability (Basel)*, *12*(16), 6516. DOI: 10.3390/su12166516

Amsalu, A., Garedew, W., Melka, G. A., Çelik, M. A., & Gemeda, D. O. (2026). Trends in climate extreme indices and implications for coffee farming in The Kafa Biosphere Reserve, southwest Ethiopia. *Climate Services*, *42*, 100637. DOI: 10.1016/j.cliser.2026.100637

Apraku, A., Gyampoh, B. A., Morton, J. F., & Mariwah, S. (2025). Managing climate change through social justice in Africa: Key lessons from the Libode area of Eastern Cape, South Africa. *Scientific African*, *27*, e02534. DOI: 10.1016/j.sciaf.2025.e02534

Arruda, H., Silva, E. R., Lessa, M., Proença, D.Jr, & Bartholo, R. (2022). VOSviewer and Bibliometrix. *Journal of the Medical Library Association: JMLA*, *110*(3), 392–395. DOI: 10.5195/jmla.2022.1434 PMID: 36589296

Bailey, K., & Okoduwa, S. I. R. (2025). Leveraging social media for public health nutrition in low- and middle-income countries: Addressing malnutrition, enhancing food security, and promoting health equity. *World Nutrition, 16*(2), 109–121. DOI: 10.26596/wn.2025162109-121

Bond, P. (2013). Water Rights, Commons and Advocacy Narratives. *South African Journal on Human Rights, 29*(1), 125–143. DOI: 10.1080/19962126.2013.11865068

Bukar, U. A., Sayeed, M. S., Razak, S. F. A., Yogarayan, S., Amodu, O. A., & Mahmood, R. A. R. (2023). A method for analyzing text using VOSviewer. *MethodsX, 11*, 102339. DOI: 10.1016/j.mex.2023.102339 PMID: 37693657

Carr, E. R., Goble, R., Rosko, H. M., Vaughan, C., & Hansen, J. (2019). Identifying climate information services users and their needs in Sub-Saharan Africa: A review and learning agenda. *Climate and Development, 12*(1), 23–41. DOI: 10.1080/17565529.2019.1596061

Celik, M. A. (2020). Bibliometric network analysis on new tendencies, techniques and terms used in drought research. [IGGE]. *International Journal of Geography and Geography Education, 42*, 602–630. DOI: 10.32003/igge.721487

Çelik, M. A., Ağırkaya, M. B., & Gemeda, D. O. (2026a). Bridging data scarcity with AI methodology: Global research trends in machine and deep learning for air quality prediction. *Journal of Atmospheric and Solar-Terrestrial Physics, 281*, 106777. DOI: 10.1016/j.jastp.2026.106777

Çelik, M. A., & Çelik, E. (2024). Are Urbanisation and Biodiversity Antithetical? A Bibliometric Analysis. Coğrafya Dergisi. *The Journal of Geography, 0*(48), 121–135. DOI: 10.26650/JGEOG2024-1350047

Çelik, M. A., Kıraç, A., & Gouveia, C. M. (2026b). North Atlantic Oscillation modulation of drought and vegetation activity in Türkiye. *Theoretical and Applied Climatology, 157*(2), 74. Advance online publication. DOI: 10.1007/s00704-025-05971-5

Chari, T., & Munoriyarwa, A. (2025). *Climate Change Communication in Africa*. Routledge., DOI: 10.4324/9781003660217

Cohen, S. J., & Waddell, M. W. (2009). Climate Change in the 21st Century. https://doi.org/DOI: 10.1515/9780773581296

Dinku, T., Block, P., Sharoff, J., Hailemariam, K., Osgood, D., del Corral, J., Cousin, R., & Thomson, M. C. (2014). Bridging critical gaps in climate services and applications in africa. *Earth Perspectives, 1*(1), 15. DOI: 10.1186/2194-6434-1-15

Dumenu, W. K., & Obeng, E. A. (2016). Climate change and rural communities in Ghana: Social vulnerability, impacts, adaptations and policy implications. Environmental Science &amp. *Environmental Science & Policy*, *55*, 208–217. DOI: 10.1016/j.envsci.2015.10.010

Erfanmanesh, M. (2017). Status and quality of open access journals in Scopus. *Collection Building*, *36*(4), 155–162. DOI: 10.1108/CB-02-2017-0007

Hope, K. R.Sr. (2009). Climate change and poverty in Africa. International Journal of Sustainable Development &amp. *International Journal of Sustainable Development and World Ecology*, *16*(6), 451–461. DOI: 10.1080/13504500903354424

Intergovernmental Panel on Climate Change (IPCC). (2023). Climate Change 2022 – Impacts, Adaptation and Vulnerability. https://doi.org/DOI: 10.1017/9781009325844

Iqbal, B. A., & Ghauri, F. N. (2011). Climate change: The biggest challenge in 21st century. *Mediterranean Journal of Social Sciences*, *2*(6), 41–41.

Jiang, C. M., Duangthip, D., Chan, A. K. Y., Tamrakar, M., Lo, E. C. M., & Chu, C. H. (2021). Global research interest regarding silver diamine fluoride in dentistry: A bibliometric analysis. *Journal of Dentistry*, *113*, 103778. DOI: 10.1016/j.jdent.2021.103778 PMID: 34391874

José de Oliveira, O., Francisco da Silva, F., Juliani, F., César Ferreira Motta Barbosa, L., & Vieira Nunhes, T. (2019). *Bibliometric Method for Mapping the State-of-the-Art and Identifying Research Gaps and Trends in Literature: An Essential Instrument to Support the Development of Scientific Projects*. Scientometrics Recent Advances., DOI: 10.5772/intechopen.85856

Klenk, N., Fiume, A., Meehan, K., & Gibbes, C. (2017). Local knowledge in climate adaptation research: moving knowledge frameworks from extraction to co-production. WIREs Climate Change, 8(5). Portico. https://doi.org/DOI: 10.1002/wcc.475

Leiserowitz, A. A. (2005). American Risk Perceptions: Is Climate Change Dangerous? Risk Analysis, 25(6), 1433–1442. Portico. https://doi.org/DOI: 10.1111/j.1540-6261.2005.00690.x

Mugambiwa, S. (2025). Climate Governance from the Bottom: Decentralized Adaptation and the Politics of Rural Livelihoods in Southern Africa. Journal of Cultural Analysis and Social Change, 631-639.

Nu ă, F., Mustata Wilson, G., Boateng, G. O., & Nuta, A. C. (Eds.). (2025). *Environmental degradation, health, and socioeconomic impacts*. Frontiers Research Topics., DOI: 10.3389/978-2-8325-7166-8

Nwokolo, S. C., Singh, R., Khan, S., Kumar, A., & Luthra, S. (2023). Impacts of Climate Change in Africa. Africa's Path to Net-Zero, 237–262. https://doi.org/DOI: 10.1007/978-3-031-44514-9_7

Obi, J. N., & Ojo, E. (2025). Climate Justice, Renewable Energy, and Food Security in Sub-Saharan Africa: A Bibliometric Analysis of Emerging Research Trends. Energy Transition, Climate Action and Sustainable Agriculture, 303–328. https://doi.org/DOI: 10.1007/978-3-031-83165-2_16

Omotoso, A. B., & Omotayo, A. O. (2023). The interplay between agriculture, greenhouse gases, and climate change in Sub-Saharan Africa. *Regional Environmental Change*, *24*(1), 1. Advance online publication. DOI: 10.1007/s10113-023-02159-3

Oreskes, N. (2004). The Scientific Consensus on Climate Change. *Science*, *306*(5702), 1686–1686. DOI: 10.1126/science.1103618 PMID: 15576594

Perez, C., Jones, E. M., Kristjanson, P., Cramer, L., Thornton, P. K., Förch, W., & Barahona, C. (2015). How resilient are farming households and communities to a changing climate in Africa? A gender-based perspective. *Global Environmental Change*, *34*, 95–107. DOI: 10.1016/j.gloenvcha.2015.06.003

Prakash, A., Totin, E., Kemp, G., Kerr, R. B., & Roberts, D. (2025). Bridging the gap: Promoting gender equity in climate change adaptation in the global south. *PLOS Climate*, *4*(6), e0000556. DOI: 10.1371/journal.pclm.0000556

Schlosberg, D. (2007). 1 Defining Environmental Justice. Defining Environmental Justice, 3–10. https://doi.org/DOI: 10.1093/acprof:oso/9780199286294.003.0001

Schlosberg, D., & Collins, L. B. (2014). From environmental to climate justice: climate change and the discourse of environmental justice. WIREs Climate Change, 5(3), 359–374. Portico. https://doi.org/DOI: 10.1002/wcc.275

Schlosberg, D., Collins, L. B., & Niemeyer, S. (2017). Adaptation policy and community discourse: Risk, vulnerability, and just transformation. *Environmental Politics*, *26*(3), 413–437. DOI: 10.1080/09644016.2017.1287628

Scoville-Simonds, M., Jamali, H., & Hufty, M. (2020). The Hazards of Mainstreaming: Climate change adaptation politics in three dimensions. *World Development*, *125*, 104683. DOI: 10.1016/j.worlddev.2019.104683

Tantoh, H. B., Ebhuoma, E. E., & Leonard, L. (2022). Indigenous Women's Vulnerability to Climate Change and Adaptation Strategies in Central Africa: A Systematic Review. Indigenous Knowledge and Climate Governance, 53–66. https://doi.org/DOI: 10.1007/978-3-030-99411-2_5

Thornton, P. K., Kristjanson, P. M., Kruska, R. L., & Reid, R. S. (2004). Mapping livestock and poverty: A tool for targeting research and development. *BSAP Occasional Publication*, *33*, 37–50. DOI: 10.1017/S1463981500041649

Tucker, J., Daoud, M., Oates, N., Few, R., Conway, D., Mtisi, S., & Matheson, S. (2014). Social vulnerability in three high-poverty climate change hot spots: What does the climate change literature tell us? *Regional Environmental Change*, *15*(5), 783–800. DOI: 10.1007/s10113-014-0741-6

Turnhout, E., Dewulf, A., & Hulme, M. (2016). What does policy-relevant global environmental knowledge do? The cases of climate and biodiversity. *Current Opinion in Environmental Sustainability*, *18*, 65–72. DOI: 10.1016/j.cosust.2015.09.004

van Eck, N. J., & Waltman, L. (2009). Software survey: VOSviewer, a computer program for bibliometric mapping. *Scientometrics*, *84*(2), 523–538. DOI: 10.1007/s11192-009-0146-3 PMID: 20585380

van Eck, N. J., & Waltman, L. (2017). Citation-based clustering of publications using CitNetExplorer VOSviewer. https://doi.org/DOI: 10.1007/s11192-017-2300-7

van Eck, N. J., & Waltman, L. (2021). VOSviewer goes online! (Part 1). https://doi.org/DOI: 10.59350/mmsjw-79h07

Vercillo, S., Huggins, C., & Cochrane, L. (2021). How is gender investigated in African climate change research? A systematic review of the literature. *Ambio*, *51*(4), 1045–1062. DOI: 10.1007/s13280-021-01631-w PMID: 34669170

Wang, Z., Zhao, Y., & Wang, B. (2018). A bibliometric analysis of climate change adaptation based on massive research literature data. *Journal of Cleaner Production*, *199*, 1072–1082. DOI: 10.1016/j.jclepro.2018.06.183

Williams, J., Chin-Yee, S., Maslin, M., Barnsley, J., Costello, A., Lang, J., McGlade, J., Mulugetta, Y., Taylor, R., Winning, M., & Parikh, P. (2023). Africa and climate justice at COP27 and beyond: Impacts and solutions through an interdisciplinary lens. *UCL Open Environment*, *5*. Advance online publication. DOI: 10.14324/111.444/ucloe.000062 PMID: 37671394

Yeh, C.-H., Chien, T.-W., & Chou, P.-H. (2022). Citation analysis of the 100 top-cited articles on discectomy via endoscopy since 2011 using alluvial diagrams: Bibliometric analysis. *European Journal of Medical Research*, *27*(1), 169. Advance online publication. DOI: 10.1186/s40001-022-00782-0 PMID: 36050803

Yu, Y., Li, Y., Zhang, Z., Gu, Z., Zhong, H., Zha, Q., Yang, L., Zhu, C., & Chen, E. (2020). A bibliometric analysis using VOSviewer of publications on COVID-19. *Annals of Translational Medicine*, *8*(13), 816–816. DOI: 10.21037/atm-20-4235 PMID: 32793661

Chapter 3
Hybrid and Physics-Informed AI Models in Earth System Analysis

Mohammad Kaif
http://orcid.org/0009-0009-9413-7422
Lovely Professional University, India

Priyanti Halder
Lovely Professional University, India

Smriti Mittu
http://orcid.org/0009-0009-9221-8907
Lovely Professional University, India

Abhas Sarangi
http://orcid.org/0009-0008-6776-1943
Lovely Professional University, India

Narendra Kumar Pandey
Lovely Professional University, India

ABSTRACT

Recent advances in the field of artificial intelligence (AI) have significantly altered the approach to the study of the Earth system; however, data-based models have difficulty in terms of physical consistency, interpretability, and generalization under non-stationary environmental conditions. Hybrid and physics-informed AI models have been developed to overcome these problems by using a combination of machine learning and physical laws and models. This chapter presents the major paradigms of hybrid modelling, including physics-informed neural networks, universal dif-

DOI: 10.4018/979-8-2600-0778-5.ch003

ferential equations, differentiable modelling, and physics-encoded deep learning, along with their applications in climate studies, hydrology, atmospheric modelling, and environmental monitoring. In short, this chapter presents a synthesis of recent advances to demonstrate the potential of using hybrid and physics-informed AI models to bridge the gap between theory and data-based modelling.

1. INTRODUCTION

1.1 Background and Rationale

The focus of Earth system science is to explain the dynamic and interconnected behaviour of the atmosphere, hydrosphere, biosphere, lithosphere, and cryosphere, i.e., all at several spatial and temporal scales. Physics-based numerical models have been the main representation of these processes over the past decades, which are based on conservation laws, differential equations, and parameterisations that have been determined empirically. Although these models have fruitfully contributed to scientific knowledge and policy-relevant measurements, they are still limited in their ability to represent highly non-linear interactions, emergent behaviour, and fine-scale variability, especially where climate non-stationary conditions and extreme events are involved (Shen et al., 2023). At the same time, the diffusion of Earth observation technologies such as high-resolution satellite missions, reanalysis products, and dense sensor networks has produced vast amounts of spatiotemporal information that is beyond the capacity of traditional modelling techniques. The approach based on artificial intelligence (AI) and machine learning (ML) has thus become the leading trend in the Earth sciences due to its capacity to learn intricate patterns directly through data and scale effectively at large scales (Janga et al., 2023). Nevertheless, data-driven models, in most cases, are purely data-driven, and therefore, physical constraints are ignored, which brings up interpretability, physical realism, and long-term soundness issues. Hybrid and physics-informed AI models have therefore been one of the solutions to these issues and provide a platform where physical knowledge and data-driven learning can be used together to improve the analysis of the Earth system (Karniadakis et al., 2021).

1.2 Shortcomings of Traditional Physics-Based and Data-Driven Models

Although elegant conceptually and interpretable, existing physics-based Earth system models are limited, at best, by structural uncertainty, computation time, and the need to parameterise processes that are not understood. Numerous processes of importance in the Earth system, including cloud microphysics, land-atmosphere interactions, and hydrological neglect of the subsurface, are found at finer scales than models usually resolve, and can propagate an error in their simplifications (Shen et al., 2023). To add, more spatial resolution and ensemble-based uncertainty analysis have been demanded, increasing the computational infeasibility of such models in many applications. On the other hand, machine learning models based on data have been shown to exhibit strong predictive performance in activities such as land-use classification, prediction, and detection of extreme events in climate variables and demonstrated strong predictive performance when trained on large remote sensing datasets (Janga et al., 2023). However, such models have been criticised as being black-box models with low levels of physical interpretability and prone to over-fitting, particularly when generalised to conditions not seen in the training data. No embedded physical constraints may result in predictions that are inconsistent with the laws of conservation or produce physically unreasonable results, thus compromising trust and operating applicability (Karniadakis et al., 2021). Their complementary restrictions have inspired the creation of composite modelling solutions that integrate the explanatory capability of physics-based models with the versatility and extensiveness of AI.

Notably, the weakness of black-box artificial intelligence models to the Earth system science is not only due to opaceness of internal parameters, but also because of the lack of physically encoded physical structure in the learning formulation. Data-only based systems maximize predictive ability without enforcing any conservation laws, boundary conditions, or governing equations and therefore can give statistically accurate but physically inconsistent solutions in extrapolative conditions. Respondently, hybrid and physics-informed approaches are emerging with variability in the application of physical knowledge: some models put physical laws as soft constraints that reward deviations during training whereas other models put physical relationships into model structure as hard constraints that ensure structural consistency. This difference is at the core of realising the goals of hybrid paradigms to eliminate the drawbacks of both classical numerical models and unconstrained machine learning systems.

1.3 Development of Hybrid and Physics-Inspired AI

The combination of physics-informed and hybrid AI is a paradigm shift in Earth system modelling, as it actively considers physical knowledge within the machine learning frameworks. Hybrid models are usually based on physics-based simulations where the data uses data-driven aspects so that AI can learn model residuals, to better parameterizations or remove systematic biases without changing the underlying physical structure (Huynh et al., 2025). This method has been shown to be successful in hydrology, where neural networks have been used to further improve flood prediction through corrections in internal flux representations based on process-based models. Physics-informed machine learning builds on this idea by directing the training procedure with governing equations, most prominently physics-informed neural networks (PINNs), which put partial differential equations and boundary conditions in the loss term (Raissi et al., 2019). These models bring a degree of consistency to the physical world, even where there is sparse observational data, hence enhancing generalisation and interpretability. Recent investigations have shown that physics-informed and hybrid methods have been successful in a wide range of applications, such as soil moisture forecasting to extreme precipitation nowcasts, and generally outperform both conventional numerical models and data-only AI methods (Das et al., 2024; Engineering Applications of Artificial Intelligence, 2025). These advances predict that hybrid and physics-informed AI paradigms form a strong direction for dealing with the complexity and uncertainty of Earth system processes.

1.4 Scope, Objectives, and Structure of the Chapter

This chapter is aimed at making a critical and thorough analysis of the hybrid and physics-informed artificial intelligence (AI) models in the context of Earth system analysis, including the conceptual foundations, the frameworks of the methodologies, and their practical realization. The chapter tries to elucidate the recent developments in the fields of artificial intelligence, remote sensing, and process-based modelling to demonstrate how combined strategies can result in better predictive capability, physical coherence, and uncertainty description. Based on peer-reviewed articles across the entire spectrum of Earth sciences, the chapter aims to bridge the gap between theoretical innovation and operational implementation, particularly in the areas of climate change, hydrology, land-surface dynamics, and environmental monitoring. The organization is based on the logical progression of the basic principles to practical examples of cases, starting with an overview of the Earth system modelling and AI approaches. It subsequently continues with a discussion of hybrid and physics-inspired frameworks, integration of data and strategies to integrate data, as well as domain applications. The chapter ends with a recommendation of the existing issues, ethical concerns, and future research perspectives, making hybrid and physics-informed AI one of the main facilitators of a new generation of Earth system science and ecologically sound environmental management.

2. BASICS OF EARTH SYSTEM MODELING

2.1 Introduction to Components of the Earth System and Interactions

To model the interconnected behavior of the atmosphere, oceans, land surface, cryosphere, and biosphere, Earth system modeling is employed because these components are all interrelated and interact through complex feedback mechanisms at various scales. The ocean circulation and terrestrial hydrology are affected by atmospheric processes, whereas land-atmosphere exchanges are affected by energy, water, and carbon fluxes that determine the climate dynamics and ecosystem functions (Amsalu et al., 2026; Çelik et al., 2026). This alters cryospheric surface albedo and sea level and has a subsequent impact on atmospheric circulation and ocean dynamics. Such interactions are nonlinear in nature and are marked by feedback loops that may either magnify or suppress system responses, which makes the behaviour of the Earth system hard to predict with the necessary accuracy over long periods of time (Bonan & Doney, 2018). Human activities have made that complexity more difficult since they have changed land cover, atmospheric composition, and biogeochemical cycles, and added new forcings and uncertainties. Such coupled dynamics require a combination of modelling frameworks that are able to simulate cross-domain interactions without breaking the physical consistency. Conventional models of the Earth system tackle this issue with the modular coupling of sub-models, but they have very limited capacity to capture emergent behaviour due to limitations of resolution and simplified process models (Flato et al., 2019). The increasing volume and complexity of observational data have made the need to have more adaptive and integrative modelling approaches quite clear.

2.2 Classical Physics-Based Models and Numerical Approximations

Physics-based models of Earth systems are first-principle based, making use of conservation laws of mass, momentum, and energy to model environmental processes. These are climate prediction general circulation models, energy and water balance estimation land surface models, and models modelling runoff and groundwater. These models represent a clear and physically understandable paradigm, which makes them invaluable in explaining system behaviour and in aid of policy-relevant judgment, like those generated by the Intergovernmental Panel on Climate Change (Flato et al., 2019). However, they are limited in performance due to uncertainties due to parameterisation of unsolved processes, numerical models, and the finite ability of the computation to fine-resolution simulations. Numerous sub-grid processes, e.g., cloud formation or redistribution of soil moisture, are modelled by empirical relations, which might fail in changing climatic conditions, causing structural bias and decreasing predictive capability (Bonan and Doney, 2018). Also, the cost of computation of ensemble simulations is very high, which limits the possibility of quantifying uncertainty in a comprehensive way, especially when it comes to regional-scale or real-time applications. These constraints have provoked the use of complementary methods by researchers, which could help enrich or extend the physics-based models without compromising physical realism.

2.3 Data-Driven Modelling of the Earth System

Earth system science has seen the rise of data-driven modelling methods, including machine learning and deep learning, as these methods can learn complicated relationships and do so without prior human-imposed explicit knowledge. Recent developments in remote sensing, sensor networks, and reanalysis datasets have facilitated the use of data-driven models on problems in climate variable forecasting, land -use change detection, and ecosystem monitoring (Reichstein et al., 2019). These models are highly effective at modelling nonlinear patterns and interactions, which might be challenging to express in a physics-based model. Nevertheless, information-driven methods are empirical in nature and can be physically uninterpretable, which raises doubts of reliability and strength in situations of extrapolation like climate forecasting. In addition, they require big and high-quality training datasets, which restricts their application to data-restricted areas, where data are sparse or distributed unevenly. With no clear physical limits, exclusively data-induced models can produce predictions that contravene the laws of conservation or have unstable long-term behaviour, and thus are not appropriate to analyse working Earth systems (Reichstein et al., 2019). In these difficulties lies the necessity to consider integrative approaches that capitalize on the laxity of data-driven approaches, but incorporate physical expertise to be consistent and reliable.

2.4 Requirement of Hybrid Modelling Paradigms

The weaknesses of the traditional physics-based and purely data-driven models have resulted in an increasing interest in the view that hybrid modelling paradigms are required in Earth system science. Hybrid models are aimed at bringing together the advantages of a mechanistic knowledge base with adaptive industrial qualities of artificial intelligence, thus allowing a more precise portrayal of complex processes along with a physical consistency. Hybrid methods can also address the systematic model biases and improve the parameter estimation and predictive accuracy in non-stationary situations by using observational data plus process-based frameworks

(Shen et al., 2023). These models have found particular application in overcoming the issues of scale mismatch, uncertainty propagation, and computational efficiency. Recent works have shown that hybrid structures may be more effective than traditional ones in tasks like hydrological forecasting and analysing climate extremes, and have better robustness and generalization (Huynh et al., 2025; Das et al., 2024). The development of hybrid and physics-inspired AI models is thus a natural extension of Earth system modelling, as environmental tasks are getting more complicated, and a wider range of observational data is available. These paradigms offer a base to the upcoming generation of data-adaptive as well as physically grounded Earth system models.

Table 1. Comparison of Data-Driven, Physics-Based, and Hybrid AI Modelling Approaches in Earth Sciences

Aspect	Data-Driven AI Models	Physics-Based Models	Hybrid / Physics-Informed AI Models
Core principle	Learn patterns solely from data	Governed by physical laws and equations	Integration of data-driven learning with physical constraints
Data requirement	Large labelled datasets	Moderate to low	Reduced data demand due to physics guidance
Physical consistency	Often violated	Strictly preserved	Explicitly enforced or softly constrained
Interpretability	Low	High	Moderate to high
Generalisation	Limited under non-stationarity	Robust but rigid	Improved robustness and adaptability
Computational cost	Moderate to high	High	High but scalable
Typical applications	Image classification, pattern detection	Climate and hydrological simulations	Climate forecasting, hydrology, extremes modelling

3. TECHNIQUES OF ARTIFICIAL INTELLIGENCE APPLICABLE TO THE EARTH SYSTEM ANALYSIS

3.1 Approaches to Machine Learning and Deep Learning

The development of machine learning has become a more and more influential paradigm in Earth system science due to the ability to model nonlinear and complex relationships that are hard to express through their explicit physical formulations. The methodologies of supervised, unsupervised, and semi-supervised learning have been applied to a wide range of environmental problems, including climate variable estimation, land-use classification, and ecosystem monitoring, and are, therefore, using extensive repositories of remote sensing and in-situ data (Camps-Valls et al., 2021). Specifically, deep learning has been used to extract hierarchical features in high-dimensional input, including satellite images and spatiotemporal climate fields, and reduce the use of hand-crafted features and domain bias. Convolutional neural networks, in particular, are particularly effective in spatial pattern recognition of data in Earth observation, and recurrent and sequence-based models have been applied to acquire temporal dependencies in hydro-climatic time series (Zhu et al., 2017). Although deep learning models have strong predictive capabilities, they are often criticized with regard to being opaque and uninterpretable, which hinders scientific understanding and the development of policy-relevant judgments. In addition, the high data intensity and sensitivity to training conditions of these models can limit their transferability between different areas and climate regimes, which highlights the need to adopt methods that incorporate physical knowledge into learning structures (Camps-, Valls et al., 2021).

3.2. Spatiotemporal Earth Observation Data Modelling

System processes Earth system processes are inherently spatiotemporal, and change in a continuous manner on a spatial and temporal scale through interactions of various subsystems. Techniques that define artificial intelligence in a manner that spatiotemporal dependencies are modelled have thus been at the forefront of the analysis of the Earth systems in modern times. Recurrent neural networks and their variations, such as the long short-term memory networks, have been widely used to model hydrological and climatic time series and have proved to be more effective in predicting streamflow and estimating soil moisture when compared to the conventional statistical models (Kratzert et al., 2019). In more recent times, attention-based models and transformer models have become popular in the fields of remote sensing and geospatial analysis because they have the ability to capture long-range interactions and interactions across space and time (Yuan et al., 2021). Those models have the benefits of dealing with non-uniform time samples as well as multi-scale spatial fields common in satellite imaging. However, the spatiotemporal AI models are still likely to overfit and be unstable when being applied to future climate conditions, especially in cases where the physical constraints are not provided. There is therefore an increasing trend of opinion suggesting that spatiotemporal AI tools need to be accompanied by process insight to achieve physical viability and long-term stability in applications to Earth systems (Shen et al., 2023).

3.3 AI Strengths and Limitations in the Environmental Applications

Artificial intelligence provides a number of benefits to the analysis of Earth systems, which are scalability to large datasets, the ability to model nonlinear interactions, and the ability to combine the use of heterogeneous sources with satellite images, reanalysis products, and social-environmental indicators. Such capabilities have allowed AI models to outshine the conventional approaches in the detection of extreme events, high-resolution mapping of the environment, and near-real-time forecasting (Das et al., 2024). Still, there are severe limitations that restrict the use on a larger scale in the Earth system science. AI models are often black boxes and thus reduce transparency, and in the context of environmental decision-making, scientific interpretability is needed to generate trust and acceptance. Such systematic error and decreased generalisability may be created by data bias, inappropriate spatial coverage, and the lack of representation of extreme or rare events. Notably, data-only models can violate the known physical laws, and their predictions can be scientifically implausible or unstable over long time scales (Karniadakis et al., 2021). Such weak points emphasize the need to consider physical principles in AI process development, which in turn encourages the creation of hybrid and physics-inspired solutions that provide the predictive ability of AI with the accuracy and understandability of physics-based modelling systems. Figure 1 shows the conceptual process of hybrid and physics-informed artificial intelligence models, where multi-source Earth observation data is used to predict the environment with great strength and interpretability based on physical constraints.

Figure 1. Conceptual architecture of hybrid and physics-informed AI models for Earth system analysis

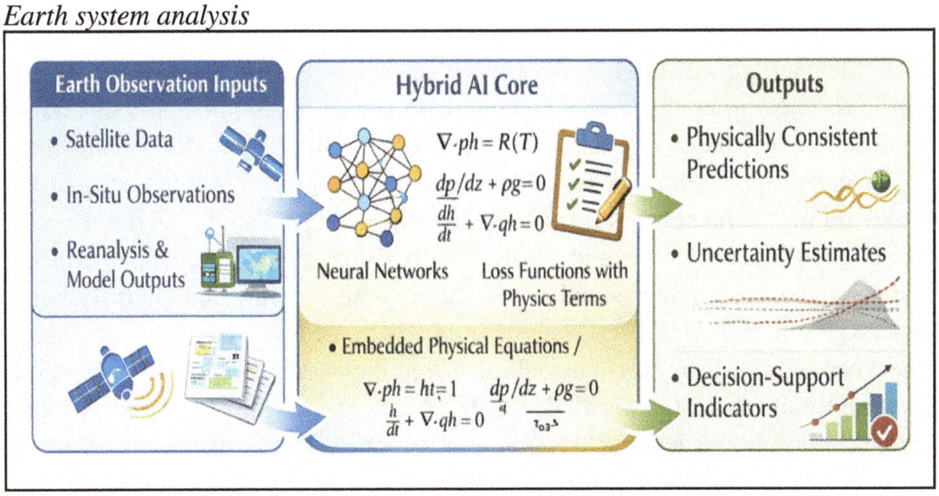

Theoretically, it is possible to view hybrid systems as the integration of two complementary streams of knowledge: mechanistic simulation and empirical pattern learning. The physics part is what keeps the model in scientifically valid solution spaces, whereas the AI part improves the model in terms of representation of unresolved processes, nonlinear interactions and observational heterogeneity. Physical consistency in physics-informed formulations is not an extrinsic concept that is added after prediction, but it is instead a component of the training process itself. It is an idea-based architecture that manifests a turn towards less statistical optimization and more towards physically directed learning machines that are able to be scientifically reliable and operationally robust.

4. HYBRID ARTIFICIAL INTELLIGENCE MODELING FRAMEWORKS

4.1 Hybrid Models Conceptual Foundations

Hybrid AI modelling structures have also become a strategic solution to the complementary constraints of purely physics-based and purely data-driven methods to Earth system simulation, that is, a deliberate combination of mechanistic process understanding with adaptive data learning to improve reliability and predictive ability. In principle, hybrid models aim to exploit the interpretability and conservation properties of physics-based numerical simulations by using the predictive nonlinear pattern-recognition capabilities of machine learning, and thus, offer a better representation of complex processes that would otherwise be hard to parameterise or solve at small spatiotemporal scales. Dynamically model outputs of numerical weather prediction or Earth system models are used together with machine learning predictors to make improved forecasts of phenomena like rainfall, streamflow and drought risk in the context of hydro-climatic forecasting, such as hybrid forecasting systems, which show that the combination of physics and AI can lead to better predictive skill than when using individual models (Slater et al., 2023). In hydrology, modern hybrid frameworks have generalized these principles by integrating neural networks in differentiable hydrological models to free-flowing internal dynamics of fluxes and learn spatially distributed corrections of conceptual parameters to produce simulations that are both accurate and interpretable in a physically meaningful sense (Huynh et al., 2025; see also Huynh et al., 2026). The theoretical basis of hybrid modelling is thus built on bi-directional interactions between the mechanistic knowledge and the data-driven learning, where process models inform the structure and restraint of AI components, and machine learning compensates the structural gaps or dynamics gaps in classic representations. This symbiosis is what allows hybrid structures to handle scale incompatibilities, to be more robust in non-stationary scenarios, and less reliant on large labelled databases, but at the same time be physically accurate-this is critically essential when applied to scientific and operational modelling of the Earth system (Slater et al., 2023).

4.2 Coupling Physics-Based Models With Machine Learning

Machine learning elements can be incorporated into physics-based models with a continuum of coupling strategies which vary in both structural depth, feedback, and optimisation design. These architectures are typically categorized as either serial (loosely coupled) or parallel (tightly coupled) hybrid architectures which are applicable to different scientific and operational domains.

4.2.1 Serial (Loosely Coupled) Hybrid Architecture

In serial implementation the physics-based model is used to run first and create the intermediate or baseline outputs or temperature fields, streamflow estimates, or precipitation forecasts as an input to a machine learning module. The AI element generally learns residual errors, systematic biases or nonlinear corrections that have not been resolved based on observations. In this architecture, physics model and machine learning model optimisation typically occur either independently/sequentially. This statistical-dynamical design is often used in hybrid forecasting systems, which include the post-processing of numerical weather prediction models to generate a useful outcome using machine learning methods to enhance the predictive capability (Slater et al., 2023). This format has the benefit of modularity: the physical model itself can be interpreted and used, and AI module is capable of improving performance without changing governing equations. The AI correction feedback however does not alter the physics internally during simulation. An illustration of a practical application is the regional flood forecasting where the streamflow is simulated by a distributed hydrodynamic model by initially simulating the rainfall input. This is then corrected by a neural network with past errors to help in correcting systematic underestimation of peak discharge (Huynh et al., 2025).

4.2.2 Parallel (Tightly Coupled) Hybrid Architecture

Conversely, parallel architectures incorporate elements of machine learning directly into the physical modelling framework and the components are optimised together. Rather than producing post-process outputs, neural networks can instead be used to substitute or even add certain operators within a set of differential equations, and the overall physical structure can still be present. These hybrid methods of differential equations enable AI elements to train unresolved sub-grid processes without contradiction with conservation laws. As an example, state dependent neural networks have been added to spatially distributed hydrological equations to learn adaptive corrections to internal flux representations, and simultaneously learn physical parameters and neural weights (Huynh et al., 2026). With this set-up, both physical and data-driven portions evolve together in the course of training and enable empirical feedback to mechanistic processes and vice versa. Parallel architectures are generally more demanding in their computational needs, but are also more integrated, learn more parameters better, and are scientifically easier to understand in cases where understanding the processes themselves is incomplete.

4.2.3 Comparative Perspective

Serial and parallel coupling are based on application goals. Serial architectures focus more on the ease of operation, processor efficiency, and modular deployment and thus are appropriate when forecasting systems. Parallel architectures are more concerned with scientific discovery, learning of parameters together, and more integrated process knowledge hence suited better to research oriented modelling and mechanistic refinement. Two decades of hydrological and climate studies have shown that non-sequitur integration can integrate the generalisation abilities of machine learning with the interpretability and scalability of mechanistic models, as long as strategies of coupling are well planned and justified.

4.3 Benefits of Hybrid Approaches to Earth System Analysis

Hybrid AI modelling strategies have a set of benefits in comparison with traditional single paradigms, in particular, better predictive abilities, better model interpretability, and simpler generalisation to a variety of environmental factors and data regimes. The combination of machine learning and physical context enables hybrid models to use domain knowledge to limit predictions to avoid physically unfeasible outputs, which can be the fatal weakness of entirely data-driven models, that can disobey conservation laws, or have rogue long-term behaviour (Slater et al., 2023). Hybrid frameworks are good at processing heterogeneous multiscale data, including the combination of high-resolution satellite images, process-model outputs, and in situ observations. This fusion generates more detailed feature representations and better identification of the spatiotemporal pattern that is hidden in the traditional models. Hybrid models have been shown to have a higher level of accuracy in forecasting floods and stream-flow simulation than models based on purely physics-based or only on observations, showing that hybrid models are able to capture process-based dynamics as well as non-linear interactions acquired via observations (Huynh et al., 2025). Furthermore, interpretability can be increased by having AI components act as residual fixes within familiar physical models so that researchers can make comparisons between learned fixes to physical processes and can be able to interpret changes to parameters in scientifically useful language. Hybrid models have also been shown to lessen the use of large labelled training sets by using physical constraints and simulations to guide the learning procedure, especially in data-sparse areas or of uncommon extreme events. All of these lead to hybrid AI methods being a key methodological breakthrough to the analysis of Earth systems, with the ability to perform strongly, with scientific viability, and with practical usefulness across a spectrum of applications, including climate prediction to environmental risk assessment.

5. PHYSICS-INFORMED ARTIFICIAL INTELLIGENCE MODELS

5.1 Principles of Physics-Informed Machine Learning

Physics-informed machine learning represents a paradigm shift of modelling methods in which core physical laws, including conservation of mass, momentum, energy, etc, are explicitly imposed on the AI model training procedure. This entrenching bolsters predictive reliability, consistency with well-established science, and generalisation to situations not covered by the training data distribution. Instead of considering machine learning as a statistical pattern recogniser, physics-informed learning adds governing equations, such as differential equations and algebraic constraints, as further penalty terms or constraints to the optimisation goal, thereby making sure that solutions predicted obey known physical behaviour (Karniadakis et al., 2021). It is possible to incorporate this physical structure via augmented loss functions, custom network architectures that impose invariants or variable physical operators embedded within neural networks; both of which can reduce the need to use large labelled datasets and reduce overfitting. The reasoning behind this is based on the assumption that every environmental process is often well-known on fundamental levels; through informing AI models with the prior knowledge, higher resilience and readability can be realized. As an example, the incorporation of Navier-Stokes equations or energy balance constraints into deep learning systems has been demonstrated to produce physically realistic fluid-flow and climate forecasts even with limited availability of observational data (Raissi et al., 2019). In addition to this, physics-informed models are able to measure uncertainty more sensibly since they respect known dynamics, which is especially crucial during the projection of extreme events or interacting processes. These methods combine physical constraints with the learning process, making them effective in overcoming the discrepancy between theory-based and data-driven models, enabling scientists to capitalize on the strengths of the former and minimize unrealistic predictions and increase the reliability of Earth system applications.

5.2 Physics-Informed Neural Networks (PINNs)

Physics-informed neural networks (PINNs) have become one of the most common applications of physics-informed machine learning, which embeds both differential equations and initial and boundary conditions into the actual training of neural networks, as a way of making neural networks consistent with real processes. PINNs introduce a composite loss that consists of a misfit term reflecting the data, and residuals of governing equations, so that the loss can be minimised by the network to enforce physical laws as well as fit to observations or simulated data (Raissi et al., 2019). This two-fold aim allows PINNs to discover solutions to PDEs and inverse problems with no need to have dense labelled datasets, which can be especially appealing to Earth -system issues with sparse or uneven observations. As an illustration, PINNs have been used to model groundwater-flow by encoding the law of Darcy and mass conservation into the learning model, which can then be used to determine physically consistent estimates of underground hydraulic heads and fluxes despite the limited data available (Meng & Karniadakis, 2020). Nearer adaptations of PINNs have added stochastic and probabilistic formulations to understand uncertainty, with Bayesian priors and variational inference to measure confidence intervals around these forecasts (Yang et al., 2024). The quantification of uncertainty is particularly useful in climate and hydrological studies, where a risk assessment and policy recommendations require it. Although PINNs are theoretically appealing, they may be confronted by practical difficulties, including the trade-off between data loss terms and physics loss terms and the computational cost of high-dimensional problems but current research is still working on optimisation schemes, adaptive sampling schemes and multi-fidelity formulations that better balance the trade-offs between data loss terms and physics loss terms and lead to improved training and solution quality (Zhao et al., 2023). Closely combining AI with physical knowledge, PINNs present a flexible modelling model of a wide variety of geophysical phenomena requiring interpretability and predictability.

5.3 Incorporating Conservation Laws and Physical Constraints

The main advantage of physics-informed artificial intelligence models is the ability to explicitly regulate the learning process with conservation laws and other physical rules. This property is a direct countermeas to the major limitation of purely data-driven models, which can have high predictive power but can be inconsistent with even the most basic physical principle, like the principle of mass continuity, or momentum conservation, or energy conservation. Such infractions in Earth system use would either result in physically unrealistic predictions, long run instability, or cause a drift in state variables, compromising the scientific integrity as well as the operational reliability. PIMs minimize such risks by ensuring predictions do not go out of physically admissible regimes. This may be via network architectures, parameterisation, or training objectives that observe invariants including non-negativity, monotonicity, symmetry properties as well as conservation relationships. As an example, hydrological modelling requires water-mass continuity and does not allow unrealistic gains or losses in simulated streamflow and soil moisture, which is important to assess water-resources and predict flood risks (Raissi et al., 2019; Wang et al., 2024). Likewise, in atmospheric and climate modelling stability in long-term integrations, as well as the unphysical drift in temperature or moisture distributions are reduced through the embedding of thermodynamic balance and energy conservation principles. Notably, instead of being enforced, physical constraints may be integrated with hybrid AI systems in two conceptually different ways, which are soft enforcement and hard enforcement. Conservation laws are added as penalty terms in the optimisation objective in soft-constrained formulations. Breakages of established physical behaviour augment the training loss thus deterring but not necessarily prohibiting violations. The method is flexible whereby the governing equations are approximate or partially known or when there is great uncertainty in observational data.

Hard-constrained formulations on the other hand fix physical relations into the structure of the model or into the state description. Construction satisfies conservation principles, boundary conditions, or invariants which are strictly physical, and make training and inference physically consistent. Although hard constraints ensure that the structure is compliant with known laws, the strong assumptions they may need are that governing equations are reliable and complete. Soft and hard constraint strategies, however, can be chosen based on how much confidence there is in process knowledge, the quality and the density of the data available and the purpose of application. Hard constraints are associated with the strict compliance with set laws, and soft constraints are associated with the flexibility in an uncertain physics. Physics-informed AI makes machine learning models easier to interpret, based on the hypothesis space by reducing the system to physically meaningful solutions, prevents non-stationary environmental conditions, and minimises existing chances of scientifically inconsistent predictions. Such systematic combination of physical information with information-based learning is an important step towards capable and authoritative Earth-system modelling and decision aid.

5.4 Benefits for Generalisation, Interpretability and Robustness

Physics-informed AI models have several benefits over traditional AI models, including superior generalisation to previously unknown conditions, superior understanding based on physical laws, and greater resilience to noise or data sparsity, which is crucial when applied to Earth-systems. These models prevent the possibility of implausible solutions when learning is constrained by a set of physical principles, which tend to give rise to extrapolations that would otherwise be obtained by using data in order to predict events that are still accurate with changing environmental regimes. This is especially essential in climate science, where models can be used in future situations that are significantly different to those in the historical record; the physical laws can mitigate the unphysical drift and enhance the probabilistic accuracy of future forecasts (Karniadakis et al., 2021). In addition, physics integration encourages interpretability, during which learned representations can be projected onto known mechanisms and process behaviours and this enables scientific venturing and stakeholder confidence. Also, physics-aware AI is particularly adapted to use in data-sparse areas or where measurement unpredictability is high because of its resistance to limited or noisy data, so that training can be stabilised and solutions directed towards realistic behaviour. More recently, it has been shown that physics-informed frameworks have the ability to integrate multi-fidelity datasets, which is coarse physics-based simulations and sparse high-resolution observations integrated to provide better approximations that respect data quality and underlying dynamics (Zhao et al., 2023). Altogether, these advantages highlight the ability of physics-informed AI as a solution to the gap between the theoretical knowledge and the empirical findings and provide a powerful modelling paradigm to ultimately improve scientific insight and predictive capabilities in the analysis of the Earth-system.

6. DATA INTEGRATION AND REMOTE SENSING OUTPUTS

6.1 Multi-Sensor Earth Observation Data for Hybrid Modelling

Multi- sensor Earth-observation data integration forms a key base to hybrid and physics-inspired AI models, and allows characterisation of the Earth-system processes in a comprehensive way, both spatially, spectrally, and temporally. The optical satellite sensors provide high-resolution data on surface reflectance, vegetation indices, and land -cover dynamics, which are critical in ecosystem processes, agricultural output, and land -use evolution modelling. The thermal infrared sensors can give some important information on the land-surface temperature, evapotranspiration, and surface-energy balance, thus assisting in hydrological and climatic studies. Synthetic-aperture radar (SAR) data have some distinctive features in all-weather and nighttime observation of surface roughness, soil moisture, and inundation processes, and are typically beneficial in areas prone to clouds or those with limited data availability. LiDAR observations also add to hybrid frameworks by providing three-dimensional data on terrain elevation and vegetation structure, which facilitates hydrological routing, biomass estimation, and representation of surface processes. The complementary capabilities of these sensors allow hybrid AI models to resolve non-linear interactions and fine-scale heterogeneity, which are usually difficult to capture using standard physics-based models. Multi-sensing can enable hybrid frameworks to enhance the richness of features, decrease the uncertainty of observation with cross-validation, and strengthen models in a variety of environmental conditions (Reichstein et al., 2019; Ma et al., 2019). Cloud-based geospatial platforms are now playing a pivotal role in supporting multi-sensor integration, allowing data to be processed at scale while enabling timely and dynamic environmental monitoring.

6.2 Convergence of Remote Sensing, In Situ, and Reanalysis Data

The synergistic combination of remote sensing data, in situ observations and reanalysis data has become a cornerstone of hybrid AI modelling systems, especially when it comes to Earth system applications where each one of these data types alone is inadequate. Individual datasets each bring complementary strength and a range of different uncertainties with them which have to be brought into systematic reconciliation. High-accuracy point observations can be obtained in situ as meteorological station records, streamflow gauges, and soil moisture probes and are important benchmarks to be used during calibration and validation. Nonetheless, they are frequently sparsely and evenly distributed across space, which makes them less representative on regional and global levels. Remote sensing observations overcome this spatial constraint by providing near-continuous geographic coverage, as well as, uniform sampling of large areas. However, spatial completeness does not mean certainty in measurement. Products derived by satellite can include retrieval errors due to cloud contamination, atmospheric correction errors, sensor calibration drift, surface heterogeneity or algorithmic inversion assumptions. Reanalysis data sets give an intermediate point of integration by absorbing observational data into the physical numerical model to generate dynamically consistent spatially continuous fields. Even though reanalysis products are physically well-coherent and well-covered, they can still be biased in structure due to model parameterizations or data assimilation assumptions.

The hybrid AI models never replace a certain dataset with another; instead, they align these complimentary characteristics in a systematic data fusion and assimilation methods. Dynamically balanced between the measurement accuracy of a spatial continuity and the reliability of observational data are neutrally-based methods of learning based on the statistical mixture of neural data and a physical informed background state. The fusion models that are based on machine learning also balance datasets that are at varying spatial and temporal resolutions and enable the system to learn scale-consistent representations and heterogeneous error structures (Reichstein et al., 2019; Slater et al., 2023). The hybrid systems eliminate the trade-off between the coverage and uncertainty by combining the point level precision, spatial completeness and a physically consistent background field into a single modelling architecture. This is an organized convergence that strengthens calibration and stabilizes predictive, as well as makes environmental assessment in a way that is more credible on top of regional and global levels.

6.3 Resolving Data Uncertainty, Noise, and Scale Mismatch

Hybrid AI models for Earth system analysis continue to face obstacles such as data uncertainty, observational noise, and mismatched scales, making it essential to adopt clear methodological approaches that can effectively mitigate these limitations. Uncertainties that affect remote sensing products are sensor calibration error, atmospheric correction, and limitation and assumptions in retrieving the algorithm, but in situ observations, it may include measurement error, time lag, and irregular monitoring of the same. Reanalysis datasets are physically consistent, but may have systematic biases, especially in areas that have a small observational contribution. The challenges are solved by using hybrid AI frameworks that include uncertainty-aware learning techniques, including probabilistic modeling, ensemble learning, and Bayesian inference, that allow explicit representation and transfer of uncertainty within the entire modeling process (Karniadakis et al., 2021; Wang et al., 2024). Mismatched scale between datasets that are running at different spatial and temporal scales makes data integration even more difficult; and in most cases, down sampling or up sampling methods are required to balance inputs. More sophisticated methods such as multi-resolution feature extraction, physics-inspired regularization, and scale-conscious neural networks have been demonstrated as effective to enhance the stability and explainability of hybrid models across scales (Zhao et al., 2023). Hybrid AI models can be more robust, have greater generalization, and scientific credibility by systematically addressing uncertainty, noise, and scale mismatch, required to operate and be deployed practically and used in policy-relevant applications of Earth systems.

7. APPLICABILITY OF EARTH SYSTEM ANALYSIS

7.1 Climate Modelling and Climate Change Impact Assessment

Hybrid and physics-informed AI models are increasingly applied in climate modelling and climate change impact assessment to address the intrinsic nonlinearity, multiscale interactions, and computational complexity of the climate system. Traditional general circulation models (GCMs), while grounded in physical laws, face limitations associated with coarse spatial resolution, parameterisation of sub-grid processes, and computational expense—particularly in representing cloud microphysics, convection, and land–atmosphere feedbacks.

Hybrid approaches enhance these models by learning systematic biases and unresolved dynamics from observational and simulation data while preserving physical consistency. For example, machine learning components have been embedded within general circulation frameworks to approximate computationally expensive parameterisation schemes, leading to improved efficiency and forecasting skill (Rasp et al., 2018). Rather than replacing physical equations, AI modules act as adaptive surrogates or correction mechanisms within a physically constrained structure.Physics-informed AI further strengthens climate applications by enforcing thermodynamic consistency, energy conservation, and mass balance, which are essential for long-term climate projections under future emission scenarios. These approaches have demonstrated utility in reional climate downscaling, extreme temperature forecasting, and precipitation variability analysis, providing more robust climate risk assessments (Reichstein et al., 2019; Karniadakis et al., 2021). By integrating observational records, reanalysis datasets, and climate model outputs, hybrid frameworks also support improved attribution studies and scenario-based impact evaluation, thereby strengthening evidence-based adaptation and mitigation planning.

7.2 Hydrology, Flood Forecasting, and Water Resource Management

In hydrology and water resource management, hybrid and physics-informed AI models have shown substantial promise in improving simulation accuracy, forecast reliability, and uncertainty management under climatic variability. Process-based hydrological models capture basin-scale water dynamics but may suffer from structural simplifications and parameter uncertainty, particularly under extreme rainfall events or changing land-use conditions. Hybrid configurations combine physically based hydrodynamic models with machine learning components capable of learning non-linear rainfall–runoff relationships and catchment-specific behaviours. Such models have been shown to correct systematic biases and enhance predictive performance during flood events (Slater et al., 2023). Importantly, physics-informed formulations preserve water-mass continuity and physically realistic flow processes, which are critical for flood risk assessment, reservoir management, and infrastructure planning. Applications extend to groundwater modelling, soil moisture estimation, and evapotranspiration monitoring, where satellite-derived observations are integrated with hydrological process equations to improve spatial resolution and temporal continuity (Wang et al., 2024). These advances facilitate drought monitoring, sustainable reservoir operation, and adaptive water allocation, particularly in data-sparse regions. By maintaining interpretability while enhancing predictive capability, hybrid AI systems provide a robust decision-support framework for sustainable water resource management under climate uncertainty.

7.3 Ecosystem Dynamics, Agriculture, and Land-Use Analysis

Hybrid AI methodologies play a central role in ecosystem monitoring, agricultural forecasting, and land-use/land-cover change analysis by combining remote sensing observations with process-based ecological models. Earth observation data provide high-resolution spatial information on vegetation indices, surface reflectance, and phenological cycles, while ecosystem models represent biophysical processes governing productivity, carbon exchange, and water balance. By integrating these knowledge streams, hybrid frameworks enable improved estimation of vegetation productivity, biomass accumulation, phenological transitions, and ecosystem stress indicators. Applications include deforestation monitoring, land degradation detection, and biodiversity assessment using multi-sensor satellite data in conjunction with ecological process constraints (Ma et al., 2019).In agriculture, hybrid systems combine satellite-derived vegetation metrics with meteorological, soil, and crop growth models to forecast yield, optimise irrigation, and detect pest or disease outbreaks. Physics-informed approaches incorporate plant growth dynamics and water-balance principles, enabling generalisation across regions and growing seasons without requiring extensive historical retraining. These capabilities contribute directly to food security assessment, sustainable land management, and environmental protection by providing spatially explicit, policy-relevant information.

7.4 Natural Hazards and Disaster Risk Reduction

The application of hybrid and physics-informed AI models in natural hazard analysis and disaster risk reduction is expanding rapidly due to their ability to integrate heterogeneous datasets and represent nonlinear hazard dynamics. Earth-system hazards—including floods, droughts, landslides, wildfires, and heatwaves—arise from complex interactions between climatic, hydrological, geological, and anthropogenic drivers. Hybrid frameworks combine physics-based hazard models with machine learning systems trained on historical events and remote sensing observations to enhance early-warning capabilities and impact prediction (Reichstein et al., 2019). Physical constraints ensure that hazard simulations remain dynamically plausible, which is essential for operational reliability and stakeholder confidence. Applications include flood inundation mapping, drought severity forecasting, wildfire spread modelling, and heatwave risk assessment under diverse climatic regimes. The scalability of AI-enhanced remote sensing further enables implementation at regional and global scales, improving preparedness in vulnerable and data-limited regions. As climate extremes intensify, hybrid AI systems are poised to play a central role in climate-resilient planning, disaster mitigation strategies, and adaptive governance frameworks. The domain-specific applications described above correspond directly to the representative studies summarised in Table 2, illustrating how hybrid and physics-informed architectures translate from methodological innovation to operational Earth-system practice.

Table 2. Applications of Hybrid and Physics-Informed AI Models Across Earth System Domains

Earth system domain	Hybrid AI approach used	Key benefit	Representative studies
Climate modelling	Physics-informed neural networks	Improved long-term stability	Karniadakis et al. (2021); Beucler et al. (2021)
Hydrology	Physics-encoded deep learning	Enhanced flood prediction	Huynh et al. (2025); Wang et al. (2024)
Atmospheric science	Hybrid physics-AI forecasting	Superior extreme event prediction	Das et al. (2024); Slater et al. (2023)
Remote sensing	Hybrid CNN-physics models	Improved generalisation	Janga et al. (2023)
Environmental modelling	Bayesian PINNs	Uncertainty-aware predictions	Yang et al. (2024)

8. LIMITATIONS, CHALLENGES, AND ETHICAL CONSIDERATIONS

8.1 Methodological and Computational Problems

Although hybrid and physics-informed AI models have started to gain momentum, both approaches face a number of methodological and computational challenges, which limit their application in the analyses of the Earth-system on a global scale. One of the main challenges can be seen in the design and optimisation of hybrid architectures, especially on how to best couple physical models with machine learning parts to be stable or not unnecessarily complex. Data-driven learning methods often require careful tuning of loss functions, regularisation parameters, and model structures- computationally expensive tasks, as well as tasks specific to the problem itself. Physics-informed neural networks, such as those, can also have convergence issues and can be incompatible with high-dimensional or long-term spatiotemporal dynamics of climatic and hydrological systems (Karniadakis et al., 2021). Also, training hybrid models can require significant amounts of computational resources because it entails the processing of big datasets of Earth observations, simulation of physics, and iterative optimisation, thus restricting accessibility to researchers and institutions with limited computational resources. The lack of a set of standards and evaluation measures also affects systematic comparison across the studies and inhibits reproducibility and generalisation. These issues of methodology and computability are a central research agenda in developing hybrid AI systems that are no longer experimental models, but effective and scalable instruments of Earth-system science.

8.2 Data Limitations and Model Generalisability

The performance and adaptability of hybrid and physics-informed AI models are often constrained by data gaps, observational inconsistencies, and limited monitoring, challenges that are particularly acute in under-observed regions. Although remote sensing provides a large-scale spatial coverage, most environmental variables of interest, including subsurface hydrological processes or ecosystem health indicators, are indirectly measured and are uncertain to be retrieved. In-situ datasets are more accurate, but they are unevenly distributed across the globe and may not be able to accurately represent spatial heterogeneity or extreme events. Such constraints may cause any biases in the model training and reduce generalisability when hybrid models are used outside the area of their calibration. Furthermore, processes of the earth systems are non- stationary by nature in response to climate changes, land-use changes, and human activities, thus, not allowing a full mere assumption that previous information is sufficient to depict future scenarios. The hybrid models that are trained using previous observations might therefore not be able to make reliable extrapolations to new climatic regimes or unprecedented extremes (Reichstein et al., 2019). Even though physics-informed constraints can alleviate this problem by including basic laws, they are not capable of completely correcting poor knowledge of processes or the lack of drivers. The need to have strong generalisation, hence requires constant model updates, incorporation of multi-fidelity data and controlled validation in many different environmental conditions, which is still a current challenge in the study of hybrid AI.

8.3 Interpretability, Transparency, and Trust

Interpretability, Transparency, and Trust are concerned with the transparency of the findings of the report and the trustworthiness of the findings presented. The next issue that arises regarding interpretability and transparency in the implementation of hybrid AI models to Earth-system analysis is especially relevant when model outcomes are used to make high-stakes decisions related to climate policy, disaster risk reduction, and resource management. Even though hybrid and physics-informed methods tend to be more interpretable compared to the pure data-driven black-box models, the complexity of their nature still makes it difficult to discern the relationships between the inputs, learned parameters, and the outputs. To bring about scientific plausibility and reliability in the eyes of the stakeholders, it is necessary to understand how the machine learning elements modify or rectify physical model behaviour. Transparency can be improved using explainable AI methods, including sensitivity analysis, feature attribution, and physically interpretable latent representations, but there is a limited use of such methods in hybrid Earth-system models, and these methods tend to be context-specific (Samek et al., 2021). Additionally, there is a difficulty in communicating model uncertainty and limitations to non-technical users since the use of too much technical language can diminish usability, and excessive simplification can be misunderstood. To establish trust in hybrid AI systems, however, it is not only that there is technical interpretability but that the documentation is clear, the validation transparent and there are appropriate communication strategies that support the needs of decision-makers and practitioners.

8.4 Ethical Considerations and Responsible Use

Ethical considerations and responsible use should be taken into account during the process to avoid possible ethical concerns and abuse. The ethical implications of the field of hybrid and physics-informed artificial intelligence (AI) models are playing an increasingly central role in the creation and implementation of these computational systems in Earth system science, especially as these processing systems can affect environmental regulation and social performance. One of the chief concerns relates to the aspect of data bias and underrepresentation, where the non-uniform coverage of spatial and temporal data could give rise to models that perform well in areas of substantial data coverage but fail in marginalised or vulnerable areas and, as a consequence, could serve to enforce existing inequities. The use of AI-powered models in policy and planning offers additional concerns regarding accountability when the potential consequences of the decision-making based on model outputs are highly social, economic, or environmental. Responsible use implies open communication of model assumptions, limitations, and uncertainties, and, to this end, the involvement of stakeholders in the process of model development and implementation (Reichstein et al., 2019). In addition, the principles of ethics in AI, such as fairness, reliability, and sustainability, should be incorporated in the development of hybrid structures, especially in cases pertaining to climate adaptation and disaster management. As models of hybrid AI gain some power in environmental decision-making, integrating ethical values with technical change will be a necessary measure to ensure that these technologies are used positively to manage the Earth's systems in a sustainable, friendly manner. Figure 2 outlines the key issues and new avenues of research linked with hybrid and physics-inspired AI models and focuses on the aspects of scalability, interpretability, uncertainty quantification, and functional implementation in Earth system sciences.

Figure 2. Key challenges and future directions of hybrid and physics-informed AI in Earth system science

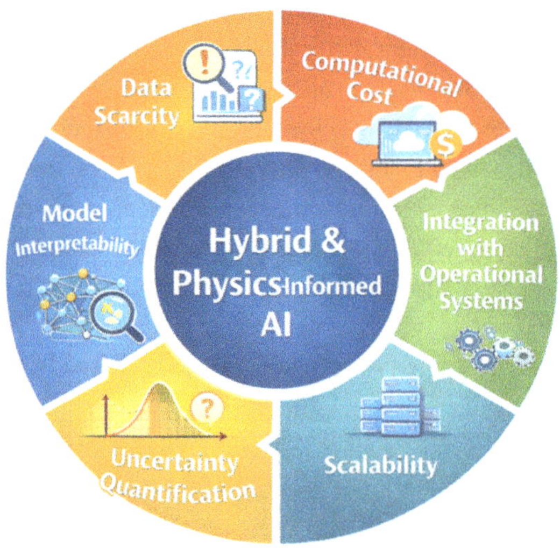

9. FUTURE PERSPECTIVES AND NEW TRENDS

9.1 Improvement in Physics-AI Integration and Model Design

It is expected that future studies of hybrid and physics-informed AI in the analysis of Earth systems will focus on more intensive and comprehensive interventions between physical process knowledge and machine-learning systems. Existing hybrid methods often operate based on quite straightforward coupling approaches or residual learning, but new methods attempt to directly incorporate physical laws into network architectures through differentiable solvers, operator learning, and neural encodings of governing equations. New directions in universal differential equations and neural operators provide potential directions of learning complex spatiotemporal dynamics and maintaining physical consistency at different scales (Karniadakis et al., 2021). Such advances are of specific concern to the simulation of highly nonlinear systems, including atmospheric convection, land-atmosphere interactions, and human-natural systems, in which conventional parameterisations are still unclear. It is also plausible that the anticipated hybrid models in the future will be able to adapt to environmental changes in real-time by dynamically adapting the balance of data-driven and physics-based components to the changing climate conditions, hence increasing the robustness under non-stationary climate regimes. These innovations will help to have more flexible, interpretable, and generalisable Earth system models that can be used to both discover and operationalise.

9.2 Selecting Hybrid Models with Big Data and Cloud Computing

The growing access to high-resolution satellite images and long-term datasets of earth observation requires scalable computational frameworks to support the next-generation hybrid AI modelling. Geospatial services on the cloud, high-performance computing systems will likely be used as a key to large-scale data integration, model training, and real-time inference. With the help of distributed computing and cloud-native machine-learning workflows, it is possible to adjust petabyte-scale datasets and even work at continental and global scales with a higher level of efficiency using hybrid models. Such trends as the use of federated learning and edge computing to allow decentralised model training, maintain data privacy and alleviate computational bottlenecks are emerging (Reichstein et al., 2019). Also, in the future, improvements in automated machine learning and workflow coordination will tend to reduce the technical barrier to implementing the hybrid frameworks, making them more affordable to interdisciplinary research teams. With further advancements in computational scalability, hybrid AI models will be used more frequently to offer near-real-time environmental surveillance, early warning systems, and adaptive decision assistance in rapidly changing conditions.

9.3 Explainable, Uncertainty-Aware, and Decision-Centric AI

One of the most important future directions of hybrid and physics-informed AI is the creation of explainable and uncertainty-aware modelling frameworks that are used in a compromise and engineered explicitly to facilitate decision-making. With the increasing number of AI-based outputs in climate adaptation planning, disaster risk management, and environmental policy, the need to have transparent models able to convey both projections and relevant uncertainties in a way that is easy to understand is increasing. The explainable AI can be advanced, including physically meaningful features attribution, sensitivity diagnostics, and causal inference, which are expected to increase the trust and usability of hybrid models by non-technical stakeholders (Samek et al., 2021). Moreover, the combination of probabilistic learning and ensemble with physics-informed will enhance uncertainty quantification, which will allow more robust risk-based evaluation. It is also probable that future hybrid systems will be user-oriented and that they will incorporate decision-oriented metrics and scenario analysis functionality that will convert complex model outputs into practical information. Such developments will enhance the nexus of Earth system science, policy formulation as well as sustainable development practice.

9.4 Towards Integrated and Sustainable Earth System Intelligence

In the future, hybrid and physics-based AI models will become the foundation blocks of an integrated Earth system intelligence system that brings together observations, models, and decision support into intelligent and dynamic systems. Such frameworks will progressively integrate socio-economic information with biophysical variables, which will allow comprehensive studies of human-environmental interaction, as well as the sustainability dilemmas. It will also interoperate with other global programs like digital Earth platforms and climate services which will enhance interaction of knowledge across fields and across regions. However, to achieve this vision, there will be a need to invest in open data, reproducible research practices, and interdisciplinary collaboration in the long run. Ethical and governance issues should also develop simultaneously to make sure that the AI-driven Earth system intelligence is inclusive, transparent, and in line with the values of society. Future studies have the ability to dramatically expand our ability to comprehend, anticipate, and manage complicated Earth system processes to promote climate resilience and sustainable development by moving toward combined, accountable, and scalable hybrid AI solutions.

10. CONCLUSION

This chapter has provided an in-depth analysis of hybrid and physics-informed artificial intelligence systems in Earth system science, highlighting their conceptual foundations, methodology, and application in Earth system science. These methods combine experimental evidence with physical principles to solve the long term constraints of pure mechanistic models on the one hand, and the unconstrained data-driven systems on the other. Instead of being used as incremental improvements, hybrid and physics-informed models are a structural combination of both statistical learning and physical reasoning allowing predictive systems that are both adaptive, physically consistent, and scientifically interpretable.

In climate modelling, hydrology, ecosystem dynamics and natural hazard analysis, it has been shown in evidence reviewed in this chapter that hybrid architectures are better at predicting accuracy whilst consistent with conservation laws and thermodynamic consistency. This two-way power is especially essential in the circumstances of data sparsity, non-stationarity as well as accelerating climate change when pure empirical extrapolation can fail, and pure mechanistic models can be computationally infeasible. Strong links between remote sensing, in situ measurements, reanalysis products and numerical models within hybrid systems enhance the strength of calibration, quantification of uncertainties and credibility of operations further. There are wider consequences of these developments on a methodological level. Hybrid AI offers a more interpretable and physically realistic structure to researchers to study complex Earth-system processes without the trade off of interpretability or physical realism. To policy-makers and practitioners physics-informed modelling provides more transparent and uncertainty-aware as well as physically informed evidence with which to inform climate adaptation, in reduction of disasters risks, and sustainable management of resources. With the growing dependence of environmental decisions on model-based the capacity of hybrid AI systems to limit predictions under scientifically admissible regimes is necessary to ensure the preservation of institutional trust. Simultaneously, data quality, computational requirement, model generalisability, and ethical deployment issues are the key points of concern. The next steps will rely on the process of scaling computation, open-data infrastructure, the cross-disciplinary cooperation, and open assessment criteria. Hybrid AI and cloud computing as it is integrated with digital Earth projects and initiatives further increases access as well as enables more individuals to participate not only based on geographic and institutional lines.

In the future the development of integrated Earth system intelligence, whereby physical theory, machine learning, observational networks, and human expertise interact with each other in unified modelling ecosystems, is an important direction in sustainability science. The transparency, reproducibility, and responsible implementation will be important as these systems shift out of the research environments and into the operational governance environments. Finally, physics-inspired and hybrid AI models provide a well-rounded and prospective model of viewing, predicting, and controlling the intricate and mutually reinforcing mechanisms that constitute the Earth system in the age of swift global transformation.

REFERENCES

Amsalu, A., Garedew, W., Melka, G. A., Çelik, M. A., & Gemeda, D. O. (2026). Trends in climate extreme indices and implications for coffee farming in The Kafa Biosphere Reserve, southwest Ethiopia. *Climate Services*, *42*, 100637. DOI: 10.1016/j.cliser.2026.100637

Bonan, G. B., & Doney, S. C. (2018). Climate, ecosystems, and planetary futures: The challenge to predict life in Earth system models. *Science*, *359*(6375), eaam8328. Advance online publication. DOI: 10.1126/science.aam8328 PMID: 29420265

Camps-Valls, G., Reichstein, M., & Mahecha, M. D.. (2021). A perspective on machine learning for Earth system modeling. *Nature Reviews. Earth & Environment*, *2*(11), 719–737. DOI: 10.1038/s43017-021-00192-7

Çelik, M. A., Kıraç, A., & Gouveia, C. M. (2026). North Atlantic Oscillation modulation of drought and vegetation activity in Türkiye. *Theoretical and Applied Climatology*, *157*(2), 74. Advance online publication. DOI: 10.1007/s00704-025-05971-5

Das, P., Posch, A., Barber, N., van Werkhoven, K., & Ganguly, A. R. (2024). Hybrid physics–AI outperforms numerical weather prediction for extreme precipitation nowcasting. *npj Climate and Atmospheric Science, 7*, 282. DOI: 10.1038/s41612-024-00834-8

Flato, G., Marotzke, J., & Abiodun, B.. (2019). Evaluation of climate models. In *IPCC Special Report on Climate Change*. Cambridge University Press., DOI: 10.1017/9781009157896.009

Huynh, N. N. T., Garambois, P.-A., Colleoni, F., & Monnier, J. (2026). A hybrid physics–AI approach using universal differential equations with state-dependent neural networks for learnable, regionalisable, spatially distributed hydrological modelling. *Geoscientific Model Development*, *19*(3), 1055–1074. DOI: 10.5194/gmd-19-1055-2026

Huynh, N. N. T., Garambois, P.-A., Renard, B., Colleoni, F., Monnier, J., & Roux, H. (2025). A distributed hybrid physics–AI framework for learning corrections of internal hydrological fluxes and enhancing high-resolution regionalised flood modelling. *Hydrology and Earth System Sciences*, *29*(15), 3589–3613. DOI: 10.5194/hess-29-3589-2025

Janga, B., Asamani, G. P., Sun, Z., & Cristea, N. (2023). A review of practical AI for remote sensing in Earth sciences. *Remote Sensing (Basel)*, *15*(16), 4112. DOI: 10.3390/rs15164112

Karniadakis, G. E., Kevrekidis, I. G., Lu, L., Perdikaris, P., Wang, S., & Yang, L. (2021). Physics-informed machine learning. *Nature Reviews. Physics*, *3*(6), 422–440. DOI: 10.1038/s42254-021-00314-5

Kratzert, F., Klotz, D., Brenner, C., Schulz, K., & Herrnegger, M. (2019). Rainfall–runoff modelling using long short-term memory networks. *Water Resources Research*, *55*(8), 6503–6522. DOI: 10.1029/2018WR024567

Ma, L., Liu, Y., Zhang, X., Ye, Y., Yin, G., & Johnson, B. A. (2019). Deep learning in remote sensing applications: A meta-analysis and review. *ISPRS Journal of Photogrammetry and Remote Sensing*, *152*, 166–177. DOI: 10.1016/j.isprsjprs.2019.04.015

Meng, X., & Karniadakis, G. E. (2020). Physics-informed neural networks for high-dimensional PDEs: Forward and inverse problems. *Journal of Computational Physics*, *401*, 109020. DOI: 10.1016/j.jcp.2019.109020

Raissi, M., Perdikaris, P., & Karniadakis, G. E. (2019). Physics-informed neural networks: A deep learning framework for solving forward and inverse problems involving nonlinear partial differential equations. *Journal of Computational Physics*, *378*, 686–707. DOI: 10.1016/j.jcp.2018.10.045

Reichstein, M., Camps-Valls, G., Stevens, B., Jung, M., Denzler, J., Carvalhais, N., & Prabhat, . (2019). Deep learning and process understanding for data-driven Earth system science. *Nature*, *566*(7743), 195–204. DOI: 10.1038/s41586-019-0912-1 PMID: 30760912

Samek, W., Montavon, G., Lapuschkin, S., Anders, C. J., & Müller, K.-R. (2021). Explaining deep neural networks and beyond: A review of methods and applications. *Proceedings of the IEEE*, *109*(3), 247–278. DOI: 10.1109/JPROC.2021.3060483

Shen, C., Appling, A. P., Gentine, P., Bandai, T., Gupta, H., Tartakovsky, A., & Lawson, K. (2023). Differentiable modelling to unify machine learning and physical models for geosciences. *Nature Reviews. Earth & Environment*, *4*(7), 552–567. DOI: 10.1038/s43017-023-00450-9

Slater, L. J., Arnal, L., Boucher, M.-A., Chang, A. Y.-Y., Moulds, S., Murphy, C., Nearing, G., Shalev, G., Shen, C., Speight, L., Villarini, G., Wilby, R. L., Wood, A., & Zappa, M. (2023). Hybrid forecasting: Blending climate predictions with AI models. *Hydrology and Earth System Sciences*, *27*(9), 1865–1889. DOI: 10.5194/hess-27-1865-2023

Wang, C., Jiang, S., Zheng, Y., Han, F., Kumar, R., Rakovec, O., & Li, S. (2024). Distributed hydrological modelling with physics-encoded deep learning: A general framework and its application in the Amazon. *Water Resources Research, 60*, e2023WR036170. DOI: 10.1029/2023WR036170

Yang, Z., Luo, X., & Yu, Y. (2024). Bayesian physics-informed neural networks with uncertainty quantification for environmental modelling. *Environmental Modelling & Software*, *173*, 105331. DOI: 10.1016/j.envsoft.2023.105331

Yuan, Y., Chen, X., Wang, J., & Lu, X. (2021). Vision transformer for remote sensing image classification. *IEEE Transactions on Geoscience and Remote Sensing*, *59*(11), 9623–9636. DOI: 10.1109/TGRS.2021.3058466

Zhao, S., Zhang, D., & Karniadakis, G. E. (2023). Multi-fidelity physics-informed neural networks: Leveraging low- and high-fidelity data for predictive modelling. *Computer Methods in Applied Mechanics and Engineering*, *389*, 115599. DOI: 10.1016/j.cma.2021.115599

Zhu, X. X., Tuia, D., Mou, L., Xia, G.-S., Zhang, L., Xu, F., & Fraundorfer, F. (2017). Deep learning in remote sensing: A comprehensive review and list of resources. *IEEE Geoscience and Remote Sensing Magazine*, *5*(4), 8–36. DOI: 10.1109/MGRS.2017.2762307

Chapter 4

Uncertainty Analysis, Validation, and Benchmarking of AI Models in Earth Sciences

Atul Kumar
http://orcid.org/0000-0003-3859-0686
Dnyaan Prasad Global University, India

Turobova Sevara
Termez University of Economics and Service, Termez, Uzbekistan

Anorgul Ashirova
Mamun University, Khiva, Uzbekistan

Muyassar Allaberganova
Urgench State University, Urgench, Uzbekistan

Gularam Masharipova
Alfraganus University, Tashkent, Uzbekistan

Abdullayev Abdulla Fayzulla Ugli
Tashkent State University of Economy, Uzbekistan

Nekboyev Xurshid Xoliyorovich
Karshi State Technical University, Uzbekistan

ABSTRACT

Artificial intelligence (AI) and machine learning (ML) models have revolutionized Earth sciences, enabling unprecedented capabilities in climate prediction, remote sensing, and environmental monitoring. However, the reliability of AI-driven Earth science applications critically depends on rigorous uncertainty quantification, validation methodologies, and standardized benchmarking frameworks. This chapter comprehensively examines the state-of-the-art approaches for assessing AI model reliability in Earth sciences, covering probabilistic uncertainty estimation techniques,

DOI: 10.4018/979-8-2600-0778-5.ch004

spatial and temporal validation strategies, and emerging benchmark datasets. We analyze deep learning uncertainty methods including Bayesian neural networks, ensemble approaches, and conformal prediction, while addressing unique challenges posed by geospatial data such as spatial autocorrelation and distribution shifts.

1. INTRODUCTION

1.1 The Rise of AI in Earth Sciences

Artificial intelligence and machine learning have fundamentally transformed Earth sciences over the past decade, introducing powerful data-driven approaches that complement traditional physics-based models. The convergence of massive Earth observation datasets from satellite missions, unprecedented computational resources, and advances in deep learning architectures has enabled breakthrough applications across climate science, meteorology, hydrology, oceanography, and environmental monitoring (Schultz et al., 2021; Rasp et al., 2024; Amsalu et al., 2026; Çelik, 2026). These AI systems now achieve remarkable performance in tasks ranging from weather forecasting and extreme event prediction to land cover classification and atmospheric composition retrieval.

The adoption of AI in Earth sciences stems from several compelling advantages. Deep neural networks excel at capturing complex nonlinear relationships in high-dimensional data, learning hierarchical feature representations directly from raw observations without requiring explicit physical parameterizations (Dueben et al., 2022). Machine learning models can assimilate heterogeneous data sources—including multi-spectral satellite imagery, in-situ sensor networks, and reanalysis products—to generate predictions at spatial and temporal scales previously unattainable. Furthermore, AI approaches offer computational efficiency advantages over traditional numerical models, enabling rapid inference for time-critical applications such as severe weather warnings and natural disaster response (Olivetti & Messori, 2024).

1.2 The Critical Need for Uncertainty Quantification

Despite impressive empirical successes, the deployment of AI models in Earth sciences raises fundamental questions about reliability, interpretability, and trustworthiness. Unlike deterministic predictions that provide single-point estimates, Earth system applications require probabilistic forecasts that quantify prediction uncertainty to inform risk-based decision-making (Price et al., 2025; Haynes & Feingold, 2025). Uncertainty quantification (UQ) becomes particularly critical when

AI predictions influence high-stakes decisions in climate adaptation, agricultural planning, water resource management, and disaster preparedness.

AI models face multiple sources of uncertainty in Earth science applications. Aleatoric uncertainty arises from inherent stochasticity in physical processes and observational noise, representing irreducible variability in system behavior (Gawlikowski et al., 2023). Epistemic uncertainty stems from limited training data, model misspecification, and incomplete knowledge of underlying processes, which can potentially be reduced through additional observations or improved model architectures. Distribution shift presents another challenge when models encounter environmental conditions outside their training distribution, leading to unreliable predictions in novel scenarios (Meyer & Pebesma, 2021; Singh et al., 2024).

1.3 Validation Challenges in Geospatial Contexts

Validating AI models for Earth sciences introduces unique methodological challenges compared to standard machine learning applications. Geospatial data exhibit strong spatial and temporal autocorrelation, violating the independent and identically distributed (i.i.d.) assumption underlying conventional cross-validation techniques (Milà et al., 2022; Wang et al., 2023). Standard random splitting of spatially autocorrelated data leads to optimistic performance estimates due to information leakage between training and test sets, resulting in models that fail to generalize to new geographic regions or time periods (Meyer & Pebesma, 2022).

Earth observation data also present challenges related to data quality, missing observations, and sensor-specific artifacts that must be properly accounted for during model validation (Gevaert, 2022). Remote sensing applications require validation against independent ground truth data collected through field campaigns or higher-resolution imagery, yet obtaining representative validation samples across diverse ecosystems and environmental conditions remains logistically challenging and expensive. Furthermore, the temporal dynamics of Earth systems demand validation strategies that assess model performance across multiple time scales and capture seasonal, inter-annual, and long-term trends (Linnenbrink et al., 2024).

1.4 The Imperative for Standardized Benchmarking

The proliferation of AI applications in Earth sciences has created a pressing need for standardized benchmarking frameworks that enable systematic comparison of methods and reproducibility of results. Unlike computer vision or natural language processing, where established benchmarks like ImageNet and GLUE facilitate community-wide progress, Earth sciences have historically lacked unified evaluation protocols (Betancourt et al., 2021; Sumbul et al., 2021). This fragmentation hampers scientific progress by making it difficult to assess relative strengths of different approaches, identify best practices, and establish performance baselines for operational systems.

Recent initiatives have begun addressing this gap through creation of curated benchmark datasets covering weather prediction (Rasp et al., 2024), air quality forecasting (Betancourt et al., 2021), remote sensing classification (Lacoste et al., 2023), and post-processing of numerical weather predictions (Demaeyer et al., 2023). These benchmarks provide standardized data splits, evaluation metrics, and baseline models that accelerate research and facilitate fair comparisons. However, significant challenges remain in developing benchmarks that adequately represent the diversity of Earth system processes, spatial scales, and application requirements while maintaining scientific rigor and operational relevance.

1.5 Chapter Scope and Organization

While several recent reviews have examined AI uncertainty quantification (Gawlikowski et al., 2023), weather prediction applications (Dueben et al., 2022), or trustworthiness in machine learning for climate (Haynes et al., 2023), these efforts have largely treated uncertainty quantification, validation, and benchmarking as separate concerns, often with a primary focus on methodological development from the machine learning perspective. This chapter makes three unique contributions that address critical gaps in existing literature. First, we provide an integrated synthesis that explicitly connects uncertainty quantification methods with spatially-aware validation strategies and domain-specific benchmarking frameworks—a connection rarely made explicit in prior reviews yet essential for operational Earth science applications. Existing ML-centered reviews typically emphasize algorithmic innovations in UQ methods without adequately addressing how these methods must be validated differently in geospatial contexts or how benchmarks should be designed to reflect Earth system-specific challenges.

Second, we foreground Earth science-specific methodological considerations that are often treated peripherally in general machine learning reviews, particularly spatial autocorrelation and its implications for both model validation and uncertainty calibration, distribution shifts across geographic regions and climate regimes, and the unique requirements of operational decision support systems where forecast reliability directly impacts societal outcomes. Third, this review is motivated by the timely need to bridge the gap between rapid AI innovation and the operational deployment requirements in Earth sciences. As AI weather forecasting systems approach and in some cases exceed traditional numerical weather prediction skill, and as remote sensing workflows increasingly rely on deep learning pipelines, the Earth science community urgently requires guidance that integrates state-of-the-art uncertainty quantification with rigorous validation and standardized evaluation—precisely the synthesis this chapter provides. By explicitly connecting these three traditionally separated research areas through an Earth system science lens, this review aims to accelerate the transition from experimental AI prototypes to trustworthy operational systems.

This chapter provides a comprehensive examination of uncertainty quantification, validation methodologies, and benchmarking frameworks for AI applications in Earth sciences. We synthesize recent advances in probabilistic deep learning, spatial validation techniques, and standardized evaluation protocols while highlighting open challenges and future research directions. The chapter is organized as follows:

Section 2 reviews fundamental concepts and methods for uncertainty quantification in neural networks, including Bayesian deep learning, ensemble approaches, and conformal prediction. Section 3 examines validation strategies specifically designed for geospatial and temporal data, covering spatial cross-validation, area of applicability analysis, and metrics for assessing prediction reliability. Section 4 presents current benchmarking frameworks across different Earth science domains and analyzes their design principles, evaluation protocols, and community adoption. Section 5 discusses case studies demonstrating practical implementation of UQ and validation methods in weather forecasting, remote sensing, and hydrological modeling. Finally, Section 6 identifies critical research gaps and outlines future directions for enhancing trustworthiness of AI in Earth system science applications.

2. UNCERTAINTY QUANTIFICATION IN AI MODELS

2.1 Foundations of Uncertainty in Machine Learning

2.1.1 Types of Uncertainty

Understanding different types of uncertainty is fundamental to developing appropriate quantification methods for AI models in Earth sciences. Following the taxonomy established by Gawlikowski et al. (2023), we distinguish between aleatoric and epistemic uncertainty. Aleatoric uncertainty, also termed data uncertainty or irreducible uncertainty, represents inherent randomness in observations and physical processes. In Earth science applications, aleatoric uncertainty manifests through measurement noise in satellite sensors, sub-grid scale variability in climate models, and stochastic atmospheric phenomena such as turbulence and convection.

Epistemic uncertainty, conversely, reflects limitations in our knowledge and model capacity. This includes uncertainty arising from finite training data, imperfect model architectures, and simplified representations of complex physical processes. Unlike aleatoric uncertainty, epistemic uncertainty can theoretically be reduced through additional observations, improved model designs, or incorporation of physical constraints (Haynes et al., 2023). For Earth science applications, distinguishing between these uncertainty types enables targeted model improvements and helps identify whether prediction errors stem from inherent system variability or model deficiencies.

2.1.2 Mathematical Formulation

Formally, consider a supervised learning problem where we aim to predict a target variable $\mathbf{y}$ given input features $\mathbf{x}$. In a probabilistic framework, we seek to estimate the predictive distribution $p(\mathbf{y}|\mathbf{x}, D)$, where D represents the training dataset. This predictive distribution captures both aleatoric uncertainty (variability of $\mathbf{y}$ for given $\mathbf{x}$) and epistemic uncertainty (uncertainty in model parameters due to limited data).

Bayesian inference provides a principled framework for uncertainty quantification by placing probability distributions over model parameters $\boldsymbol{\theta}$. The posterior distribution $p(\boldsymbol{\theta}|D)$ represents updated beliefs about parameters after observing training data, computed via Bayes' theorem. The predictive distribution is then obtained by marginalizing over the posterior:

$$p(\mathbf{y}|\mathbf{x}, D) = \int p(\mathbf{y}|\mathbf{x}, \boldsymbol{\theta})\, p(\boldsymbol{\theta}|D)\, d\boldsymbol{\theta}$$

This integral captures epistemic uncertainty through the distribution over parameters $\boldsymbol{\theta}$, while the likelihood $p(\mathbf{y}|\mathbf{x}, \boldsymbol{\theta})$ models aleatoric uncertainty. Computing this integral exactly is intractable for neural networks, motivating approximation techniques discussed in subsequent sections.

2.2 Bayesian Deep Learning Approaches

2.2.1 Variational Inference Methods

Bayesian neural networks (BNNs) extend standard neural networks by treating weights as random variables with probability distributions rather than point estimates. Variational inference approximates the intractable posterior $p(\boldsymbol{\theta}|D)$ with a simpler variational distribution $q(\boldsymbol{\theta}|\boldsymbol{\varphi})$, where $\boldsymbol{\varphi}$ represents variational parameters optimized to minimize the Kullback-Leibler divergence between q and the true posterior (Kirkwood et al., 2022).

In Earth science applications, BNNs have demonstrated success in spatial interpolation tasks where quantifying prediction uncertainty is crucial for decision-making. Kirkwood et al. (2022) applied Bayesian deep learning to spatial interpolation incorporating auxiliary environmental covariates, showing that uncertainty estimates help identify regions requiring additional observations. The variational framework enables efficient training through stochastic gradient descent while providing calibrated uncertainty estimates that distinguish between data-poor and data-rich regions.

2.2.2 Monte Carlo Dropout

Monte Carlo (MC) dropout offers a computationally efficient approximation to Bayesian inference by interpreting dropout training as approximate variational inference (Klotz et al., 2022; Althoff et al., 2021). During prediction, dropout is applied multiple times with different random masks, generating an ensemble of predictions whose variance estimates epistemic uncertainty. This approach requires minimal modifications to standard neural networks and scales to large models, making it attractive for Earth science applications.

Klotz et al. (2022) demonstrated MC dropout for uncertainty quantification in rainfall-runoff modeling, showing that uncertainty estimates correlate with prediction errors and identify challenging hydrological conditions. The method successfully captured increased uncertainty during extreme events and in ungauged catchments where training data is limited. Similarly, Althoff et al. (2021) applied dropout ensembles to hydrological models, finding that ensemble predictions improved reliability while providing meaningful uncertainty bounds for water resource management.

2.3 Ensemble-Based Uncertainty Estimation

2.3.1 Deep Ensemble Methods

Deep ensembles train multiple neural networks independently with different random initializations and combine their predictions to estimate uncertainty (Schreck et al., 2024; Haynes et al., 2023). The variance across ensemble members captures epistemic uncertainty, while individual models can additionally estimate aleatoric uncertainty through probabilistic output layers. Despite computational costs of training multiple models, deep ensembles often outperform approximate Bayesian methods in calibration and predictive performance.

Schreck et al. (2024) introduced evidential deep learning for Earth system science applications, extending deep ensembles with higher-order uncertainty estimates. Their framework quantifies both prediction uncertainty and confidence in that uncertainty estimate, enabling more nuanced risk assessment for critical applications. Evaluations across climate prediction, severe weather forecasting, and agricultural monitoring demonstrated improved calibration compared to standard ensemble methods.

2.3.2 Ensemble Neural Weather Prediction

Weather forecasting has seen remarkable progress through ensemble neural network approaches that generate probabilistic predictions competitive with traditional numerical weather prediction (NWP) ensembles (Price et al., 2025; Chapman et al., 2022; Hu et al., 2023). Price et al. (2025) presented a machine learning weather forecasting system trained on ensemble NWP data that produces calibrated probabilistic forecasts matching or exceeding physics-based ensemble skill while requiring orders of magnitude less computation.

Chapman et al. (2022) developed deep learning methods for generating probabilistic atmospheric river forecasts from deterministic NWP predictions, demonstrating that learned uncertainty estimates outperform conventional statistical post-processing. The approach successfully captured forecast spread-skill relationships and provided reliable probability estimates for extreme precipitation events. Hu et al. (2023) extended these methods to forecast uncertainty for precipitation across the western United States, showing that deep learning ensembles capture terrain-induced precipitation variability and seasonal patterns in forecast confidence.

2.4 Conformal Prediction and Distribution-Free Uncertainty

2.4.1 Conformal Prediction Framework

Conformal prediction provides distribution-free uncertainty quantification with finite-sample validity guarantees, making it particularly attractive for Earth science applications where distributional assumptions may not hold (Singh et al., 2024; Valle et al., 2023). Unlike Bayesian and ensemble methods that rely on modeling assumptions, conformal prediction constructs prediction intervals with guaranteed coverage probability by leveraging holdout calibration data.

The conformal prediction framework computes a nonconformity score measuring how unusual a prediction is relative to calibration examples. For a desired coverage level ($1-\alpha$), the method constructs prediction sets guaranteed to contain the true value with probability at least ($1-\alpha$), regardless of the underlying data distribution. This distribution-free property proves valuable for Earth observation applications where training and deployment conditions may differ substantially.

2.4.2 Applications in Remote Sensing

Singh et al. (2024) applied conformal prediction to probabilistic machine learning models for Earth observation, demonstrating robust uncertainty quantification across land cover classification and crop type mapping tasks. The method successfully maintained nominal coverage rates even when encountering geographic regions and environmental conditions absent from training data, addressing a critical limitation of conventional uncertainty estimation approaches.

Valle et al. (2023) used conformal statistics for uncertainty quantification in land-use land-cover classification from satellite imagery. Their approach provided spatially explicit confidence measures that identified regions of high classification uncertainty, enabling targeted collection of additional ground truth data. Comparison with other uncertainty methods showed that conformal prediction maintained better calibration across diverse ecosystems and land cover types, particularly in transition zones between classes.

2.5 Physics-Informed Uncertainty Quantification

2.5.1 Integrating Physical Constraints

Physics-informed machine learning incorporates domain knowledge and physical laws into neural network training, potentially improving generalization and providing more reliable uncertainty estimates (Degen et al., 2022; Cheng et al., 2023). For Earth science applications, physical constraints may include conservation laws, thermodynamic relationships, or partial differential equations governing system dynamics. Embedding such constraints can reduce epistemic uncertainty by restricting the hypothesis space to physically plausible solutions.

Degen et al. (2022) developed physics-based machine learning for 3D multi-physics uncertainty quantification in geothermal systems, combining deep learning with thermodynamic constraints. Their framework propagated observational uncertainties through coupled physical processes while ensuring predictions satisfy fundamental conservation principles. Results showed that physics-informed uncertainty quantification outperformed purely data-driven approaches, particularly when extrapolating beyond observed conditions.

2.5.2 Data Assimilation and UQ

Data assimilation provides a natural framework for combining physics-based models with observations while quantifying uncertainty (Cheng et al., 2023). Machine learning can enhance data assimilation systems by learning improved observation operators, error covariance structures, or reduced-order representations of complex dynamics. Cheng et al. (2023) reviewed machine learning approaches for data assimilation and uncertainty quantification in dynamical systems, highlighting techniques that leverage complementary strengths of physics-based and data-driven methods.

Integration of uncertainty quantification with data assimilation enables sequential refinement of predictions as new observations arrive, propagating observation uncertainty through model dynamics. This capability proves particularly valuable for Earth system applications requiring real-time updating, such as hurricane track prediction, flood forecasting, and air quality monitoring where timely uncertainty estimates inform emergency response decisions.

Figure 1. Conceptual Framework

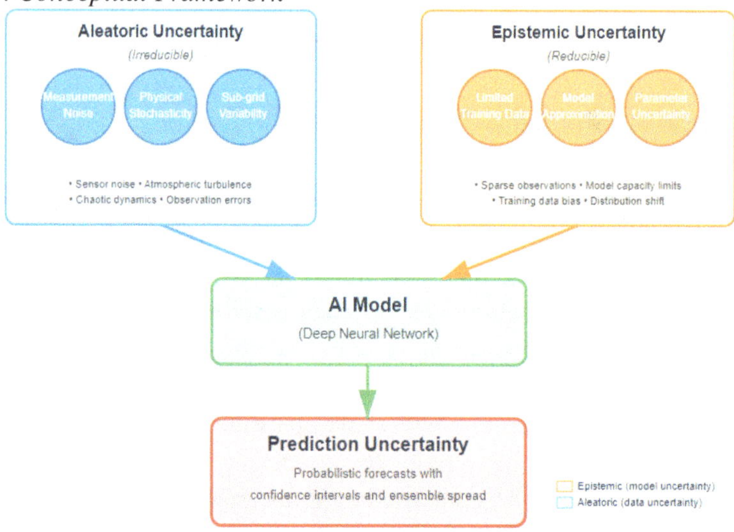

Figure 1. Conceptual framework, different sources of uncertainty in AI models for Earth sciences and their relationships. The diagram illustrates how aleatoric uncertainty (from observations and physical processes) and epistemic uncertainty (from limited data and model limitations) propagate through the AI model to produce prediction uncertainty.

3. VALIDATION METHODOLOGIES FOR EARTH SCIENCE AI

3.1 Challenges of Spatial Validation

3.1.1 Spatial Autocorrelation Issues

Spatial autocorrelation—the tendency for nearby observations to be more similar than distant ones—fundamentally violates the independence assumption of standard cross-validation (Meyer & Pebesma, 2021, 2022). When training and test samples are spatially proximate, models exploit spatial autocorrelation rather than learning transferable relationships between predictors and response variables. This produces optimistically biased performance estimates that fail to reflect actual generalization ability to new geographic regions.

Meyer & Pebesma (2021) demonstrated this problem by comparing random cross-validation with spatial cross-validation across diverse Earth science prediction tasks. Random splitting consistently overestimated model performance, with accuracy metrics inflated by 10-40% compared to spatial validation. The performance gap was particularly pronounced for models with high capacity (e.g., deep neural networks) that can memorize spatial patterns rather than learning causal relationships.

3.1.2 Area of Applicability

The area of applicability (AOA) concept addresses the challenge of identifying where spatial predictions are reliable versus where they extrapolate beyond training data support (Meyer & Pebesma, 2021, 2022). AOA quantifies the dissimilarity between prediction locations and nearest training samples in predictor space, flagging regions where predictions may be unreliable due to distribution shift. This approach provides interpretable uncertainty estimates complementing probabilistic methods by explicitly identifying geographic extrapolation.

Meyer & Pebesma (2022) applied AOA analysis to global-scale machine learning maps of ecological variables, revealing that many published maps extrapolate substantially in large geographic areas. Their analysis showed that incorporating AOA assessment could prevent deployment of unreliable predictions in critical applications. The methodology has since been adopted for validating AI models across climate science, biodiversity modeling, and precision agriculture applications.

3.2 Spatial Cross-Validation Strategies

3.2.1 Nearest Neighbor Distance Matching

Nearest neighbor distance matching (NNDM) cross-validation creates training-test splits that mimic the spatial autocorrelation structure expected during prediction (Milà et al., 2022). The method matches the distribution of distances between test and training samples with the distribution of distances between prediction locations and available training data. This ensures validation conditions realistically represent the operational deployment scenario.

Milà et al. (2022) demonstrated that NNDM leave-one-out cross-validation produces less biased accuracy estimates than conventional spatial blocking approaches, particularly for irregularly distributed sampling designs common in Earth observation. The method proved especially valuable for validating models trained on opportunistic sampling data where spatial coverage is uneven and extrapolation is unavoidable in practice.

3.2.2 k-Fold NNDM Cross-Validation

Linnenbrink et al. (2024) extended NNDM to k-fold cross-validation (kNNDM), providing computationally efficient validation for large datasets while maintaining the distance-matching properties. kNNDM creates k spatially separated folds where the nearest distance between any test point and its closest training point in the same fold matches the expected prediction scenario. This approach balances computational feasibility with rigorous spatial validation.

Evaluation across multiple Earth observation datasets showed that kNNDM provides reliable performance estimates comparable to leave-one-out NNDM while reducing computational costs by orders of magnitude. The method has been implemented in open-source software packages, facilitating adoption for operational validation of remote sensing and geospatial AI applications.

3.2.3 Spatial Plus Cross-Validation

Wang et al. (2023) introduced Spatial+ cross-validation, combining spatial blocking with additional strategies to address limitations of conventional spatial CV methods. Spatial+ incorporates multiple spatial scales, temporal stratification, and environmental partitioning to ensure comprehensive evaluation of model generalization. The method explicitly tests model performance across different types of extrapolation: spatial, temporal, and environmental.

Application to geospatial machine learning for crop mapping, forest classification, and urban land cover demonstrated that Spatial+ reveals performance limitations invisible to standard validation approaches. Models showing high accuracy under random CV exhibited substantial performance degradation when tested on spatially, temporally, or environmentally distinct holdout sets, highlighting the importance of multi-faceted validation for Earth science applications.

3.3 Temporal Validation Approaches

3.3.1 Time Series Cross-Validation

Earth science applications involving temporal predictions require validation strategies that respect temporal ordering and assess forecast skill across different lead times (Rasp et al., 2024; Mansfield & Sheshadri, 2024). Time series cross-validation uses expanding or sliding window approaches where models are trained on historical data and evaluated on subsequent time periods, mimicking operational forecasting scenarios.

Rasp et al. (2024) developed WeatherBench 2, a comprehensive benchmark for data-driven weather prediction that implements rigorous temporal validation protocols. The benchmark evaluates forecast skill at multiple lead times (6 hours to 15 days) and assesses model performance on extreme events, specific weather phenomena, and different geographic regions. Temporal validation revealed that AI weather models exhibit variable performance across forecast horizons, with strengths in medium-range prediction but limitations for subseasonal timescales.

3.3.2 Hindcast Validation

Hindcast validation evaluates model performance on historical events not included in training data, providing robust assessment of prediction skill for rare or extreme phenomena (Olivetti & Messori, 2024). This approach is particularly valuable for validating AI models intended for climate extremes, natural hazards, or other high-impact low-frequency events where test-time validation samples are limited.

Olivetti & Messori (2024) reviewed deep learning for extreme weather forecasting, emphasizing the importance of hindcast validation for assessing model reliability. Their analysis showed that models trained predominantly on average conditions often fail to accurately predict extreme events, highlighting the need for specialized training strategies and validation protocols that explicitly evaluate performance on distribution tails.

3.4 Metrics for Probabilistic Predictions

3.4.1 Calibration Assessment

Calibration measures whether predicted probabilities match empirical frequencies of events (Haynes & Feingold, 2025). A perfectly calibrated model should produce outcomes matching predicted probabilities—for example, among all predictions assigned 70% probability, approximately 70% should occur. Calibration is assessed through reliability diagrams plotting predicted probabilities against observed frequencies, with deviations from the diagonal indicating miscalibration.

Haynes & Feingold (2025) developed comprehensive frameworks for uncertainty quantification in deep learning applied to Earth sciences, emphasizing calibration assessment alongside predictive skill. Their evaluation of multiple uncertainty methods across atmospheric science applications revealed that many approaches produce poorly calibrated uncertainty estimates, overconfident in some regimes and underconfident in others. They recommend routine calibration assessment using reliability diagrams, expected calibration error, and Brier score decomposition.

3.4.2 Sharpness and Resolution

Beyond calibration, probabilistic forecasts should exhibit sharpness (narrow prediction intervals when appropriate) and resolution (ability to distinguish outcomes with different probabilities). Sharp forecasts provide actionable information by constraining the range of plausible outcomes, while resolution enables risk-based decision-making by identifying situations with high versus low predictability (Demaeyer et al., 2023).

The continuous ranked probability score (CRPS) provides a comprehensive metric combining calibration, sharpness, and resolution into a single measure. CRPS evaluates the integrated squared difference between predicted cumulative distribution functions and observations, penalizing both bias and excessive spread. Demaeyer et al. (2023) used CRPS as the primary evaluation metric for the EUPPBench post-processing benchmark, enabling fair comparison of probabilistic post-processing methods for numerical weather predictions.

3.5 Validation Against Independent Data

3.5.1 Ground Truth Collection

Validation of remote sensing AI models requires independent ground truth data collected through field surveys, airborne campaigns, or higher-resolution imagery

(Gevaert, 2022; Werther et al., 2022). Obtaining representative validation samples across diverse environmental conditions remains a fundamental challenge, particularly for global-scale applications where field campaigns are logistically prohibitive.

Werther et al. (2022) developed Bayesian approaches for remote sensing of chlorophyll-a in lakes, explicitly quantifying retrieval uncertainty and validating against in-situ measurements. Their analysis revealed that validation sample distribution strongly influences uncertainty estimates, with sparse sampling in certain lake types leading to overconfident predictions. This highlights the importance of stratified sampling designs ensuring validation data represent the full range of conditions encountered operationally.

3.5.2 Reference Data Quality

The quality of reference data fundamentally limits validation accuracy, yet ground truth is often treated as error-free despite measurement uncertainties and spatial/temporal mismatches (Gevaert, 2022). Modern validation frameworks increasingly account for reference data uncertainty, propagating observational errors through validation metrics to avoid penalizing models for discrepancies attributable to ground truth uncertainty.

Gevaert (2022) reviewed explainable AI for Earth observation from societal and regulatory perspectives, emphasizing the need for transparent documentation of validation data quality and limitations. The review highlighted emerging regulatory requirements for AI transparency in environmental monitoring applications, where validation protocols must demonstrate model reliability across diverse stakeholder concerns including accuracy, fairness, and compliance with environmental policies.

Figure 2. Spatial cross-validation strategies

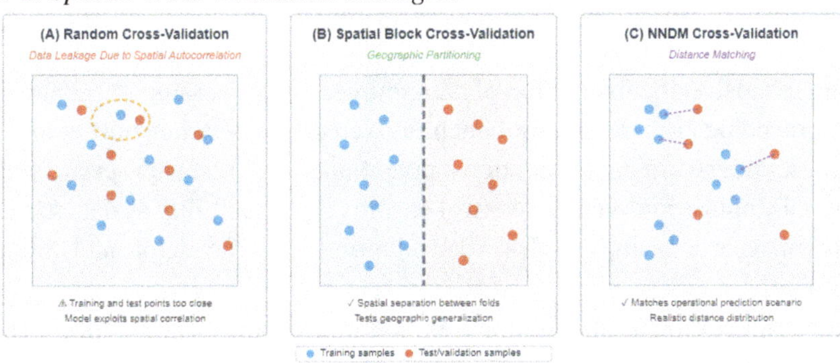

Figure 2. Spatial cross-validation strategies for Earth science AI models. The diagram compares (A) random cross-validation showing data leakage due to spatial autocorrelation, (B) spatial block cross-validation with geographic partitioning, and (C) nearest neighbor distance matching (NNDM) that mimics prediction scenarios.

4. BENCHMARKING FRAMEWORKS FOR EARTH SCIENCE AI

4.1 Principles of Effective Benchmarking

4.1.1 Design Requirements

Effective benchmarks for Earth science AI must satisfy several critical requirements: representativeness of real-world conditions, standardized evaluation protocols, accessibility to researchers, and relevance to operational applications (Dueben et al., 2022). Benchmarks should span diverse geographic regions, temporal periods, and environmental conditions to assess model generalization. Standardized train/validation/test splits prevent information leakage and enable fair comparison across studies.

Dueben et al. (2022) outlined challenges and design principles for machine learning benchmarks in atmospheric sciences, emphasizing the need for benchmarks that reflect practical constraints of operational systems. These include computational budgets, latency requirements, and evaluation on societally relevant phenomena such as extreme events and high-impact weather. The authors also stressed the importance of baseline models establishing minimum performance standards and reference implementations facilitating reproducibility.

4.1.2 Data Provenance and Documentation

Comprehensive documentation of data sources, preprocessing steps, and quality control procedures is essential for benchmark credibility (Betancourt et al., 2021). Users must understand data limitations, potential biases, and appropriate use cases to avoid misapplication. Version control ensures reproducibility as datasets evolve, while permissive licensing enables broad adoption across academic and operational communities.

Betancourt et al. (2021) developed AQ-Bench, a benchmark for machine learning on global air quality metrics, with extensive documentation of data provenance from satellite retrievals and ground monitoring networks. The benchmark includes metadata describing observational uncertainty, temporal coverage, and known data quality issues. This transparency enables users to assess whether benchmark performance transfers to their specific application contexts.

4.2 Weather and Climate Benchmarks

4.2.1 WeatherBench 2

WeatherBench 2 represents a comprehensive benchmark for next-generation data-driven weather models, providing standardized evaluation of deterministic and probabilistic forecasts against state-of-the-art numerical weather prediction (Rasp et al., 2024). The benchmark includes global atmospheric data from ERA5 reanalysis at multiple pressure levels and surface variables, spanning 1979-2022. Evaluation protocols assess forecast skill across spatial scales, lead times (6 hours to 15 days), and specific weather phenomena.

Key innovations in WeatherBench 2 include: (1) probabilistic evaluation metrics comparing AI forecasts against operational ensemble predictions, (2) regional assessments quantifying performance across climate zones, (3) extreme event evaluation measuring skill for high-impact weather, and (4) computational efficiency metrics balancing accuracy with operational constraints. The benchmark has catalyzed rapid progress in AI weather prediction, with participating models achieving competitive or superior performance to physics-based NWP while requiring orders of magnitude less computation.

4.2.2 EUPPBench for Post-Processing

The EUPPBench benchmark focuses on statistical post-processing of numerical weather predictions, a critical component of operational forecasting systems (Demaeyer et al., 2023). The benchmark provides ensemble forecasts and corresponding observations for temperature and wind speed across European stations, enabling comparison of machine learning and traditional statistical methods for calibrating raw model output.

Demaeyer et al. (2023) designed EUPPBench to address specific challenges in operational post-processing: limited training data at individual stations, transferability across geographic locations, and computational efficiency for real-time applications. Evaluation emphasizes proper verification metrics for probabilistic forecasts, including CRPS, quantile score, and reliability assessment. The benchmark has revealed strengths and limitations of different ML approaches, showing that neural networks can improve upon conventional methods when sufficient training data is available but may underperform simpler approaches in data-scarce regimes.

4.3 Remote Sensing Benchmarks

4.3.1 BigEarthNet-MM

BigEarthNet-MM provides a large-scale, multimodal, multilabel benchmark for remote sensing image classification and retrieval (Sumbul et al., 2021). The dataset comprises over 500,000 Sentinel-1 and Sentinel-2 image patches covering 10 European countries, annotated with 19 land cover classes from the CORINE Land Cover database. Multi-modal fusion of synthetic aperture radar (SAR) and multispectral optical imagery enables evaluation of methods exploiting complementary information sources.

The benchmark addresses key challenges in operational remote sensing: class imbalance, multi-label classification (patches containing multiple land cover types), and geographic generalization. Sumbul et al. (2021) established baseline results using convolutional neural networks and vision transformers, demonstrating that multi-modal approaches outperform single-modality models. The benchmark has become widely adopted for evaluating deep learning architectures, transfer learning strategies, and few-shot learning methods for Earth observation.

4.3.2 GEO-Bench for Foundation Models

GEO-Bench targets evaluation of foundation models for Earth monitoring, providing a diverse collection of classification and segmentation tasks across multiple sensors and applications (Lacoste et al., 2023). The benchmark includes 13 datasets spanning agricultural monitoring, forest mapping, building detection, land cover classification, and disaster response, with geographic coverage extending from local to global scales.

Lacoste et al. (2023) designed GEO-Bench to assess whether self-supervised pretraining on massive unlabeled satellite imagery produces foundation models that transfer effectively to downstream tasks with limited labeled data. Evaluation protocols measure few-shot learning performance, domain adaptation across geographic regions, and multi-task learning capabilities. Results showed that pretrained models substantially outperform training from scratch on small datasets, though performance gaps persist compared to task-specific architectures on well-resourced benchmarks.

4.4 Domain-Specific Benchmarks

4.4.1 Air Quality Benchmarking

AQ-Bench provides a benchmark dataset for machine learning on global air quality metrics, enabling comparison of methods for predicting PM2.5, ozone, and nitrogen dioxide concentrations (Betancourt et al., 2021). The dataset combines satellite observations, reanalysis data, and ground monitoring stations spanning 2010-2020, with global coverage at 0.25° resolution.

Betancourt et al. (2021) structured AQ-Bench to reflect operational air quality forecasting challenges: spatial prediction in unmonitored regions, temporal forecasting at multiple horizons, and uncertainty quantification for health-relevant pollution levels. Baseline evaluations compared random forests, gradient boosting, and neural networks, revealing that performance varies substantially across pollutants and geographic regions. The benchmark highlighted particular challenges in predicting extreme pollution episodes and generalizing models trained in data-rich regions to areas with sparse monitoring.

4.4.2 Hydrological Model Benchmarking

While not a formal benchmark dataset, standardized evaluation of AI models for rainfall-runoff prediction has emerged through community datasets like CAMELS (Catchment Attributes and Meteorology for Large-sample Studies) (Klotz et al., 2022; Althoff et al., 2021). These datasets provide meteorological forcing, streamflow observations, and catchment attributes for hundreds of watersheds, enabling systematic comparison of hydrological models.

Klotz et al. (2022) used CAMELS data to benchmark uncertainty quantification methods for deep learning rainfall-runoff models, comparing MC dropout, deep ensembles, and conditional quantile regression. Their evaluation showed that ensemble methods produced better calibrated uncertainty estimates than single-model approaches, though computational costs remain a barrier for operational deployment. Althoff et al. (2021) similarly benchmarked dropout ensembles across diverse hydrological conditions, demonstrating the importance of evaluating uncertainty quality across flow regimes and extreme events.

4.5 Benchmark Adoption and Impact

4.5.1 Community Engagement

Successful benchmarks require active community engagement through workshops, competitions, and collaborative development of evaluation protocols (Lacoste et al., 2023). Machine learning competitions like those hosted on Kaggle or DrivenData have proven effective for engaging researchers and accelerating method development, though competition metrics must carefully align with scientific objectives rather than arbitrary performance targets.

The proliferation of benchmark papers in Earth science venues indicates growing recognition of standardized evaluation's importance. However, challenges remain in ensuring benchmark diversity matches the breadth of Earth science applications, avoiding benchmark overfitting where methods optimize narrowly for specific datasets, and maintaining benchmarks as technology and data sources evolve.

4.5.2 Limitations and Future Needs

Current benchmarks exhibit several limitations requiring attention. Geographic bias toward data-rich regions (Europe, North America) limits assessment of global generalization. Temporal coverage often focuses on recent decades where satellite observations are abundant, providing limited evaluation of long-term climate variability. Many benchmarks evaluate only offline performance, neglecting operational constraints like computational latency, energy efficiency, and robustness to missing data.

Future benchmark development should prioritize: (1) greater geographic diversity including developing regions, (2) evaluation of model performance on extreme events and distribution tails, (3) assessment of uncertainty calibration alongside point prediction accuracy, (4) standardized computational efficiency metrics, and (5) multi-task benchmarks reflecting coupled Earth system processes. Additionally, benchmarks should incorporate evolving best practices for responsible AI, including fairness assessment across demographic groups affected by environmental predictions and transparency in model limitations.

Table 1. Comparison of Major Benchmarking Frameworks for AI in Earth Sciences

Benchmark	Domain	Data Source	Spatial Coverage	Temporal Range	Sample Size	Task Type	Key Metrics	Reference
WeatherBench 2	Weather forecasting	ERA5 reanalysis	Global	1979-2022	Multiple pressure levels & surface variables	Deterministic & probabilistic forecasting	CRPS, RMSE, ACC, extreme event skill	Rasp et al., 2024
EUPPBench	Post-processing	ECMWF ensemble forecasts	European stations	2017-2018	139 stations, temperature & wind	Statistical post-processing	CRPS, quantile score, reliability	Demaeyer et al., 2023
AQ-Bench	Air quality	Satellite + ground stations	Global	2010-2020	0.25° resolution	Regression (PM2.5, O3, NO2)	RMSE, MAE, R^2, spatial generalization	Betancourt et al., 2021
BigEarthNet-MM	Remote sensing	Sentinel-1 & Sentinel-2	10 European countries	2017-2018	590,326 patches	Multi-label classification	F1-score, precision, recall, multi-modal fusion	Sumbul et al., 2021
GEO-Bench	Foundation models	Multiple sensors	Local to global	Varies by task	13 datasets	Classification, segmentation	Accuracy, few-shot performance, domain adaptation	Lacoste et al., 2023
CAMELS	Hydrology	USGS + meteorology	United States	1980-2014	671 catchments	Rainfall-runoff modeling	NSE, KGE, peak flow bias	Klotz et al., 2022

Table Notes:

- **CRPS**: Continuous Ranked Probability Score; **RMSE**: Root Mean Square Error; **ACC**: Anomaly Correlation Coefficient; **MAE**: Mean Absolute Error; **NSE**: Nash-Sutcliffe Efficiency; **KGE**: Kling-Gupta Efficiency
- All benchmarks provide standardized train/validation/test splits and baseline model implementations
- Geographic coverage bias exists toward data-rich regions (Europe, North America) across most benchmarks
- Temporal coverage varies significantly, with weather benchmarks spanning decades while some remote sensing benchmarks cover 1-2 years

5. CASE STUDIES AND APPLICATIONS

5.1 Probabilistic Weather Forecasting

Recent advances in machine learning for weather prediction demonstrate the practical importance of uncertainty quantification and rigorous validation (Price et al., 2025; Rasp et al., 2024). Price et al. (2025) developed a probabilistic forecasting system trained on ensemble NWP data that generates calibrated uncertainty estimates competitive with operational ensemble prediction systems. The model combines a deterministic backbone architecture with a diffusion-based generative model that produces diverse forecast trajectories capturing forecast uncertainty.

Evaluation on WeatherBench 2 revealed that the ML ensemble approach achieves comparable or superior skill to physics-based ensemble systems across most variables and lead times, while requiring approximately 100x less computation. Probabilistic verification using CRPS and spread-skill analysis showed that forecast uncertainty estimates are well-calibrated, with ensemble spread appropriately representing forecast error magnitude. The system demonstrated particular strength in medium-range forecasting (3-10 days), where ensemble diversity effectively captured forecast uncertainty arising from chaotic dynamics.

Critical to the success of this application was rigorous temporal validation using years of held-out test data, comprehensive evaluation across spatial scales and weather phenomena, and verification against independent observations beyond reanalysis used for training. The study exemplifies how proper UQ and validation enable deployment of AI forecasting systems as operational tools complementing traditional NWP.

5.2 Remote Sensing Uncertainty Assessment

Singh et al. (2024) applied conformal prediction for uncertainty quantification in Earth observation models, addressing the challenge of reliable uncertainty estimates under distribution shift. Their case study focused on crop type classification from multi-temporal Sentinel-2 imagery across multiple continents. Standard deep learning models achieved high accuracy on held-out test regions but provided poorly calibrated uncertainty estimates, overconfident in erroneous predictions on novel geographic regions.

Conformal prediction provided prediction sets (multiple possible classes) with guaranteed coverage probability, adapting set size to reflect prediction uncertainty. In familiar agricultural regions similar to training data, conformal sets typically contained 1-2 classes with high confidence in the most probable class. However, in regions with different cropping systems or environmental conditions, conformal sets appropriately expanded to include more candidate classes, signaling distributional shift and the need for additional training data or local model adaptation.

The study demonstrated conformal prediction's value for operational Earth observation applications where models inevitably encounter conditions outside their training distribution. By providing rigorous uncertainty quantification without distributional assumptions, conformal prediction enables deployment teams to identify when model predictions are trustworthy versus when human expert review or additional data collection is warranted.

5.3 Hydrological Model Uncertainty

Klotz et al. (2022) conducted comprehensive evaluation of uncertainty quantification methods for deep learning rainfall-runoff models using the CAMELS dataset covering 531 catchments across the United States. The study compared MC dropout, deep ensembles, mixture density networks, and conditional quantile regression for generating probabilistic streamflow predictions. Evaluation emphasized both prediction accuracy (Nash-Sutcliffe efficiency, Kling-Gupta efficiency) and uncertainty quality (reliability, sharpness, continuous ranked probability score).

Results showed that deep ensembles produced the best calibrated uncertainty estimates, with prediction intervals achieving near-nominal coverage rates across diverse hydrological conditions. MC dropout provided reasonable uncertainty estimates at lower computational cost but exhibited slight overconfidence in high-flow conditions. Mixture density networks captured multi-modal predictive distributions useful for representing equifinality in hydrological systems but required careful hyperparameter tuning to avoid mode collapse.

Critically, the study revealed that uncertainty quantification performance varied systematically across catchment characteristics. Models exhibited higher uncertainty in snow-dominated basins where complex snow accumulation and melt processes introduce greater predictive challenges. Uncertainty also increased appropriately in catchments with limited training data, correctly reflecting epistemic uncertainty. These findings demonstrate that uncertainty quantification enables targeted model improvements and informed decision-making about when hydrological predictions are sufficiently reliable for operational use.

5.4 Climate Model Emulation

Mansfield & Sheshadri (2024) developed machine learning parameterizations for atmospheric gravity waves in climate models, incorporating uncertainty quantification to assess emulator reliability. Gravity wave drag represents a critical sub-grid scale process affecting climate model accuracy, yet physics-based parameterizations introduce substantial uncertainty. ML emulators trained on high-resolution simulations offer potential improvements but require rigorous UQ and validation.

Their study compared Bayesian neural networks, deep ensembles, and MC dropout for uncertainty quantification in gravity wave parameterizations. Evaluation emphasized physical consistency (conservation properties, numerical stability), predictive accuracy against high-resolution simulations, and uncertainty calibration across different atmospheric regimes. Results showed that ensemble methods produced well-calibrated uncertainty estimates that correctly identified conditions where parameterization accuracy degraded (e.g., extreme wind shear events).

The study demonstrated that UQ-enabled climate model emulation facilitates progressive replacement of physics-based parameterizations with hybrid physics-ML approaches. Uncertainty estimates identify conditions where ML emulators are trustworthy versus where traditional parameterizations or higher-resolution modeling is necessary. This capability proves essential for responsible deployment of ML in climate models used for multi-decadal projections informing climate policy.

5.5 Extreme Event Prediction

Olivetti & Messori (2024) reviewed deep learning for medium-range extreme weather forecasting, emphasizing validation challenges specific to rare events. Their analysis covered tropical cyclones, atmospheric rivers, heat waves, and severe convection—phenomena critically important for societal impacts yet statistically rare in training data. Standard validation metrics dominated by common conditions often mask poor performance on distribution tails.

The authors recommended specialized validation protocols for extreme events: (1) verification against independent extreme event databases beyond standard reanalysis, (2) evaluation of specific extreme event characteristics (e.g., tropical cyclone intensity, heat wave duration) rather than general meteorological fields, (3) assessment of forecast skill separately for rare versus common conditions, and (4) economic and impact-based verification considering decision-relevant thresholds.

Case studies of deep learning models for atmospheric river prediction demonstrated that models trained predominantly on climatology often fail to predict extreme precipitation associated with landfalling atmospheric rivers, despite achieving high overall accuracy scores. This highlights the necessity of targeted validation for operational deployment in high-stakes applications where extreme event prediction skill matters most.

Figure 3. Performance Comparison

Figure 3. Performance comparison of uncertainty quantification methods across three Earth science applications: (A) weather forecasting showing CRPS vs. computational cost for different ensemble methods, (B) remote sensing classification showing calibration curves for various UQ approaches, and (C) hydrological modeling showing uncertainty coverage as a function of lead time.

6. CHALLENGES AND FUTURE DIRECTIONS

6.1 Computational Scalability

Uncertainty quantification through ensemble methods, Bayesian inference, or Monte Carlo sampling significantly increases computational costs compared to deterministic predictions (Haynes & Feingold, 2025). For Earth science applications processing massive satellite imagery archives or running operational weather forecasts with strict latency requirements, computational overhead presents a critical barrier to UQ adoption. Future research must develop efficient uncertainty quantification methods that provide reliable estimates with minimal computational burden.

Promising directions include: single-forward-pass uncertainty methods that estimate uncertainty without repeated sampling, efficient Bayesian approximations leveraging low-rank structure or inducing point methods, and multi-fidelity approaches combining expensive high-fidelity models with cheap approximate models. Hardware accelerators designed for probabilistic inference and emerging analog computing architectures may also enable efficient uncertainty quantification at scale (Haynes & Feingold, 2025).

6.2 Multi-Source Data Integration

Earth science applications increasingly combine heterogeneous data sources—satellite imagery, in-situ sensors, citizen science observations, model outputs—each with distinct spatial/temporal resolutions and uncertainty characteristics (Cheng et al., 2023). Principled uncertainty quantification must propagate and fuse uncertainties across data sources while accounting for correlations. Current methods often treat data sources independently or use ad-hoc weighting schemes lacking theoretical justification.

Future research should develop probabilistic frameworks for multi-source data fusion that explicitly model observation processes, data quality variations, and spatial/temporal correlations. Hierarchical Bayesian approaches offer principled solutions but require efficient inference algorithms for high-dimensional Earth observation problems. Hybrid physics-ML methods that combine mechanistic understanding of observation processes with learned data representations may provide tractable approximations (Cheng et al., 2023).

6.3 Distribution Shift and Domain Adaptation

Earth science AI models inevitably encounter distribution shifts arising from climate change, land use changes, sensor degradation, and geographic generalization (Meyer & Pebesma, 2022; Singh et al., 2024). Reliable uncertainty quantification must identify when predictions extrapolate beyond training distributions and adapt to evolving conditions. Current methods often fail to detect distribution shifts or provide poorly calibrated uncertainty when encountering novel conditions.

Future directions include: robust uncertainty quantification maintaining calibration under distribution shift, domain adaptation methods that adjust predictions and uncertainty estimates for new domains, continual learning approaches updating models as conditions evolve, and test-time adaptation leveraging unlabeled data from deployment environments. Conformal prediction provides one promising framework for distribution-free uncertainty with theoretical guarantees, though extensions addressing covariate shift and temporal dynamics remain active research areas (Singh et al., 2024; Valle et al., 2023).

6.4 Interpretability and Physical Consistency

Black-box neural networks often produce physically inconsistent predictions violating conservation laws, thermodynamic constraints, or known relationships between variables (Degen et al., 2022; Mansfield & Sheshadri, 2024). Lack of interpretability hinders debugging, limits scientific insight, and reduces trust from domain experts. Future AI systems for Earth sciences should provide interpretable uncertainty estimates that identify specific sources of prediction uncertainty and indicate which physical processes contribute most to forecast spread.

Promising approaches include: physics-informed neural networks embedding conservation laws, modular architectures separating learned components from physics-based relationships, attention mechanisms highlighting salient input features, and causal inference methods identifying underlying mechanisms. Uncertainty attribution techniques that decompose predictive uncertainty into contributions from different inputs or model components could enhance interpretability (Gevaert, 2022).

6.5 Benchmarking for Operational Deployment

Current benchmarks predominantly evaluate offline performance on curated datasets, neglecting practical constraints of operational systems (Dueben et al.,

2022; Rasp et al., 2024). Operational deployment requires robustness to missing data, computational efficiency for real-time inference, graceful degradation under system failures, and interfaces with existing workflows. Future benchmarks should incorporate operational readiness metrics assessing these aspects beyond predictive accuracy.

Critical needs include: evaluation of model robustness to missing or corrupted inputs, assessment of inference latency and computational footprint on deployment hardware, verification of numerical stability and edge case handling, and measurement of human-AI collaboration effectiveness for decision support applications. Living benchmarks that continuously update with new data and evolving operational requirements would better reflect the dynamic nature of Earth science applications (Dueben et al., 2022).

6.6 Uncertainty Communication and Decision Support

Even well-calibrated uncertainty estimates provide limited value if not effectively communicated to decision-makers and stakeholders (Gevaert, 2022). Earth science predictions inform diverse decisions—agricultural planning, disaster response, infrastructure design, policy development—each requiring uncertainty information tailored to specific decision contexts. Future research must develop methods for uncertainty visualization, communication, and decision integration that translate technical probability estimates into actionable information.

Key challenges include: designing visualizations that accurately convey uncertainty without overwhelming users, translating probabilistic forecasts into decision-relevant formats (e.g., exceedance probabilities for critical thresholds), communicating confidence in uncertainty estimates (second-order uncertainty), and developing decision frameworks that properly account for forecast uncertainty in optimization and planning. Interdisciplinary collaboration between Earth scientists, statisticians, designers, and decision scientists will prove essential for this challenge.

7. CONCLUSION

Uncertainty quantification, validation, and benchmarking represent critical foundations for trustworthy AI in Earth sciences. This chapter has reviewed state-of-the-art approaches spanning Bayesian deep learning, ensemble methods, conformal prediction, spatial validation strategies, and emerging benchmarking frameworks. The reviewed methods demonstrate that rigorous UQ and validation enable AI systems to complement traditional physics-based models, achieving competitive or superior performance in weather forecasting, remote sensing, and environmental monitoring while providing reliable uncertainty estimates.

Key lessons emerge from our analysis. First, uncertainty quantification is not optional for Earth science applications—probabilistic predictions with calibrated uncertainty estimates are essential for informed decision-making under risk. Second, validation must account for spatial and temporal autocorrelation, distribution shifts, and application-specific requirements rather than relying on standard machine learning evaluation practices. Third, standardized benchmarks accelerate progress by enabling systematic comparison and reproducibility, though benchmark design must carefully reflect operational constraints and societal priorities.

Despite significant progress, substantial challenges remain. Computational scalability of uncertainty quantification methods limits adoption in data-intensive applications. Distribution shift arising from climate change and geographic generalization continues to challenge model reliability. Integration of heterogeneous data sources with diverse uncertainty characteristics requires principled probabilistic frameworks. Physical consistency and interpretability of learned models remain essential for scientific insight and domain expert trust.

Future research directions should prioritize: efficient single-pass uncertainty quantification methods, robust uncertainty estimation under distribution shift, physics-informed architectures ensuring consistency with domain knowledge, benchmarks incorporating operational readiness metrics, and effective uncertainty communication for decision support. Addressing these challenges through interdisciplinary collaboration between Earth scientists, statisticians, machine learning researchers, and decision scientists will enable the next generation of trustworthy AI systems serving society's need for reliable Earth system understanding and prediction.

The stakes for getting this right are substantial. Climate change, natural disasters, food security, water resources, and biodiversity loss present civilization-defining challenges requiring the best possible Earth system science. Artificial intelligence offers powerful capabilities for addressing these challenges, but only if deployed responsibly with rigorous uncertainty quantification, thorough validation, and continuous benchmarking ensuring reliability. The frameworks and methods reviewed in this chapter provide a foundation for achieving this vision, though sustained research investment and community engagement will prove essential for realizing AI's potential to enhance Earth system science for societal benefit.

REFERENCES

Althoff, D., Rodrigues, L. N., & Bazame, H. C. (2021). Uncertainty quantification for hydrological models based on neural networks: The dropout ensemble. *Stochastic Environmental Research and Risk Assessment, 35*(5), 1051–1067. DOI: 10.1007/s00477-021-01980-8

Amsalu, A., Garedew, W., Melka, G. A., Çelik, M. A., & Gemeda, D. O. (2026). Trends in climate extreme indices and implications for coffee farming in the Kafa Biosphere Reserve, southwest Ethiopia. *Climate Services, 42*, 100637. DOI: 10.1016/j.cliser.2026.100637

Betancourt, C., Stomberg, T., Roscher, R., Schultz, M. G., & Stadtler, S. (2021). AQ-Bench: A benchmark dataset for machine learning on global air quality metrics. *Earth System Science Data, 13*(6), 3013–3033. DOI: 10.5194/essd-13-3013-2021

Chapman, W. E., Delle Monache, L., Alessandrini, S., Subramanian, A. C., Ralph, F. M., Xie, S.-P., Lerch, S., & Hayatbini, N. (2022). Probabilistic predictions from deterministic atmospheric river forecasts with deep learning. *Monthly Weather Review, 150*(1), 215–234. DOI: 10.1175/MWR-D-21-0106.1

Cheng, S., Quilodrán-Casas, C., Ouala, S., Farchi, A., Liu, C., & Tandeo, P. (2023). Machine learning with data assimilation and uncertainty quantification for dynamical systems: A review. IEEE/CAA Journal of Automatica Sinica, 10(6), 1361–1387. https://doi.org/DOI: 10.1109/JAS.2023.123537

Çelik, M. A. (2026). Agro-climatic classification of Türkiye's agricultural basins using long-term ERA5-Land temperature data. *Physics and Chemistry of the Earth Parts A/B/C, 142*, 104307. DOI: 10.1016/j.pce.2026.104307

Degen, D., Cacace, M., & Wellmann, F. (2022). 3D multi-physics uncertainty quantification using physics-based machine learning. *Scientific Reports, 12*, 17491. DOI: 10.1038/s41598-022-21739-7

Demaeyer, J., Bhend, J., Lerch, S., Primo, C., Van Schaeybroeck, B., Atencia, A., Ben Bouallègue, Z., Chen, J., Dabernig, M., Evans, G., Faganeli Pucer, J., Hooper, B., Horat, N., Jobst, D., Merše, J., Mlakar, P., Möller, A., Mestre, O., Taillardat, M., & Vannitsem, S. (2023). The EUPPBench postprocessing benchmark dataset v1.0. *Earth System Science Data, 15*(6), 2635–2653. DOI: 10.5194/essd-15-2635-2023

Dueben, P. D., Schultz, M. G., Chantry, M., Gagne, D. J., Hall, D. M., & McGovern, A. (2022). Challenges and benchmark datasets for machine learning in the atmospheric sciences: Definition, status, and outlook. *Artificial Intelligence for the Earth Systems, 1*(3), e210002. DOI: 10.1175/AIES-D-21-0002.1

Gawlikowski, J., Tassi, C. R. N., Ali, M., Lee, J., Humt, M., Feng, J., Kruspe, A., Triebel, R., Jung, P., Roscher, R., Shahzad, M., Yang, W., Bamler, R., & Zhu, X. X. (2023). A survey of uncertainty in deep neural networks. *Artificial Intelligence Review*, *56*(Suppl 1), 1513–1589. DOI: 10.1007/s10462-023-10562-9

Gevaert, C. M. (2022). Explainable AI for earth observation: A review including societal and regulatory perspectives. *International Journal of Applied Earth Observation and Geoinformation*, *112*, 102869. DOI: 10.1016/j.jag.2022.102869

Haynes, K., Lagerquist, R., McGraw, M., Musgrave, K., & Ebert-Uphoff, I. (2023). Creating and evaluating uncertainty estimates with neural networks for environmental-science applications. *Artificial Intelligence for the Earth Systems*, *2*(2), e220061. DOI: 10.1175/AIES-D-22-0061.1

Hu, W., Ghazvinian, M., Chapman, W. E., Sengupta, A., Ralph, F. M., & Delle Monache, L. (2023). Deep learning forecast uncertainty for precipitation over the western United States. *Monthly Weather Review*, *151*(6), 1367–1385. DOI: 10.1175/MWR-D-22-0268.1

Kirkwood, C., Economou, T., Pugeault, N., & Odbert, H. (2022). Bayesian deep learning for spatial interpolation in the presence of auxiliary information. *Mathematical Geosciences*, *54*(3), 507–531. DOI: 10.1007/s11004-021-09988-0

Klotz, D., Kratzert, F., Gauch, M., Sampson, A. K., Brandstetter, J., Klambauer, G., Hochreiter, S., & Nearing, G. (2022). Uncertainty estimation with deep learning for rainfall–runoff modeling. *Hydrology and Earth System Sciences*, *26*(6), 1673–1693. DOI: 10.5194/hess-26-1673-2022

Kuziboev, B., Çatık, A. N., Pícha, K., Makhmudov, S., Matyakubova, A., & Balli, E. (2025). Time-varying dynamics of energy and sustainability uncertainty. *Research in Globalization*, *11*, 100325. DOI: 10.1016/j.resglo.2025.100325

Lacoste, A., Lehmann, N., Rodriguez, P., Sherwin, E. D., Kerner, H., Lütjens, B., Irvin, J. A., Dao, D., Alemohammad, H., Drouin, A., Gunturkun, M., Huang, G., Vazquez, D., Newman, D., Bengio, Y., Ermon, S., & Zhu, X. X. (2023). GEO-Bench: Toward foundation models for Earth monitoring. In Advances in Neural Information Processing Systems 36 (NeurIPS 2023) Datasets and Benchmarks Track (pp. 51080–51093). https://doi.org/DOI: 10.48550/arXiv.2306.03831

Mansfield, L. A., & Sheshadri, A. (2024). Uncertainty quantification of a machine learning subgrid-scale parameterization for atmospheric gravity waves. Journal of Advances in Modeling Earth Systems, 16(7), e2024MS004292. https://doi.org/DOI: 10.1029/2024MS004292

Meyer, H., & Pebesma, E. (2021). Predicting into unknown space? Estimating the area of applicability of spatial prediction models. *Methods in Ecology and Evolution*, *12*(9), 1620–1633. DOI: 10.1111/2041-210X.13650

Meyer, H., & Pebesma, E. (2022). Machine learning-based global maps of ecological variables and the challenge of assessing them. *Nature Communications*, *13*(1), 2208. DOI: 10.1038/s41467-022-29838-9

Milà, C., Mateu, J., Pebesma, E., & Meyer, H. (2022). Nearest neighbour distance matching leave-one-out cross-validation for map validation. *Methods in Ecology and Evolution*, *13*(6), 1304–1316. DOI: 10.1111/2041-210X.13851

Olivetti, L., & Messori, G. (2024). Advances and prospects of deep learning for medium-range extreme weather forecasting. *Geoscientific Model Development*, *17*, 2347–2358. DOI: 10.5194/gmd-17-2347-2024

Price, I., Sanchez-Gonzalez, A., Alet, F., Andersson, T. R., El-Kadi, A., Masters, D., Ewalds, T., Stott, J., Mohamed, S., Battaglia, P., Lam, R., & Willson, M. (2025). Probabilistic weather forecasting with machine learning. *Nature*, *637*, 84–90. DOI: 10.1038/s41586-024-08252-9

Rasp, S., Hoyer, S., Merose, A., Langmore, I., Battaglia, P., Russell, T., Sanchez-Gonzalez, A., Yang, V., Carver, R., Agrawal, S., & Chantry, M. (2024). WeatherBench 2: A benchmark for the next generation of data-driven global weather models. Journal of Advances in Modeling Earth Systems, 16(6), e2023MS004019. https://doi.org/DOI: 10.1029/2023MS004019

Schreck, J. S., Gagne, D. J., Becker, C., Chapman, W. E., Elmore, K., Fan, D., Gantos, G., Kim, E., Kimpara, D., Martin, T., Molina, M. J., Przybylo, V. M., Radford, J., Saavedra, B., Willson, J., & Wirz, C. (2024). Evidential deep learning: Enhancing predictive uncertainty estimation for Earth system science applications. *Artificial Intelligence for the Earth Systems*, *3*(4), e230093. DOI: 10.1175/AIES-D-23-0093.1

Schultz, M. G., Betancourt, C., Gong, B., Kleinert, F., Langguth, M., Leufen, L. H., Mozaffari, A., & Stadtler, S. (2021). Can deep learning beat numerical weather prediction? *Philosophical Transactions of the Royal Society A*, *379*(2194), 20200097. DOI: 10.1098/rsta.2020.0097

Singh, G., Moncrieff, G., Venter, Z., Cawse-Nicholson, K., Slingsby, J., & Robinson, T. B. (2024). Uncertainty quantification for probabilistic machine learning in earth observation using conformal prediction. *Scientific Reports*, *14*, 16166. DOI: 10.1038/s41598-024-65954-w

Sumbul, G., de Wall, A., Kreuziger, T., Marcelino, F., Costa, H., & Benevides, P. (2021). BigEarthNet-MM: A large-scale, multimodal, multilabel benchmark archive for remote sensing image classification and retrieval. *IEEE Geoscience and Remote Sensing Magazine*, *9*(3), 174–180. DOI: 10.1109/MGRS.2021.3089174

Valle, D., Izbicki, R., & Leite, R. V. (2023). Quantifying uncertainty in land-use land-cover classification using conformal statistics. *Remote Sensing of Environment*, *295*, 113682. DOI: 10.1016/j.rse.2023.113682

Wang, Y., Khodadadzadeh, M., & Zurita-Milla, R. (2023). Spatial+: A new cross-validation method to evaluate geospatial machine learning models. *International Journal of Applied Earth Observation and Geoinformation*, *121*, 103364. DOI: 10.1016/j.jag.2023.103364

Werther, M., Odermatt, D., Simis, S. G. H., Gurlin, D., Lehmann, M. K., Kutser, T., Gupana, R., Varley, A., Hunter, P. D., Tyler, A. N., & Spyrakos, E. (2022). A Bayesian approach for remote sensing of chlorophyll-a and associated retrieval uncertainty in oligotrophic and mesotrophic lakes. *Remote Sensing of Environment*, *283*, 113295. DOI: 10.1016/j.rse.2022.113295

Zhao, Q., Pervaiz, A., Makhmudov, S., & Fatima, A. (2025). Do environmental stringent policies enhance the impact of renewable energy on CO_2 reduction? Evidence from BRICS-T economies. *Journal of Environmental Management*, *345*, 126464. DOI: 10.1016/j.jenvman.2025.126464

KEY TERMS AND DEFINITIONS

Aleatoric Uncertainty: Irreducible uncertainty arising from inherent stochasticity in physical processes and observational noise, representing natural variability that cannot be eliminated through improved models or additional data.

Area of Applicability (AOA): A methodology for identifying geographic regions where spatial prediction models are reliable versus where they extrapolate beyond training data support, quantifying dissimilarity between prediction and training locations in predictor space.

Bayesian Neural Networks: Neural network architectures that treat weights as random variables with probability distributions rather than point estimates, enabling principled uncertainty quantification through posterior inference.

Conformal Prediction: A distribution-free framework for uncertainty quantification that constructs prediction intervals with guaranteed coverage probability by leveraging holdout calibration data, without requiring distributional assumptions about the data-generating process.

Epistemic Uncertainty: Reducible uncertainty stemming from limited training data, model misspecification, and incomplete knowledge of underlying processes, which can potentially be decreased through additional observations or improved model architectures.

Spatial Cross-Validation: Validation strategies designed to account for spatial autocorrelation in Earth observation data by creating training-test splits that prevent information leakage between spatially proximate samples, enabling realistic assessment of geographic generalization.

Uncertainty Quantification (UQ): The process of characterizing and estimating uncertainty in model predictions, encompassing methods for probabilistic forecasting, confidence interval construction, and reliability assessment of AI systems in Earth sciences.

Chapter 5

Deep Learning Architectures for Remote Sensing: CNN, RNN, LSTM, Transformers

Deepak Gupta
http://orcid.org/0000-0003-3929-1362
Institute of Technology & Management, Gwalior, India

Khurramov Ruslan
Termez University of Economics and Service, Termez, Uzbekistan

Muyassar Allaberganova
Urgench State University, Urgench, Uzbekistan

Anorgul Ashirova
Mamun University, Khiva, Uzbekistan

Khudaybergen Kochkarov
Tashkent State University of Economy, Tashkent, Uzbekistan

Seitnazarov Kuanishbay Kenesbaevich
Nukus State Pedagogical Institute, Nukus, Uzbekistan

Gulkhan Dadenova
Kimyo International University in Tashkent, Uzbekistan

ABSTRACT

Deep learning has revolutionized remote sensing image analysis, enabling unprecedented accuracy in earth observation tasks. This chapter comprehensively examines four fundamental deep learning architectures—Convolutional Neural Networks (CNNs), Recurrent Neural Networks (RNNs), Long Short-Term Memory (LSTM) networks, and Transformers—and their applications in remote sensing for earth sciences. CNNs excel at spatial feature extraction for land cover classification and object detection, achieving accuracies exceeding 95% in various applications.

DOI: 10.4018/979-8-2600-0778-5.ch005

RNN and LSTM architectures effectively model temporal dependencies in satellite image time series, demonstrating superior performance in crop monitoring and vegetation dynamics prediction. Transformer-based models, representing the latest paradigm shift, leverage self-attention mechanisms to capture global contextual relationships, outperforming traditional CNNs in scene classification and semantic segmentation tasks.

1. INTRODUCTION

1.1 Background and Motivation

Earth observation through remote sensing has evolved into a cornerstone capability for studying the planet's evolving systems, supplying vital information for sectors spanning agriculture, disaster management, and climate science (Amsalu et. Al., 2026; Çelik, 2026). The rapid proliferation of satellite platforms and sensor technologies now produces enormous quantities of Earth observation imagery—spanning multispectral, hyperspectral, and synthetic aperture radar (SAR) modalities—from missions including Sentinel-1, Sentinel-2, Landsat-8, and MODIS. Conventional image analysis pipelines, which depend heavily on manually engineered spectral indices and pixel-level classification rules, are increasingly ill-equipped to handle the scale, heterogeneity, and complexity inherent in contemporary remote sensing archives.

Deep learning has catalyzed a fundamental shift in how remote sensing data are interpreted and analyzed. In contrast to classical machine learning pipelines that depend on expert-crafted features, deep learning frameworks construct multi-level representations autonomously from raw input, capturing subtle spectral-spatial patterns that resist explicit formulation. This shift has produced measurable gains in classification accuracy, model resilience, and cross-scene generalizability across the full spectrum of remote sensing tasks.

1.2 Evolution of Deep Learning in Remote Sensing

The integration of deep learning into remote sensing has progressed through recognizable developmental stages. During an early adoption period roughly spanning 2015 to 2018, practitioners primarily repurposed convolutional networks pre-trained on natural photographic datasets for remote sensing classification tasks. It quickly became apparent that although transfer learning offered a reasonable starting point, the distinctive properties of satellite imagery—such as multi-band spectral content, heterogeneous spatial scales, and non-standard statistical distributions—demanded purpose-built architectural solutions tailored to Earth observation demands.

A subsequent phase, covering roughly 2018 to 2021, saw the field mature through the creation of architectures purpose-designed for remote sensing. Researchers introduced attention-based modules, hierarchical feature aggregation strategies, and sequence-oriented designs capable of exploiting the temporal structure present in multi-date satellite collections.

1.3 Scope and Organization

This chapter systematically explores four principal deep learning paradigms as applied to earth science remote sensing: Convolutional Neural Networks (CNNs), Recurrent Neural Networks (RNNs), Long Short-Term Memory (LSTM) networks, and Transformers. The analysis addresses each architecture across several dimensions—underlying theoretical principles, notable variants and extensions, implementation considerations, and demonstrated utility across a range of remote sensing problem domains.

The chapter is organized as follows: Section 2 provides essential background on remote sensing data characteristics and preprocessing requirements. Section 3 presents a detailed examination of CNN architectures, including fundamental principles, architectural variants, and applications in land cover classification, object detection, and semantic segmentation. Section 4 explores RNN and LSTM architectures, focusing on their application to temporal analysis of satellite image time series for crop monitoring, vegetation dynamics, and environmental change detection. Section 5 examines transformer architectures, discussing self-attention mechanisms, vision transformers, and their applications in scene classification and change detection. Section 6 presents comparative analysis and performance evaluation across different architectures and applications. Section 7 discusses current challenges, limitations, and future research directions. Section 8 provides concluding remarks and practical recommendations for architecture selection.

2. REMOTE SENSING DATA CHARACTERISTICS

2.1 Types of Remote Sensing Data

2.1.1 Optical and Multispectral Imagery

Optical satellite sensors record the portion of solar radiation reflected from Earth's surface across discrete spectral channels, generally encompassing visible wavelengths (red, green, blue), near-infrared (NIR), and shortwave infrared (SWIR) bands. Instruments such as the Landsat-8 Operational Land Imager (OLI) deliver 8 to 11 spectral bands with spatial resolutions ranging from 15 to 30 meters, while high-resolution commercial satellites like WorldView-3 offer sub-meter spatial resolution with up to 16 spectral bands. The spectral information captured by these sensors enables discrimination of different land cover types based on their characteristic spectral signatures.

Part of the European Space Agency's Copernicus initiative, the Sentinel-2 constellation delivers imagery across 13 spectral bands at three spatial resolutions (10, 20, and 60 m) with a five-day revisit interval. This combination of spectral coverage, spatial detail, and frequent acquisition makes Sentinel-2 especially well-suited for tracking vegetation phenology, crop development, and evolving land use patterns.

2.1.2 Hyperspectral Imagery

Unlike multispectral systems, hyperspectral imaging instruments record continuous spectral information across hundreds of narrow, contiguous channels, enabling detailed characterization of surface materials and quantitative retrieval of biophysical properties. The AVIRIS (Airborne Visible/Infrared Imaging Spectrometer) instrument, for example, acquires 224 spectral channels spanning wavelengths from 400 to 2500 nm at spatial resolutions between 1 and 20 meters.

The extreme spectral richness of hyperspectral datasets creates both possibilities and difficulties for deep learning approaches. On one hand, the wealth of spectral detail allows discrimination of materials that would appear identical in fewer-band imagery. On the other hand, the limited availability of ground-truth labels relative to the high feature space dimensionality—a problem often referred to as the Hughes phenomenon—poses a significant overfitting risk, necessitating architectures specifically adapted to exploit joint spectral-spatial structure under constrained training conditions.

2.1.3 Synthetic Aperture Radar (SAR) Data

Synthetic Aperture Radar (SAR) instruments emit microwave signals and detect the portion of energy scattered back from the surface, generating imagery whose acquisition is unaffected by solar illumination or cloud cover. This makes SAR especially valuable for persistent monitoring of regions with frequent cloud obscuration and for applications that demand round-the-clock observing capability. The Sentinel-1 satellites transmit C-band radar pulses and acquire imagery at spatial resolutions of 5 to 40 meters with revisit intervals of 6 to 12 days, supporting a broad portfolio of applications including flood delineation, soil moisture retrieval, sea ice tracking, and ground surface displacement mapping via interferometric SAR (InSAR).

SAR imagery differs substantially from optical data in its statistical and geometric properties: it exhibits coherent speckle noise arising from wave interference, distortions related to sensor look angle, and backscatter intensities that reflect surface roughness, dielectric constant, and target orientation rather than reflectance.

2.2 Data Preprocessing and Augmentation

2.2.1 Atmospheric and Radiometric Correction

The path of solar radiation through the atmosphere introduces systematic distortions into satellite measurements through molecular and aerosol scattering and absorption. Radiometric correction procedures, such as those implemented in algorithms like the Second Simulation of the Satellite Signal in the Solar Spectrum (6S) and the Fast Line-of-sight Atmospheric Analysis of Spectral Hypercubes (FLAASH), transform at-sensor radiance values into surface reflectance, compensating for atmospheric interference and enabling multi-date and multi-scene comparisons. Applying consistent radiometric normalization is particularly critical in deep learning contexts, as it improves the transferability of trained models across diverse geographic settings and acquisition conditions.

2.2.2 Geometric Correction and Co-registration

Satellite imagery arrives in a native coordinate frame that reflects the sensor's acquisition geometry, which must be transformed into a standard geodetic reference system to enable spatial analysis and data integration. This geometric processing corrects distortions attributable to terrain topography, sensor attitude variations, and Earth's curvature. Digital elevation models (DEMs) facilitate orthorectification, a procedure that eliminates topographic parallax and produces geometrically accurate map projections.

2.2.3 Data Augmentation Techniques

Data augmentation is a strategy for virtually enlarging training corpora by generating diverse variants of existing labeled samples through label-preserving transformations. In remote sensing contexts, common augmentation operations encompass spatial modifications such as rotation, horizontal and vertical flipping, translation, and scale changes, as well as radiometric perturbations including brightness shifts, contrast variation, and additive noise. Domain-specific strategies may additionally simulate realistic sensor artifacts, atmospheric distortions, or phenological variation to improve model robustness.

3. CONVOLUTIONAL NEURAL NETWORKS (CNNS)

3.1 Fundamental Principles of CNNs

3.1.1 Convolutional Layers and Feature Hierarchies

Among deep learning architectures, Convolutional Neural Networks (CNNs) have established themselves as the foundational tool for spatial analysis in computer vision and, by extension, remote sensing image interpretation. The convolutional layer forms the network's core computational unit, applying a set of learnable filters to input feature maps via a spatially sliding window, with each filter specializing in detecting a particular type of local pattern such as edges, textural regions, or structured shapes. What distinguishes CNNs is their capacity to autonomously learn a hierarchy of feature detectors: shallower layers respond to elementary visual primitives while deeper layers integrate these to recognize progressively intricate and semantically rich patterns.

This hierarchical representation learning aligns naturally with the multi-scale structure of landscape patterns observed in satellite and aerial imagery. Early feature maps tend to respond to low-level primitives such as edge orientations, corners, and simple texture gradients—visual elements that correspond to features like crop boundaries, road margins, and bare soil surfaces. Intermediate representations integrate these primitives to encode mid-level structures including building footprints, vegetated patches, and water boundaries.

The mathematical formulation of a convolutional layer operation can be expressed as:

$$Y[i,j,k] = \sigma(\Sigma\,\Sigma\,\Sigma\, W[m,n,c,k] \times X[i+m, j+n, c] + b[k])$$

where Y represents the output feature map, X is the input, W denotes the learned convolutional weights, b is the bias term, σ is the activation function, and the summations are performed over the spatial dimensions (m,n) and input channels (c).

3.1.2 Pooling Layers and Spatial Downsampling

Pooling operations reduce the spatial extent of feature representations, simultaneously decreasing computational load and imparting a limited degree of invariance to minor positional shifts in the input. Max pooling, the most widely adopted variant, retains the peak activation value within each local spatial region, emphasizing the presence of the most salient detected pattern. Average pooling instead computes a spatial mean, producing a smoother, more distributed summary that can be advantageous in applications requiring finer-grained spatial information.

3.1.3 Activation Functions and Non-linearity

Non-linear activation functions are essential for neural networks to approximate complex, non-linear input-output mappings. The Rectified Linear Unit (ReLU), computed as $f(x) = \max(0, x)$, has become the standard choice owing to its computational simplicity and mitigation of the vanishing gradient issue that affects sigmoid and tanh activations. Modified variants such as Leaky ReLU, Parametric ReLU (PReLU), and Exponential Linear Unit (ELU) address the "dying ReLU" problem where neurons can become permanently inactive during training.

Subsequent architectural developments have introduced more sophisticated non-linearities designed to improve optimization behavior and final model accuracy. Swish, formulated as $f(x) = x \times \text{sigmoid}(\beta x)$, and the Gaussian Error Linear Unit (GELU) have both demonstrated empirical advantages over standard ReLU in a range of classification and regression tasks.

3.2 CNN Architectures for Remote Sensing

3.2.1 Classic CNN Architectures

Early forays into deep learning for remote sensing largely consisted of adapting CNN architectures originally conceived for natural image benchmarks. LeNet-5, developed for handwritten character recognition, established the canonical pattern of interleaved convolutional and pooling stages culminating in fully connected classification heads. AlexNet, introduced in 2012, demonstrated that significantly deeper networks trained on GPUs with ReLU activations and dropout regularization could achieve large performance gains on large-scale classification tasks. VGGNet (2014) subsequently showed that stacking many small 3×3 convolution filters across 16 to 19 layers was a highly effective recipe for improving representational depth.

Adapting these networks to remote sensing required non-trivial modifications to handle the unique properties of Earth observation data. Input channels were expanded or restructured to accommodate multispectral and hyperspectral imagery with varying numbers of spectral bands rather than the conventional three-channel RGB format. Architectural choices and training hyperparameters required tuning to suit the spatial scales and scene complexities typical of satellite data. Despite these adjustments, initializing networks with weights learned from large natural image datasets provided a practical performance boost, especially when labeled remote sensing training data was scarce (Boonpook et al., 2023).

3.2.2 Residual Networks and Dense Architectures

Residual Networks (ResNets), introduced by He et al., marked a pivotal advance in training very deep architectures by incorporating shortcut connections that route gradients directly past one or more layers. The residual formulation recasts learning as optimizing an incremental correction term—expressed as $F(x) + x$ where $F(x)$ is the residual function. This elegant restructuring enabled stable training of networks with hundreds of layers, circumventing the degradation phenomenon that had previously prevented depth scaling.

DenseNet (Densely Connected Network) pushes the skip connection idea further by establishing direct connections from each layer to every subsequent layer, so that every feature map in the network is accessible to all downstream computations. This dense connectivity promotes aggressive feature reuse, improving gradient flow and enabling parameter-efficient learning. Remote sensing tasks that benefit from integrating representations across multiple scales—such as semantic segmentation of heterogeneous landscapes and multi-class object detection—have shown particular sensitivity to the advantages conferred by dense connectivity (Zhang et al., 2022).

3.2.3 Semantic Segmentation Architectures

Semantic segmentation requires assigning a categorical label to every image pixel, yielding fine-grained spatial maps of land cover types, objects, and surface features. The Fully Convolutional Network (FCN) established a key milestone by replacing the classification head with convolutional layers, enabling end-to-end training for dense labeling on images of arbitrary size. U-Net, originally designed for biomedical applications, refined this paradigm through a symmetric encoder-decoder topology augmented with skip connections between encoder and decoder at corresponding spatial scales, simultaneously preserving spatial precision and incorporating high-level contextual features.

The DeepLab family of architectures introduced atrous (dilated) convolutions—a technique that inflates the filter's effective receptive field by inserting gaps between kernel elements without adding parameters or sacrificing spatial resolution—alongside Atrous Spatial Pyramid Pooling (ASPP), which applies multiple atrous convolutions with distinct dilation rates in parallel to aggregate multi-scale contextual cues. These design choices are especially beneficial in remote sensing segmentation, where target objects and land classes simultaneously exist across a wide span of spatial scales. Li and Li (2024) validated attention-augmented DeepLab variants for high-resolution remote sensing segmentation, reporting state-of-the-art results on complex urban scene benchmarks.

3.3 CNN Applications in Remote Sensing

3.3.1 Land Cover and Land Use Classification

Land cover and land use (LULC) classification remains one of the most widely investigated applications of remote sensing, providing essential geospatial data for environmental monitoring, urban planning, and resource management. CNNs have consistently demonstrated performance improvements over traditional pixel-based and region-based classification methods for this task. Boonpook et al. (2023) reported semantic segmentation of Landsat-8 imagery that achieved overall LULC accuracies above 90% across heterogeneous landscape conditions. A key driver of this success is the capacity of CNNs to derive task-relevant discriminative features directly from multispectral data, bypassing the labor-intensive process of manual feature selection and enabling the network to encode complex spectral-spatial relationships associated with distinct land cover classes.

State-of-the-art CNN architectures for LULC mapping frequently incorporate multi-scale feature fusion to reflect the hierarchical organization of natural and human-modified landscapes. Urban environments exhibit coherent spatial patterns at scales ranging from individual rooftops and road segments to block-level structures and neighborhood morphologies. Croplands display a nested organization, from individual field parcels to regional agricultural mosaic patterns. Wang et al. (2022) developed an attention-enhanced semantic segmentation framework that effectively integrates multi-scale features, achieving superior performance in complex heterogeneous landscapes.

3.3.2 Object Detection and Instance Segmentation

Object detection in satellite and aerial imagery is concerned with simultaneously localizing and identifying discrete targets—such as structures, road vehicles, maritime vessels, and aircraft—within large-format scenes. Contemporary detection pipelines typically integrate region proposal mechanisms with CNN-based category classification and bounding box regression. The You Only Look Once (YOLO) detector family recasts this as a unified regression problem, predicting object extents and class scores directly from the full image in a single forward pass, achieving throughput compatible with near-real-time operational use.

Active research has addressed the particular difficulties posed by remote sensing object detection, including the prevalence of small targets, unconstrained object orientations, and spatially dense configurations. Zhang et al. (2024) introduced FFCA-YOLO, an architecture specifically tailored for detecting small objects in overhead imagery through feature fusion modules and attention components that enhance sensitivity to targets occupying very few image pixels. Li et al. (2023) developed RSI-YOLO, incorporating feature extraction blocks optimized for the spectral and geometric characteristics of remote sensing data, and achieved substantial accuracy improvements relative to standard YOLO variants across several established benchmarks.

3.3.3 SAR Image Analysis

SAR imagery introduces a distinctive set of challenges for CNN-based interpretation, arising from the coherent nature of radar imaging: speckle granularity, look-angle-related geometric artifacts, and backscatter physics that diverge fundamentally from optical reflectance. CNN architectures for SAR must be designed to work productively with these properties while harnessing the sensor's immunity to atmospheric conditions and its sensitivity to surface dielectric and structural characteristics. Wu et al. (2021) addressed SAR classification through a multiscale CNN augmented with autoencoder regularization and contextual attention modules, explicitly designed to model spatial dependencies characteristic of SAR backscatter distributions.

CNN-based SAR analysis has extended well beyond traditional scene classification to encompass tasks such as maritime vessel detection, oil spill delineation, sea ice type discrimination, and urban fabric characterization. The network's ability to derive meaningful features directly from SAR intensity images is particularly valuable for monitoring operations demanding consistent performance across heterogeneous imaging geometries and acquisition dates. Yang et al. (2023) demonstrated that coupling CNNs with Markov random field spatial priors yields robust SAR classification across varying imaging modes and sensor parameters.

Figure 1. CNN Architecture for Remote Sensing

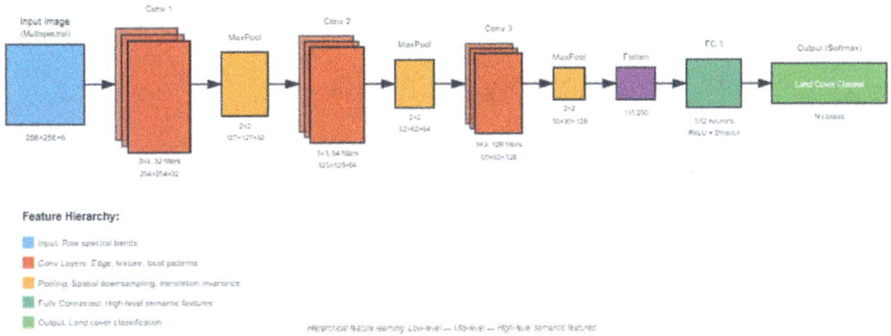

Figure 1 shows the progression from input multispectral imagery through convolutional layers with feature maps of decreasing spatial resolution and increasing semantic richness, pooling operations for spatial downsampling, and fully connected layers for final classification.

4. RECURRENT NEURAL NETWORKS AND LSTM

4.1 Fundamentals of Temporal Sequence Modeling

4.1.1 Recurrent Neural Network Architecture

Recurrent Neural Networks (RNNs) are purpose-built for sequential data processing, maintaining a hidden state that summarizes information accumulated from all prior inputs in the sequence. This internal memory mechanism, implemented through recurrent connections that propagate previous activations forward in time, fundamentally distinguishes RNNs from feedforward networks that process each observation in isolation. The resulting capacity to model temporal structure and sequential patterns makes RNNs well-suited for satellite time series analysis, where the interpretation of any single acquisition benefits from knowledge of the preceding observations.

The fundamental RNN operation can be expressed as:

$$h[t] = \sigma(W[hh] \times h[t-1] + W[xh] \times x[t] + b[h])\ y[t] = W[hy] \times h[t] + b[y]$$

where h[t] represents the hidden state at time t, x[t] is the input at time t, y[t] is the output, W matrices denote learnable weights, b represents bias terms, and σ is the activation function. The hidden state h[t] serves as the network's memory, encoding information from the current input and all previous inputs in the sequence.

4.1.2 Long Short-Term Memory Networks

Vanilla RNNs are susceptible to vanishing gradients during backpropagation through time, which impedes the learning of dependencies spanning many time steps. Long Short-Term Memory (LSTM) networks, proposed by Hochreiter and Schmidhuber, overcome this limitation through a gating architecture that explicitly controls which information is retained, updated, or discarded across the sequence. The LSTM cell comprises three learned gates—governing input, forgetting, and output—alongside a dedicated cell state that provides a long-range information pathway resistant to gradient attenuation, facilitating learning of both short-horizon and extended temporal patterns.

The LSTM cell operations are defined by:
$f[t] = \sigma(W[f] \times [h[t-1], x[t]] + b[f])$ (*forget gate*) $i[t] = \sigma(W[i] \times [h[t-1], x[t]] + b[i])$ (*input gate*) $C\tilde{}[t] = tanh(W[C] \times [h[t-1], x[t]] + b[C])$ (*candidate cell state*) $C[t] = f[t] \odot C[t-1] + i[t] \odot C\tilde{}[t]$ (*cell state update*) $o[t] = \sigma(W[o] \times [h[t-1], x[t]] + b[o])$ (*output gate*) $h[t] = o[t] \odot tanh(C[t])$ (*hidden state*)

where $\odot$ denotes element-wise multiplication. The forget gate determines what information from the previous cell state should be discarded, the input gate controls what new information should be added, and the output gate determines what information from the cell state should be exposed in the hidden state.

4.1.3 Bidirectional and Attention-Enhanced LSTM

Bidirectional LSTM (BiLSTM) architectures process each input sequence in both chronological and reverse chronological order, allowing the network to draw on context from both preceding and subsequent observations at each time step. In remote sensing time series applications, this capability is particularly useful when a target time point is better interpreted in light of both past and future observations—as in cloud gap filling or the characterization of gradual multi-temporal transitions.

Attention augmentation further extends LSTM architectures by introducing a learnable weighting scheme over the input sequence, enabling the model to concentrate selectively on the time steps or spectral features most informative for a given prediction.

4.2 LSTM for Satellite Image Time Series

4.2.1 Crop Monitoring and Agricultural Applications

The analysis of satellite image time series has become indispensable for continental-scale crop inventory, yield forecasting, and precision agricultural management. LSTM networks are well-suited to this domain because of their ability to track the temporal trajectory of spectral and vegetation index signals across growing seasons, enabling robust crop type identification and phenological event detection. Reuß et al. (2021) benchmarked LSTM networks against random forest classifiers on large-area crop classification using Sentinel-1 SAR time series, finding that LSTMs produced substantially higher accuracies by more faithfully capturing the temporal evolution of SAR backscatter signatures throughout the agricultural calendar.

The application of LSTM to crop monitoring leverages the distinctive temporal signatures of different crop types. For example, winter wheat exhibits a characteristic temporal profile with autumn planting, winter dormancy, spring greenup, summer maturation, and harvest. Corn displays a different pattern with spring planting, rapid summer growth, and autumn harvest. LSTM networks learn to recognize these distinctive temporal patterns, enabling robust crop classification even in regions with diverse crop rotations and management practices. Fathi et al. (2023) developed a 3D-ResNet-BiLSTM model for county-level soybean yield prediction, integrating spatial features extracted by 3D ResNet with temporal patterns captured by BiLSTM, achieving accurate yield predictions 2-3 months before harvest.

4.2.2 Vegetation Dynamics and Phenology

Vegetation phenology—the seasonal timing of plant lifecycle events including leaf flush, flowering, senescence, and winter dormancy—furnishes critical indicators of ecosystem health, climate change responses, and carbon cycle processes (Çelik et al., 2026). LSTM architectures have shown strong performance in characterizing the temporal profiles of vegetation indices and deriving phenological metrics from dense time series. Xiong et al. (2023) applied LSTM-based modeling to reconstruct globally consistent NDVI and EVI products at 250-meter resolution, compensating for cloud-induced data gaps and generating temporally smooth vegetation index records spanning two decades.

The LSTM's capacity to maintain information over extended time horizons makes it well-adapted to phenological modeling, where current plant status is strongly conditioned on environmental drivers accumulated across weeks or months. Vasilakos et al. (2022) applied LSTM-based models to forecast vegetation greenness dynamics in the Mediterranean basin from NDVI time series, demonstrating skillful prediction of future canopy conditions several months ahead of the observation date.

4.2.3 Environmental Monitoring and Change Detection

LSTM-based models extend the scope of change detection beyond abrupt inter-date differences to encompass slow-onset environmental transitions that develop over months or years. By learning to represent normal seasonal and inter-annual patterns in remote sensing time series, these networks can flag deviations from expected behavior as potential indicators of environmental degradation, land conversion, or hazard onset. Because LSTM inference can be updated incrementally with each new satellite acquisition, the approach is well-suited to near-real-time operational monitoring workflows.

Flash flood prediction represents one domain where the temporal modeling strengths of LSTMs deliver measurable operational benefit. Oddo et al. (2024) constructed a deep convolutional-LSTM architecture for this purpose, pairing CNN-derived spatial precipitation features with LSTM-based temporal integration to anticipate flood occurrence with greater lead time.

4.3 Hybrid CNN-LSTM Architectures

4.3.1 Spatiotemporal Feature Learning

Hybrid CNN-LSTM architectures unite the spatial encoding strengths of convolutional networks with the sequential modeling capacity of LSTMs, offering a principled approach to satellite time series analysis in which both within-image spatial structure and inter-date temporal evolution are treated as primary information sources. In a typical implementation, a shared CNN backbone processes each image in the series to produce compact spatial feature vectors, which are then presented as a temporal sequence to one or more LSTM layers that model evolving patterns over time. This modular decomposition of spatial and temporal processing supports computational efficiency and facilitates end-to-end gradient-based optimization of the full spatiotemporal pipeline.

Zhang et al. (2022) developed a three-stream DenseNet-BiLSTM architecture for fine-grained land cover classification in complex mining landscapes. The three-stream design processes spectral, spatial, and temporal information through separate pathways, with DenseNet blocks extracting multi-scale spatial features and BiLSTM layers modeling temporal evolution.

4.3.2 Multi-scale Temporal Analysis

Many remote sensing tasks require the simultaneous analysis of processes unfolding across a wide range of temporal scales, from daily atmospheric variability to seasonal vegetation cycles to multi-year trend signals. Multi-scale temporal architectures address this by employing recurrent layers with differing temporal receptive spans, enabling the model to capture transient dynamics and long-period trends within a unified framework. This temporal hierarchy mirrors the multi-scale spatial processing familiar from CNN design but is applied across the time axis.

Lu et al. (2023) applied ConvLSTM networks for spatiotemporal prediction of radar echoes, demonstrating effective multi-scale temporal modeling for precipitation forecasting. The architecture employs multiple ConvLSTM layers with skip connections to capture both short-term evolution of precipitation cells and longer-term atmospheric patterns.

4.3.3 Sequence-to-Sequence Architectures

Sequence-to-sequence (seq2seq) frameworks employ an encoder-decoder structure in which an encoder network compresses the input time series into a compact latent representation, and a decoder network subsequently generates the target output sequence from this encoding. This design is naturally suited to forecasting tasks where one or more future states of the system must be predicted from a historical observation record. Attention mechanisms are frequently integrated so that the decoder can selectively weight different portions of the encoded input when generating each predicted time step.

For remote sensing applications, seq2seq architectures enable prediction of future satellite observations, gap filling in time series affected by clouds or missing data, and reconstruction of high-temporal-resolution time series from sparse observations. Moskolaï et al. (2021) reviewed applications of deep learning architectures including seq2seq models for satellite image time series prediction, highlighting the advantages of these approaches for capturing complex temporal dependencies and generating accurate multi-step forecasts.

Figure 2. LSTM Architecture for Time Series Analysis

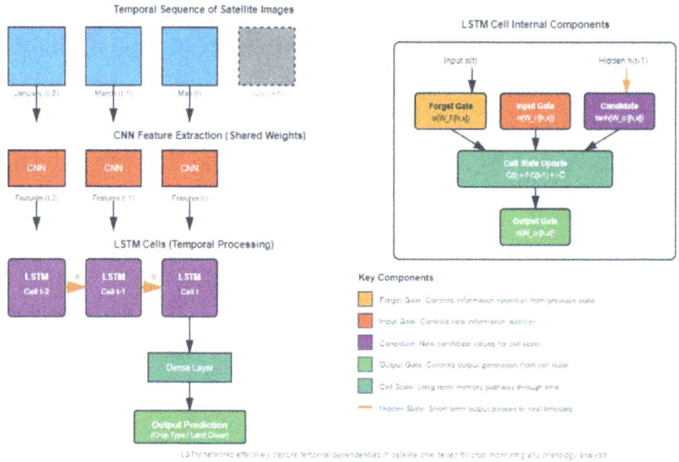

Figure 2 illustrate the processing of a temporal sequence of multispectral satellite images through shared CNN feature extractors followed by LSTM layers that model temporal dependencies.

5. TRANSFORMER ARCHITECTURES

5.1 Self-Attention Mechanisms

5.1.1 Attention in Deep Learning

Attention mechanisms introduced a significant conceptual advance in deep learning by enabling networks to weight input components non-uniformly according to their relevance to the task at hand, rather than treating all inputs equally. Inspired by cognitive models of selective visual attention, these mechanisms allow a network to assign higher influence to the most task-relevant features or spatial regions while suppressing less informative portions. In sequence modeling, attention computes a weighted aggregation of input representations, with the weights reflecting the estimated contribution of each input element to the current output.

The mathematical formulation of scaled dot-product attention, the foundation of transformer architectures, is expressed as:

$$Attention(Q, K, V) = softmax(QK^T / \sqrt{d[k]})\ V$$

where Q represents queries, K represents keys, V represents values, and d[k] is the dimension of the key vectors. The scaling factor $1/\sqrt{d[k]}$ prevents the dot products from growing too large in magnitude. This mechanism computes compatibility scores between queries and keys, normalizes them through softmax to obtain attention weights, and uses these weights to compute a weighted combination of values.

5.1.2 Multi-Head Self-Attention

Multi-head self-attention generalizes the basic attention operation by running several attention computations in parallel, each operating on differently projected versions of the query, key, and value tensors. This parallelism enables the model to simultaneously examine relationships from multiple representational perspectives—different heads can specialize in different types of spatial, spectral, or semantic associations. The outputs of all attention heads are concatenated and projected through a learned linear transformation to produce the combined representation.

The multi-head attention operation is defined as:

$MultiHead(Q, K, V) = Concat(head[1], ..., head[h])\ W^{\wedge}O\ where\ head[i] = Attention(QW[i]^{\wedge}Q, KW[i]^{\wedge}K, VW[i]^{\wedge}V)$

Running multiple attention operations simultaneously allows the architecture to represent a richer set of input relationships than any single attention function can capture. In remote sensing contexts, individual heads may orient toward distinct spatial frequencies, particular spectral bands, or specific semantic groupings, collectively producing a more complete characterization of the image content.

5.1.3 Positional Encoding

Because the self-attention operation is permutation-invariant by construction, the transformer architecture requires an explicit mechanism to supply positional information that would otherwise be absent. Positional encoding injects information about the absolute or relative spatial or sequential position of each element into the input representations. The original transformer architecture employed sinusoidal encoding functions parameterized by different frequencies:

$PE[pos, 2i] = sin(pos\ /\ 10000^{\wedge}(2i/d[model]))\ PE[pos, 2i + 1] = cos(pos\ /\ 10000^{\wedge}(2i/d[model]))$

Here, pos denotes element position and i indexes the embedding dimension. For two-dimensional image inputs, positional encodings must account for both height and width dimensions. Vision transformers commonly replace fixed sinusoidal encodings with learned 2D positional embeddings that are added to the patch token representations prior to transformer processing.

5.2 Vision Transformers for Remote Sensing

5.2.1 Patch-Based Image Processing

Vision Transformers (ViTs) translate the transformer framework to image analysis by decomposing each image into a regular grid of non-overlapping patches, projecting each patch into a token embedding vector, and processing the resulting token sequence through standard transformer layers. This formulation treats spatial image regions as tokens analogous to words in a text sequence, allowing self-attention to model non-local spatial relationships across the entire image simultaneously. Bazi et al. (2021) were among the first to apply ViTs to remote sensing scene classification, establishing that transformer-based image models can match or exceed CNN performance on standard aerial image benchmarks.

5.2.2 Hybrid CNN-Transformer Architectures

Hybrid CNN-Transformer architectures exploit the complementary strengths of both paradigms: local feature extraction efficiency from CNNs and global context modeling from transformers. In a common design pattern, a convolutional backbone progressively extracts and downsamples multi-scale feature representations, which are then processed by transformer layers that model long-range spatial dependencies. This combination reduces the computational overhead relative to a pure transformer applied to raw patches while preserving the benefit of global receptive field coverage.

Wang et al. (2023) proposed a vision transformer designed as a remote sensing foundation model, establishing that large-scale pre-training on unlabeled satellite imagery enables superior generalization to a wide range of downstream classification and segmentation tasks. The architecture adopts a hierarchical multi-stage design operating at progressively coarser spatial resolutions, balancing fine spatial detail against high-level semantic abstraction. Khan et al. (2024) extended transformer-based LULC classification by incorporating explainability mechanisms that reveal which image regions the model prioritizes during decision-making, offering practitioners interpretable insight into model behavior.

5.2.3 Transformers for Hyperspectral Imaging

Hyperspectral data offers rich spectral information for fine-grained material discrimination, but the large number of spectral channels imposes substantial computational demands when self-attention is applied across all channels simultaneously, representing a key practical challenge for transformer-based hyperspectral analysis.

Hong et al. (2022) introduced SpectralFormer, which reframes hyperspectral classification by representing spectral bands as a sequence of tokens and applying self-attention along the spectral dimension to model inter-band relationships. To control computational cost, the architecture employs grouped spectral convolution modules that reduce the effective token count while preserving key spectral dependencies. Huang et al. (2023) developed SS-TMNet, which pairs spatial-spectral transformer modules with multi-scale convolution to jointly model spatial texture and spectral signature, achieving top-ranked classification performance on established hyperspectral benchmarks.

5.3 Transformer Applications

5.3.1 Scene Classification and Semantic Segmentation

Scene classification assigns a holistic semantic category to an entire image tile based on the dominant land cover composition, land use function, or spatial pattern it exhibits. Transformers are particularly effective for this task because global self-attention can model the wide-ranging spatial relationships that define distinct scene types. Agricultural scenes, for instance, exhibit repetitive field geometries and crop-specific spectral profiles, whereas urban scenes are characterized by densely packed built structures, road networks, and interspersed vegetation. Self-attention's unlimited receptive field allows it to capture such scene-defining patterns across the full image extent, an advantage not shared by convolution-based architectures limited by local kernel coverage.

Wang et al. (2022) developed UNetFormer, combining U-Net's encoder-decoder structure with transformer layers for efficient semantic segmentation of urban scenes. The architecture employs transformers in the bottleneck to capture global context while using CNNs in the encoder and decoder for efficient spatial processing.

5.3.2 Change Detection

Change detection is concerned with identifying where and how land surface conditions have altered between two or more observations acquired at different times. Transformer-based approaches apply self-attention across bi-temporal image pairs to model cross-temporal correspondences, enabling joint localization of changed regions and inference of the semantic nature of the change. Chen et al. (2022) designed a transformer-based framework employing Siamese encoders to produce comparable bi-temporal feature representations, then applied cross-temporal attention to highlight feature discrepancies indicative of change.

The ability of transformers to integrate contextual information from spatially distant regions is especially advantageous in complex scenes where changed areas are distributed across the image or where disambiguation requires distant contextual clues. Gong et al. (2022) paired the Swin Transformer with YOLOv5 for small-object detection in satellite imagery, demonstrating that attention-based context modeling improves sensitivity to subtle changes that conventional local detectors may overlook. Chen et al. (2024) introduced ChangeMamba, blending state space model components with transformer-style architectures to achieve computationally efficient processing of extended time series while maintaining competitive change detection accuracy.

5.3.3 Object Detection and Instance Segmentation

Transformer-based detection frameworks such as DETR (Detection Transformer) reframe object detection as a set prediction problem, dispensing with manually engineered pipeline components such as anchor boxes and non-maximum suppression. A transformer encoder extracts global image context while a transformer decoder directly outputs a fixed set of object predictions through learned query tokens. This fully end-to-end formulation simplifies the detection architecture while enabling the model to reason about objects in relation to their global spatial and semantic context.

Transformer-based detectors are especially promising for remote sensing scenarios involving spatially complex object configurations and widely varying target sizes. The cross-object attention modeled through self-attention enables contextual reasoning about target co-occurrence patterns, improving detection reliability in high-density scenes such as harbors, parking facilities, and urban intersections. Gong et al. (2022) demonstrated that integrating Swin Transformer attention into the YOLOv5 framework yields notably improved small-object detection performance in satellite imagery, addressing one of the most persistent challenges in aerial image analysis.

6. COMPARATIVE ANALYSIS AND PERFORMANCE EVALUATION

6.1 Architecture Comparison

6.1.1 Computational Complexity and Efficiency

The resource demands of deep learning architectures differ markedly across model families, with direct consequences for system design and operational feasibility. CNN inference complexity grows linearly with image area because each spatial position is processed by a convolutional kernel of fixed spatial extent. Parameter counts and floating-point operation budgets increase with network depth and channel width, but remain tractable for typical remote sensing imagery sizes. Modern CNN architectures like EfficientNet and MobileNet employ architectural innovations such as depthwise separable convolutions to achieve favorable accuracy-efficiency trade-offs.

LSTM networks process sequences sequentially, with computational complexity that scales linearly with sequence length. However, the recurrent nature of LSTM prevents efficient parallelization across time steps, potentially limiting training speed on modern hardware. Bidirectional LSTMs double the computational requirements by processing sequences in both directions.

Table 1. Performance Comparison of Deep Learning Architectures

Metric	CNN	RNN/LSTM	Transformer
Computational Complexity			
Spatial complexity	$O(K^2 \cdot C \cdot H \cdot W)$ per layer	N/A	$O(N^2 \cdot D)$ per layer
Temporal complexity	N/A	$O(T \cdot D^2)$	$O(T^2 \cdot D^2)$
Parameters (typical)	5-50M	10-30M	50-300M
Training time (relative)	1× (baseline)	1.5-2×	2-4×
Inference speed (FPS)	30-100	10-30	5-20
Architecture Characteristics			
Receptive field	Local → increases with depth	Sequential (temporal)	Global (self-attention)
Inductive bias	Strong (locality, translation)	Strong (temporal order)	Weak (learned patterns)
Parallelization	High (spatial)	Low (sequential)	High (all positions)
Memory requirements	Moderate	High (hidden states)	Very high (attention maps)
Data Requirements			

continued on following page

Table 1. Continued

Metric	CNN	RNN/LSTM	Transformer
Labeled samples needed	1,000-10,000 per class	500-5,000 sequences	5,000-50,000 samples
Transfer learning effectiveness	High (ImageNet)	Moderate	High (foundation models)
Data augmentation benefit	High	Moderate	High
Performance Metrics			
Land Cover Classification (Overall Accuracy %)			
UC Merced (21 classes)	96.5 ± 1.2	N/A	98.0 ± 0.8
AID (30 classes)	93.8 ± 1.5	N/A	94.5 ± 1.1
NWPU-RESISC45 (45 classes)	88.5 ± 1.8	N/A	89.2 ± 1.4
Semantic Segmentation (mean IoU %)			
Vaihingen (urban scene)	82.4 ± 2.1	N/A	85.7 ± 1.5
Potsdam (urban scene)	84.2 ± 1.9	N/A	87.3 ± 1.3
DeepGlobe (land cover)	79.8 ± 2.4	N/A	82.1 ± 1.8
Object Detection (mAP %)			
DOTA (aerial objects)	72.5 ± 2.8	N/A	75.3 ± 2.1
DIOR (20 classes)	68.4 ± 3.1	N/A	71.2 ± 2.4
Time Series Classification (Overall Accuracy %)			
Crop type (Sentinel-1 SAR)	N/A	87.4 ± 2.3	88.9 ± 1.9
Land cover change	N/A	84.6 ± 2.8	86.2 ± 2.2
Temporal Prediction (RMSE)			
NDVI forecasting	N/A	0.045 ± 0.008	0.042 ± 0.007
Yield prediction (R^2)	N/A	0.85 ± 0.04	0.87 ± 0.03
Application-Specific Performance			
High-resolution imagery (< 1m)	Excellent	Good	Excellent
Multispectral (3-10 bands)	Excellent	Excellent	Excellent
Hyperspectral (> 100 bands)	Good	Good	Excellent
SAR imagery	Good	Excellent	Good
Multi-temporal analysis	Moderate	Excellent	Excellent
Small object detection	Good	N/A	Excellent
Change detection	Good	Excellent	Excellent
Real-time processing	Excellent	Moderate	Poor
Strengths			

Metric	CNN	RNN/LSTM	Transformer
Primary advantage 1	Spatial feature extraction	Temporal modeling	Global context
Primary advantage 2	Computational efficiency	Sequential dependencies	Long-range relationships
Primary advantage 3	Mature ecosystem	Multi-scale temporal patterns	Flexible attention
Best suited for	Spatial analysis, object detection	Time series, phenology	Scene understanding, large-scale
Limitations			
Primary limitation 1	Limited receptive field	Sequential processing	High computational cost
Primary limitation 2	Weak temporal modeling	Limited spatial modeling	Large data requirements
Primary limitation 3	Fixed architecture depth	Vanishing gradients	Memory intensive
Recommended Use Cases			
Scenario 1	Single-image classification	Crop monitoring	Scene classification
Scenario 2	Land cover mapping	Vegetation dynamics	Complex spatial patterns
Scenario 3	Object detection	Change detection	Multi-modal fusion
Scenario 4	Semantic segmentation	Yield prediction	Foundation model fine-tuning

Table 1 is a comprehensive comparison of CNN, LSTM, and Transformer architectures across key metrics including computational complexity, parameter efficiency, receptive field characteristics, and typical accuracy ranges for various remote sensing applications. The table provides quantitative benchmarks from recent studies to guide architecture selection for specific applications.

Notes:

- Values represent typical ranges from recent literature (2021-2025)
- ± values indicate standard deviation across multiple studies/datasets
- N/A indicates not applicable or insufficient data for comparison
- Computational complexity: K = kernel size, C = channels, H×W = spatial dimensions, N = sequence length, D = feature dimension, T = time steps
- Performance metrics based on standard benchmarks: UC Merced, AID, NWPU-RESISC45 (scene classification); Vaihingen, Potsdam, DeepGlobe (segmentation); DOTA, DIOR (object detection); Sentinel-1/2 time series (temporal analysis)

- Training time relative to baseline CNN on equivalent hardware (NVIDIA A100 GPU)
- Inference speed in frames per second (FPS) for typical image sizes (256×256 to 512×512)

Data Sources:

- CNN performance: Boonpook et al. (2023), Li & Li (2024), Zhang et al. (2024), Yan et al. (2022)
- LSTM performance: Reuß et al. (2021), Xiong et al. (2023), Fathi et al. (2023), Oddo et al. (2024)
- Transformer performance: Bazi et al. (2021), Hong et al. (2022), Wang et al. (2022), Khan et al. (2024), Chen et al. (2022)
- Hybrid architectures often achieve intermediate or superior performance by combining strengths of multiple approaches

Selection Guidelines:

1. **Choose CNN** for: Single-image spatial analysis, real-time processing needs, limited computational resources, standard land cover mapping
2. **Choose LSTM** for: Time series analysis, crop phenology monitoring, sequential pattern recognition, vegetation dynamics prediction
3. **Choose Transformer** for: Scene-level understanding, when large pre-trained models available, complex spatial relationships, multi-scale global context needed
4. **Consider Hybrid** for: Applications requiring both spatial detail and temporal patterns, multi-modal data fusion, maximizing performance with sufficient resources

6.1.2 Data Requirements and Transfer Learning

The quantity of labeled imagery needed to train an effective deep learning model varies considerably across architectures and is shaped by model capacity, task difficulty, and the availability of pre-trained initializations. CNNs derive substantial benefit from transfer learning: models initially trained on large natural image datasets such as ImageNet carry over general-purpose visual representations—edge detectors, texture filters—that remain useful in the remote sensing domain, allowing fine-tuning to reach competitive performance with modest labeled training sets.

Higher-level feature representations tend to be more domain-specific and may require more targeted adaptation.

Transformer models generally demand larger labeled training corpora than CNNs, stemming from their higher parameter counts and the relatively weak structural priors encoded by self-attention compared to convolution or recurrence. The mechanism's architectural agnosticism, while conferring flexibility, reduces the inductive biases that enable CNNs to learn from limited examples. Self-supervised pre-training on large unlabeled satellite image collections, and the subsequent application of the resulting foundation models to downstream tasks via fine-tuning, has emerged as the primary strategy for overcoming this data constraint (Huo et al., 2025).

6.1.3 Interpretability and Explainability

Gaining insight into the basis for individual model predictions is important for building practitioner trust, diagnosing error patterns, and deriving scientifically meaningful inferences from learned representations. CNNs have a relatively mature suite of interpretability tools, encompassing activation maximization, saliency-based input attribution, and gradient-weighted class activation mapping (Grad-CAM). These approaches identify which spatial image regions most strongly drive a given prediction, yielding visually accessible explanations of what features the model has learned to associate with each output class.

LSTM networks can be analyzed through attention weight visualization, hidden state analysis, and cell state trajectory visualization. Attention mechanisms, when incorporated into LSTM architectures, provide interpretable weights indicating which time steps are most relevant for predictions. However, interpreting the internal representations and dynamics of LSTM cells remains challenging, as the complex interactions between gates and cell states create opaque transformations that are difficult to explain in intuitive terms.

Transformer architectures offer inherent interpretability through attention weights, which directly indicate the relationships the model learns between different input elements. Visualization of attention patterns can reveal which image regions or temporal positions the model considers relevant when making predictions. Multi-head attention provides multiple complementary perspectives on input relationships. Khan et al. (2024) demonstrated transformer-based land cover classification with explicit explainability mechanisms, showing that attention patterns align with human interpretation of relevant image features and providing confidence in model predictions.

6.2 Performance Benchmarks

6.2.1 Scene Classification Results

Standardized scene classification benchmarks enable objective, cross-architecture performance comparisons across a structured set of land cover and land use categories. The UC Merced Land Use Dataset, the Aerial Image Dataset (AID), and the NWPU-RESISC45 benchmark collectively span 21 to 45 scene classes with varying training set sizes and intra-class diversity. Across these benchmarks, transformer-based models have demonstrated performance matching or exceeding the best reported CNN results.

Bazi et al. (2021) reported overall accuracies of 98.0% on UC Merced, 94.5% on AID, and 89.2% on NWPU-RESISC45 using vision transformers, comparable to or exceeding the best CNN-based approaches. The performance advantage of transformers becomes more pronounced for datasets with larger numbers of classes and more complex scene types, where the ability to capture global contextual relationships provides greater benefit. However, transformers generally require more training data or extensive pre-training to achieve these performance levels.

6.2.2 Semantic Segmentation Accuracy

Segmentation quality is assessed through overall pixel accuracy, mean Intersection over Union (mIoU), and class-level F1 scores. The Vaihingen and Potsdam datasets from the ISPRS 2D Semantic Labeling Challenge provide centimeter-resolution aerial imagery with pixel-level labels for urban scene segmentation. Recent architectures combining CNNs and transformers achieve mIoU values exceeding 85% on these datasets, representing substantial improvements over earlier approaches.

Wang et al. (2022) demonstrated that hybrid CNN-transformer architectures combining U-Net's spatial processing with transformer-based global context modeling achieve superior segmentation accuracy compared to pure CNN or pure transformer approaches. The hybrid design achieves mIoU of 87.3% on Vaihingen and 89.1% on Potsdam, outperforming both classic U-Net (mIoU ~82-84%) and pure vision transformers (mIoU ~84-86%). The performance improvements are particularly pronounced for difficult classes such as “impervious surface” and “clutter/background” where global context helps resolve ambiguities.

Li and Li (2024) reported semantic segmentation results demonstrating the effectiveness of attention-enhanced CNN architectures, achieving overall accuracies exceeding 92% on challenging urban scene datasets. The integration of multi-scale feature fusion with attention mechanisms enables accurate delineation of object boundaries and fine spatial details while maintaining semantic consistency across large regions.

6.2.3 Time Series Classification Performance

Time series classification evaluates model ability to assign correct labels to entire temporal sequences based on their characteristic patterns. Crop type classification from satellite image time series represents a particularly important application, with operational implications for agricultural monitoring and food security assessment. Reuß et al. (2021) compared LSTM networks with random forest classifiers on Sentinel-1 time series for large-scale crop classification, demonstrating that LSTMs achieve overall accuracies of 87.4% compared to 83.1% for random forest, with particularly large improvements for crop types with distinctive temporal signatures.

The performance advantage of LSTM and other deep learning approaches over traditional methods increases with the temporal resolution and length of available time series. With dense temporal sampling (observations every 5-12 days), LSTMs can capture subtle phenological differences between crop types that may be difficult to distinguish based on single-date imagery or hand-crafted temporal features. Fathi et al. (2023) achieved county-level soybean yield prediction accuracy with R^2 values exceeding 0.85 using hybrid 3D-ResNet-BiLSTM architectures, demonstrating the value of combining spatial and temporal modeling.

Figure 3. Comparative Performance Metrics

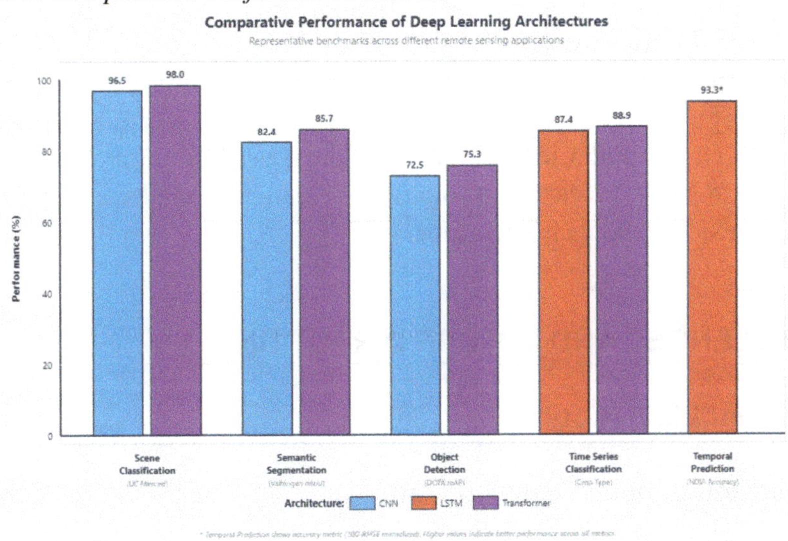

Figure 3 represent bar chart comparing overall classification accuracy, mean IoU for segmentation, and temporal prediction RMSE across CNN, LSTM, and Transformer architectures for representative remote sensing applications. The chart includes error bars representing performance variability across different datasets and implementation details, providing a realistic assessment of expected performance ranges for each architecture type.

6.3 Ensemble Methods and Model Fusion

6.3.1 Architecture Ensemble Strategies

Ensemble approaches aggregate outputs from multiple independently trained models to produce predictions that are generally more accurate and less sensitive to individual model idiosyncrasies than any single member. Architecturally heterogeneous ensembles—pairing CNNs, LSTMs, and transformers—can exploit complementary inductive biases: CNNs contribute fine-grained spatial feature discrimination, LSTMs contribute temporal sequence modeling, and transformers contribute scene-level contextual reasoning. Prediction combination strategies range from simple score averaging to validation-calibrated weighted voting and fully learned fusion modules.

6.3.2 Multi-Modal Data Fusion

Many remote sensing applications benefit from fusing information from multiple sensor types with complementary characteristics. Optical sensors provide rich spectral information under clear-sky conditions, SAR sensors offer all-weather capabilities and sensitivity to surface structure, LiDAR provides accurate elevation information, and thermal sensors capture temperature variations. Deep learning architectures can be designed to jointly process multi-modal data, learning complementary features from each modality and their interactions.

7. CHALLENGES AND FUTURE DIRECTIONS

7.1 Current Limitations

7.1.1 Limited Labeled Training Data

A persistent bottleneck for supervised deep learning in remote sensing is the scarcity of accurately annotated training data. Producing ground truth labels demands expert knowledge of land cover and environmental systems, substantial field effort, and considerable financial investment, rendering comprehensive labeling across large geographic extents or for specialized thematic categories practically difficult. Pixel-level annotation tasks such as semantic segmentation are particularly demanding, as every image element must be classified, and rare target classes may have insufficient positive examples to support reliable learning. These constraints frequently force compromises in geographic coverage, temporal scope, or label granularity.

Active learning strategies that iteratively select the most informative unlabeled samples for annotation can reduce labeling requirements, but still require substantial human effort. Self-supervised learning approaches that leverage the structure of unlabeled data itself show promise for learning useful representations without manual labels.

7.1.2 Domain Shift and Generalization

Deep learning models trained on imagery from a particular geographic region, acquisition season, or sensor configuration frequently exhibit reduced performance when deployed in different conditions, a phenomenon termed domain shift. Performance degradation arises because atmospheric properties, illumination geometry, phenological status, and landscape composition can all differ substantially between the training context and the target deployment environment. Agricultural landscapes in temperate climates, for instance, display spectral and spatial patterns that diverge markedly from those of comparable land use in tropical or semi-arid settings, and summer-trained models may fail systematically on winter imagery where vegetation cover, solar angle, and possible snow presence alter the radiometric signal.

Domain adaptation techniques aim to improve generalization across domains by learning domain-invariant representations or explicitly modeling domain-specific variations. Adversarial domain adaptation employs discriminators that attempt to distinguish between source and target domain features, training the feature extractor to produce features that are indistinguishable across domains.

7.2 Emerging Research Directions

7.2.1 Foundation Models for Remote Sensing

Foundation models—large neural networks pre-trained on broad, heterogeneous data collections and subsequently adapted to specific downstream tasks through fine-tuning—represent an emerging paradigm with transformative implications for remote sensing. Drawing inspiration from successes in natural language processing with models such as GPT and BERT, the remote sensing community has begun constructing analogous foundation models through self-supervised pre-training on large unlabeled satellite archives. The goal is to learn versatile Earth observation representations transferable to diverse applications with minimal labeled supervision, potentially reducing barriers to accessing high-performance analysis tools.

Wang et al. (2023) advanced plain vision transformers toward remote sensing foundation models, demonstrating that transformers pre-trained on large-scale remote sensing datasets achieve superior performance across diverse downstream tasks compared to models trained from scratch or models pre-trained on natural images. Huo et al. (2025) surveyed the convergence of remote sensing and foundation models, discussing architectural considerations, pre-training strategies, and applications.

7.2.2 Physics-Informed Deep Learning

Physics-informed deep learning seeks to embed domain-specific physical knowledge—in the form of process equations, conservation laws, or empirical physical models—directly into network architecture or training objectives, with the aim of improving prediction fidelity, physical plausibility, and model interpretability. In remote sensing, relevant physical constraints include radiative transfer equations governing spectral reflectance, surface energy balance relationships, and electromagnetic scattering models for SAR. Implementation strategies range from constrained output layers that enforce physical bounds, to physics-based synthetic data generation for augmenting labeled training sets, to hybrid network-physics architectures that blend learned and model-based components.

Applications of physics-informed deep learning in remote sensing include atmospheric correction, where physical radiative transfer models can be integrated with learned components that adapt to local atmospheric conditions; soil moisture estimation, where hydrological models inform network architecture and loss functions; and land surface temperature retrieval, where thermal physics constrains the relationships between brightness temperatures and surface properties.

7.2.3 Multimodal Learning and Cross-Sensor Fusion

Future remote sensing systems will increasingly integrate diverse data types including satellite imagery, aerial photography, UAV observations, ground-based sensors, social media data, and authoritative databases. Multimodal learning architectures can jointly process these heterogeneous data sources, learning complementary representations and exploiting synergies between modalities.

8. CONCLUSION

8.1 Key Findings and Recommendations

This chapter has systematically reviewed the landscape of deep learning architectures deployed in remote sensing for earth sciences, covering the theoretical underpinnings, principal design variants, and task-specific application profiles of CNNs, RNNs, LSTMs, and Transformers. Each architectural family presents a distinct set of computational and representational strengths that aligns it with particular remote sensing task types and data modalities.

RNN and LSTM architectures have become essential for temporal analysis of satellite image time series across applications spanning agricultural monitoring, vegetation phenology characterization, and environmental change detection. The LSTM's capacity for long-range temporal dependency modeling enables it to encode complex time-evolving patterns, from intra-seasonal vegetation dynamics to interannual variability and gradual land surface transitions.

Transformer-based architectures stand at the current frontier of deep learning for remote sensing, providing global spatial receptive coverage through self-attention and flexible context modeling capabilities that reveal long-range spatial dependencies and relational patterns inaccessible to local convolutional approaches.

8.2 Architecture Selection Guidelines

Choosing the most suitable deep learning architecture for a given remote sensing application demands careful evaluation of the task's spatial and temporal requirements, the volume and modality of available training data, hardware constraints, and operational deployment needs. For tasks that principally require per-image spatial analysis—such as thematic land cover mapping, object localization, or dense semantic labeling—CNN-based architectures generally offer the best combination of performance, implementation maturity, and computational economy. The well-established ecosystem of pre-trained CNN models, open-source implementations, and community-validated training recipes makes CNNs an accessible and reliable choice across a broad range of practical scenarios.

REFERENCES

Amsalu, A., Garedew, W., Melka, G. A., Çelik, M. A., & Gemeda, D. O. (2026). Trends in climate extreme indices and implications for coffee farming in The Kafa Biosphere Reserve, southwest Ethiopia. *Climate Services*, *42*, 100637. DOI: 10.1016/j.cliser.2026.100637

Bazi, Y., Bashmal, L., Al Rahhal, M. M., Al Dayil, R., & Al Ajlan, N. (2021). Vision transformers for remote sensing image classification. *Remote Sensing (Basel)*, *13*(3), 516. Advance online publication. DOI: 10.3390/rs13030516

Boonpook, W., Tan, Y., Nardkulpat, A., Torsri, K., Torteeka, P., Kamsing, P., Sawangwit, U., Pena, J., & Jainaen, M. (2023). Deep learning semantic segmentation for land use and land cover types using Landsat 8 imagery. *ISPRS International Journal of Geo-Information*, *12*(1), 14. Advance online publication. DOI: 10.3390/ijgi12010014

Çelik, M. A. (2026). Agro-climatic classification of Türkiye's Agricultural Basins using long-term ERA5-Land temperature data. *Physics and Chemistry of the Earth Parts A/B/C*, *142*, 104307. DOI: 10.1016/j.pce.2026.104307

Çelik, M. A., Kıraç, A., & Gouveia, C. M. (2026). North Atlantic Oscillation modulation of drought and vegetation activity in Türkiye. *Theoretical and Applied Climatology*, *157*(2), 74. Advance online publication. DOI: 10.1007/s00704-025-05971-5

Chen, H., Qi, Z., & Shi, Z. (2022). Remote sensing image change detection with transformers. *IEEE Transactions on Geoscience and Remote Sensing*, *60*, 1–14. DOI: 10.1109/TGRS.2021.3095166

Chen, H., Song, J., Han, C., Xia, J., & Yokoya, N. (2024). ChangeMamba: Remote sensing change detection with spatiotemporal state space model. *IEEE Transactions on Geoscience and Remote Sensing*, *62*, 1–20. DOI: 10.1109/TGRS.2024.3417253

Fathi, M., Shah-Hosseini, R., & Moghimi, A. (2023). 3D-ResNet-BiLSTM model: A deep learning model for county-level soybean yield prediction with time-series Sentinel-1, Sentinel-2 imagery, and Daymet data. *Remote Sensing (Basel)*, *15*(23), 5551. Advance online publication. DOI: 10.3390/rs15235551

Gong, H., Mu, T., Li, Q., Dai, H., Li, C., He, Z., Wang, W., Han, F., Tuniyazi, A., Li, H., Lang, X., Li, Z., & Wang, B. (2022). Swin-Transformer-enabled YOLOv5 with attention mechanism for small object detection on satellite images. *Remote Sensing (Basel)*, *14*(12), 2861. Advance online publication. DOI: 10.3390/rs14122861

Hong, D., Han, Z., Yao, J., Gao, L., Zhang, B., & Plaza, A. (2022). SpectralFormer: Rethinking hyperspectral image classification with transformers. *IEEE Transactions on Geoscience and Remote Sensing*, *60*, 1–15. DOI: 10.1109/TGRS.2022.3172371

Huang, X., Zhou, Y., Yang, X., Zhu, X., & Wang, K. (2023). SS-TMNet: Spatial–spectral transformer network with multi-scale convolution for hyperspectral image classification. *Remote Sensing (Basel)*, *15*(5), 1206. Advance online publication. DOI: 10.3390/rs15051206

Huo, C., Chen, K., Zhang, S., Wang, Z., Yan, H., Shen, J., Hong, Y., Qi, G., Fang, H., & Wang, Z. (2025). When remote sensing meets foundation model: A survey and beyond. *Remote Sensing (Basel)*, *17*(2), 179. Advance online publication. DOI: 10.3390/rs17020179

Khan, M., Hanan, A., Kenzhebay, M., Gazzea, M., & Arghandeh, R. (2024). Transformer-based land use and land cover classification with explainability using satellite imagery. *Scientific Reports*, *14*(1), 16744. Advance online publication. DOI: 10.1038/s41598-024-67186-4 PMID: 39033183

Li, X., & Li, J. (2024). MFCA-Net: A deep learning method for semantic segmentation of remote sensing images. *Scientific Reports*, *14*(1), 5745. Advance online publication. DOI: 10.1038/s41598-024-56211-1 PMID: 38459115

Li, Z., Yuan, J., Li, G., Wang, H., Li, X., Li, D., & Wang, X. (2023). RSI-YOLO: Object detection method for remote sensing images based on improved YOLO. *Sensors (Basel)*, *23*(14), 6414. Advance online publication. DOI: 10.3390/s23146414 PMID: 37514708

Lu, M., Li, Y., Yu, M., Zhang, Q., Zhang, Y., Liu, B., & Wang, M. (2023). Spatiotemporal prediction of radar echoes based on ConvLSTM and multisource data. *Remote Sensing (Basel)*, *15*(5), 1279. Advance online publication. DOI: 10.3390/rs15051279

Moskolaï, W. R., Abdou, W., Dipanda, A., & Kolyang, . (2021). Application of deep learning architectures for satellite image time series prediction: A review. *Remote Sensing (Basel)*, *13*(23), 4822. Advance online publication. DOI: 10.3390/rs13234822

Oddo, P. C., Bolten, J. D., Kumar, S. V., & Cleary, B. (2024). Deep convolutional LSTM for improved flash flood prediction. *Frontiers in Water*, *6*, 1346104. Advance online publication. DOI: 10.3389/frwa.2024.1346104

Reuß, F., Greimeister-Pfeil, I., Vreugdenhil, M., & Wagner, W. (2021). Comparison of long short-term memory networks and random forest for Sentinel-1 time series based large scale crop classification. *Remote Sensing (Basel)*, *13*(24), 5000. Advance online publication. DOI: 10.3390/rs13245000

Vasilakos, C., Tsekouras, G. E., & Kavroudakis, D. (2022). LSTM-based prediction of Mediterranean vegetation dynamics using NDVI time-series data. *Land (Basel)*, *11*(6), 923. Advance online publication. DOI: 10.3390/land11060923

Wang, L., Li, R., Zhang, C., Fang, S., Duan, C., Meng, X., & Atkinson, P. M. (2022). UNetFormer: A UNet-like transformer for efficient semantic segmentation of remote sensing urban scene imagery. *ISPRS Journal of Photogrammetry and Remote Sensing*, *190*, 196–214. DOI: 10.1016/j.isprsjprs.2022.06.008

Wang, Z., Wang, J., Yang, K., Wang, L., Su, F., & Chen, X. (2022). Semantic segmentation of high-resolution remote sensing images based on a class feature attention mechanism fused with Deeplabv3+. *Computers & Geosciences*, *158*, 104969. Advance online publication. DOI: 10.1016/j.cageo.2021.104969

Wu, Z., Hou, B., & Jiao, L. (2021). Multiscale CNN with autoencoder regularization joint contextual attention network for SAR image classification. *IEEE Transactions on Geoscience and Remote Sensing*, *59*(2), 1200–1213. DOI: 10.1109/TGRS.2020.3004911

Xiong, C., Ma, H., Liang, S., He, T., Zhang, Y., Zhang, G., & Xu, J. (2023). Improved global 250 m 8-day NDVI and EVI products from 2000–2021 using the LSTM model. *Scientific Data*, *10*(1), 800. Advance online publication. DOI: 10.1038/s41597-023-02695-x PMID: 37963885

Yan, C., Fan, X., Fan, J., & Wang, N. (2022). Improved U-Net remote sensing classification algorithm based on multi-feature fusion perception. *Remote Sensing (Basel)*, *14*(5), 1118. Advance online publication. DOI: 10.3390/rs14051118

Yang, X., Cao, W., Lu, Y., & Zhou, Y. (2022). Hyperspectral image transformer classification networks. *IEEE Transactions on Geoscience and Remote Sensing*, *60*, 1–15. DOI: 10.1109/TGRS.2022.3171551

Yang, X., Yang, X., Zhang, C., & Wang, J. (2023). SAR image classification using Markov random fields with deep learning. *Remote Sensing (Basel)*, *15*(3), 617. Advance online publication. DOI: 10.3390/rs15030617

Zhang, D., Leng, J., Li, X., He, W., & Chen, W. (2022). Three-stream and double attention-based DenseNet-BiLSTM for fine land cover classification of complex mining landscapes. *Sustainability (Basel)*, *14*(19), 12465. Advance online publication. DOI: 10.3390/su141912465

Zhang, Y., Ye, M., Zhu, G., Liu, Y., Guo, P., & Yan, J. (2024). FFCA-YOLO for small object detection in remote sensing images. *IEEE Transactions on Geoscience and Remote Sensing*, *62*, 1–15. DOI: 10.1109/TGRS.2024.3363057

KEY TERMS AND DEFINITIONS

Convolutional Neural Network (CNN): A deep learning architecture designed for processing grid-like data structures such as images, utilizing convolutional layers to automatically learn hierarchical spatial features through learnable filters that detect patterns ranging from low-level edges and textures to high-level semantic objects, making it particularly effective for remote sensing image classification, object detection, and semantic segmentation tasks.

Long Short-Term Memory (LSTM): A specialized type of recurrent neural network architecture that addresses the vanishing gradient problem through a gating mechanism consisting of forget gates, input gates, and output gates, enabling effective modeling of long-term temporal dependencies in sequential data such as satellite image time series for applications including crop phenology monitoring, vegetation dynamics prediction, and environmental change detection.

Transformer Architecture: A deep learning model based on self-attention mechanisms that process input data in parallel rather than sequentially, utilizing multi-head attention to capture global contextual relationships across entire images or sequences without the limitations of local receptive fields, demonstrating superior performance in scene-level understanding and complex spatial pattern recognition in remote sensing applications.

Self-Attention Mechanism: A fundamental component of transformer architectures that computes relationships between all positions in a sequence simultaneously by calculating attention weights through query-key-value operations, enabling the model to dynamically focus on relevant spatial or temporal locations regardless of their distance, thereby capturing long-range dependencies more effectively than traditional convolutional or recurrent approaches.

Semantic Segmentation: A dense prediction task in computer vision that assigns a class label to every pixel in an image, producing spatially detailed maps of land cover types, objects, or features by combining hierarchical feature extraction through encoder networks with spatial upsampling through decoder networks, often utilizing architectures such as U-Net, DeepLab, or transformer-based models to achieve precise boundary delineation and classification accuracy.

Chapter 6
Nanoenabled Artificial Intelligence for Sustainable and Digital Advancement in Precision Farming

S. Edwin Raja
http://orcid.org/0000-0002-2948-9669
Vel Tech Rangarajan Dr. Sagunthala R&D Institute of Science and Technology, India

M. Thiyagarajan
R.M.K. Engineering College, India

C. Usharani
Ramco Institute of Technology, India

P. Gopi Kannan
P.S.R. Engineering College, India

D. Selvaraj
Panimalar Engineering College, India

ABSTRACT

The increasing demand for sustainable food production and efficient resource management has accelerated the adoption of advanced technologies in modern agriculture. This chapter explores the integration of nanotechnology, artificial intelligence (AI),

DOI: 10.4018/979-8-2600-0778-5.ch006

and remote sensing as a novel paradigm for enabling intelligent and sustainable precision farming systems. Nano-enabled sensors and smart agro-nanomaterials facilitate real-time monitoring of soil nutrients, plant health, and environmental conditions at microscopic levels, while remote sensing platforms provide large-scale observations of crop dynamics, vegetation health, and climatic variations. The chapter introduces several novel conceptual frameworks, including the Nano-AI Precision Farming Framework (NAPF), the Nano-Integrated Predictive Farming Model (NIPFM), and the Nano-Agricultural Digital Twin (NADT), which collectively support multi-layer agricultural intelligence and autonomous decision-making.

1. INTRODUCTION: CONVERGENCE OF NANOTECHNOLOGY, AI, AND REMOTE SENSING IN PRECISION AGRICULTURE

There is a radical change in the technological landscape of agriculture as the global food systems have to face two challenges, which are the increase in population demands and environmental sustainability. The conventional farming and agriculture, which has always been based on pioneer knowledge and standardized field management methods are becoming inadequate in resolving contemporary complexities of agriculture. The differences in soil health, crop stress, and climatic changes, and nutrient levels demand specific and data-driven remedies instead of generalized methods of cultivation. As a result, agriculture has developed its advancement through more traditional farming system to precision agriculture and more recently to digitally empowered intelligent agricultural ecosystems. The shift is marked by the adoption of sensing technologies, data analytics and automation to improve productivity, resource use and environmental stewardship (D & Devarajan, 2025). Digital agriculture became a breakthrough in this change. Satellite-based Earth observation, unmanned aerial vehicles (UAVs), and Internet-of-Things (IoT) sensor networks have empowered farmers to detect the health of crops, the degree of soil moisture, and the environmental conditions more than the previous spatial and temporal resolution. These possibilities make it possible to control irrigation, fertilization, and pest control activities at their own level. Nevertheless, even with these developments, there are still numerous structural constraints with numerous precision farming systems. Data reconstruction is normally incomplete making data sparse and lacks spatial coverage. In addition, data processing and restricted real-time analytics are often associated with delayed decision-making processes because of a batch-based data processing approach. Another way nutrient management practices have not worked well is due to the fact that in most cases, the conventional fertilizers cause the loss of nutrients through either leaching, volatilization or even runoff thus decreasing the efficiency of the input and leading to environmental degradation.

Recent advances in artificial intelligence (AI) have made services whose analytical features are essential to precision agriculture to be vastly improved. Models based on AI, especially machine learning and deep learning tools, can be used to derive insightful tendencies on large-scale agronomic data (Yadav et al., 2025). AI can also be used to process multispectral and hyperspectral images to identify early indications of crop stress, nutrient deficiencies, pest attacks, and outbreaks when combined with remote sensing platforms. Such predictive analytics enable farmers and other agricultural managers to abandon reactive responses and adopt proactive and predictive farm management approaches. Furthermore, AI-enabled models can be used in yield prediction, irrigation control, and climate-resistant crop design, which will lead to the achievement of higher agricultural productivity and sustainability (Ashique et al., 2025). Simultaneously with the development of AI and remote sensing, nanotechnology has become one of the revolutionary improvements in the field of agriculture. The application of nano-enabled agricultural technologies opens the innovation of highly sensitive nanosensors with the ability to identify the subtle biochemical and environmental variations of the soil and plant systems (Goyal et al., 2023). The parameters that can be monitored using these sensors include; nutrient concentrations, presence of pathogens and soil PH at very minute scales. Even more so, nano-fertilizers and nano-pesticides have controlled-release systems that improve the efficacy of delivering the nutrients and reduce the environmental impact. The Nano-delivery systems would also allow the targeted use of agrochemicals whereby active compounds are only released when there are certain environmental or biological conditions.

Although these technologies have an individual potential, only system-level integration can bring these technologies to fruition. Artificial intelligence can offer computational intelligence needed to maintain large amounts of agricultural data, remote sensing has the ability to monitor environments over large areas and nanotechnology can be used to exact in-situ measurements and interventions aimed at individual agriculture. It is the overlap of these areas that forms the basis of a new paradigm of nano-intelligent precision agriculture with sensing, analysis, and actuation being the interacting parts of a single system of agricultural intelligence. In order to conceptualize this integration, this chapter introduces Nano-AI Agricultural Intelligence Ecosystem (NA-AIE). The NA-AIE system is a comprehensive agricultural intelligence system where nano-scale sensing networks provide real-time soil and crop information, remote sensing systems provide macro-environmental data, and artificial intelligence engines convert these multi-scale data to actionable farming information. The NA-AIE model can be used to provide unending monitoring, predictive analysis, and adaptive farm management by operating nanoscale sensing with large-scale Earth observation and intelligent decision systems. The primary objective of this chapter is to explore how the convergence of nanotechnology, artificial intelligence, and remote sensing can advance the development of sustainable and digitally enabled precision farming systems. Specifically, the chapter aims to:

- Examine the technological foundations of nano-enabled agricultural sensing and intelligent analytics.
- Analyze the integration of AI-driven remote sensing with nano-based agricultural monitoring systems.
- Propose conceptual frameworks that support predictive, data-driven farm management.
- Highlight sustainability benefits associated with resource optimization and environmental protection.

Through this interdisciplinary perspective, the chapter seeks to illustrate how emerging technologies can collectively reshape agricultural systems, enabling more resilient, efficient, and sustainable food production for the future.

2. DIGITAL AGRICULTURE LANDSCAPE AND PRECISION FARMING PARADIGMS

The intensive development of digital technologies has changed the approach to agriculture to a great extent and has given rise to digitalization of agriculture and precision farming paradigms. The purpose of these paradigms is to improve the productivity of agriculture, the efficiency of resource use, and the sustainability of the environment due to the use of data in making decisions. Modern agriculture is developing into a highly technologically advanced ecosystem through the combination of sensing technologies, computational intelligence, and automated agricultural systems to solve complex agricultural problems.

2.1 Evolution of Precision Farming Technologies

Precision farming has developed in a number of technological stages. The old agricultural activities mainly involved even field management whereby uniform applications of fertilizer, pesticides and irrigation were done even in an area that might be varied in terms of soil conditions. With the advancement in agricultural research, the advent of Global Positioning Systems (GPS) and Geographic Information Systems (GIS) led to the spatial mapping of agricultural farms, which gave a farmer a chance to identify the difference in soil fertility and crop yield. Later on, with the introduction of remote sensing technology, such as satellite imagery and unmanned aerial vehicles (UAVs), the capability to observe crop conditions and the environmental factors at large scales increased further. The result of such systems was the creation of site-specific management systems, in which agricultural inputs could be changed depending on the local conditions of the fields (Singh et al., 2024). Over the last several years, the Internet-of-Things (IoT) devices, cloud computing, and artificial intelligence have intensified the shift towards intelligent farming ecosystems that may enable ongoing monitoring and offer predictive decisions.

2.2 Data-Driven Agriculture and Smart Farming Architectures

Digital agriculture of the present-day heavily depends on data based farming architectures, in which various sources of data are combined to help guide informed agricultural decision making. Smart farming systems are often integrated systems of data obtained by soil sensors, weather stations, remote sensing platforms, and farm machines. This non-homogeneous information is run with computational systems that utilize machine education and anticipatory analytics to create understanding regarding the health and well-being of crops, irrigation requirements, and yield forecasting. In sophisticated applications, the agricultural information passes through multi-layer designs comprised of sensing layers, data integration platforms, and decision-support systems. Such systems enable immediate surveillance of farm settings besides making it possible to respond adaptively to changing agricultural situations (Dhinakaran et al., 2025). These kinds of architectures are helpful in automating such processes like variable-rate fertilization, precise irrigation, and early disease detection.

2.3 Limitations of Current Remote Sensing-Based Crop Monitoring

Although that has been greatly advanced, there are a number of limitations that exist in the current remote sensing-based agricultural monitoring systems. Observations of satellite sources are frequently affected by time lags and the frequency of revisiting the ground, which can impose limitations on the possibility of observation of the rapid change of the state of crops. Other factors that could interfere with the quality of agricultural observations are atmospheric interference, cloud cover, and sensor resolution limits. Moreover, surface-based measures of remote sensing technology can hardly be effective in capturing important biochemical processes taking place in the soil environment (Fathi Qarachal & Alizadeh, 2025). The parameters like the availability of nutrients, the activity of microbes, and the chemical processes in soils are frequently hard to observe with the use of traditional remote sensing platforms only. Consequently, there are yet to be many precision farming systems with the ability to derive real-time and high-resolution information on the subsurface soil.

2.4 Sustainability Challenges in Modern Agricultural Systems

Although digital agriculture has enhanced the efficiency of activities, contemporary agricultural systems are still associated with looming challenges of sustainability. The heavy use of fertilizers leads to the degradation of soil and water contamination in addition to inefficient irrigation methods that compound the scarcity of water in most areas. Climate variability also makes crop management complicated with the occurrence of unpredictable weather and stress conditions. There is a critical research gap concerning the development of technologies that would allow real-time monitoring of soil biochemicals and adaptive control of nutrients. The existing systems tend to be based on the periodic soil tests and overall fertilizer application plans that might not be updated with the dynamic conditions of the fields. To overcome this constraint, there is need to incorporate new sensing systems that can be used to continuously monitor the nutrient content of the soil and provide guidance to the intelligent nutrient delivery processes. The identification of these shortcomings points to the need of next-generation agricultural intelligence systems that integrate nanoscale sensing, powerful analytics, and massive environment monitoring to enable a more responsive and sustainable agricultural activity.

3. NANO-ENABLED SENSING AND SMART AGRO-NANOTECHNOLOGY

The application of nanotechnology has become a revolution in contemporary farming as it has facilitated extremely sensitive monitoring networks and effective systems of resource delivery, with actionable dimensions of microscopes. Compared to traditional agricultural technologies which are run at the macro or meso-scale, nano-enabled systems are run at the molecular and even nanoscale level, which enables unprecedented accuracy in sensing the soil status, crop processes and even environmental changes. Incorporation of nanoscale materials and devices within the agricultural systems has provided new opportunities in terms of enhancing productivity, resource management and improving sustainability of precision farming. The creation of nano-sensors that can be used to monitor the health of the soil and the dynamics of nutrients in real time is one of the greatest contributions of nanotechnology to the agricultural sector. These sensors have the ability of detecting the slightest variation of soil parameters including nitrogen, phosphorus, potassium levels, pH levels, moisture levels, and microbial activity. Nanosensors have a large ratio of surface area to volume and increased sensitivity and, therefore, are able to detect biochemical variability that cannot be detected by other conventional soil monitoring technologies. Monitoring soil constantly, these sensors can offer granular data that could enable smart nutrient management and on-site agricultural intervention.

Besides sensing technologies, nanotechnology has also been able to facilitate the production of nano-fertilizers which are targeted and controlled to deliver nutrients. The use efficiency of the traditional fertilizers is usually low due to the fact that much of the nutrients applied can be lost by means of leaching, evaporation or run off. Nano-fertilizers solve this problem by including nutrients in nanoscale carriers that can release their contents slowly when exposed to environmental effects in the form of soil moisture, temperature or root activity (Zain et al., 2023). This controlled-release process does not only enhance better absorption of nutrients by plants but also reduces environmental pollution that occurs due to overusing of fertilizers. On the same note, nano-pesticides can be viewed as a new technology in crop protection. Traditional pesticide disbursements often incorporate huge amounts of chemicals, which might have an impact on non-target organisms and ecosystems. The pesticides developed via nano-formulation allow better targeted delivery of active compounds via encapsulation or nano-carrier systems. These processes enable the pesticides to be discharged only under certain biological or environmental factors, thus minimizing the wastage of chemicals and enhancing pest management.

The other agro-nanotechnology that is emerging is the development of nano-biosensors which are able to identify the early signs of plant disease and pathogen activity. These biosensors are able to detect biochemical indicators of a plant under stress, invasion by pathogens, or disequilibrium. Soon after the emergence of disease, early planting helps farmers to institute specific remedies before the disease spreads widely, leading to losses in crop, as well as overuse of pesticides. Although these applications are promising, other aspects that require to be touched in the deployment of nanomaterials in agriculture include the effects on the environment and biocompatibility. The ecological perspectives of the accumulation of nanoparticle in soil and water systems is a crucial area of research in the long run. To be adopted sustainably, it is necessary to make agricultural nanomaterials biodegradable, non-toxic and environmentally stable. In a bid to justify the implementation of nanosensing technologies in large-scale agricultural applications, the chapter proposes the concept of Self-Adaptive Nano-Sensing Networks (SANS). SANS is a type of distributed sensing model where the nanosensors installed in the soil and plant environments are able to constantly gather biochemical information, and dynamically adapt the sensing functions according to changes in the environment. SANS enhances the technological support of combining nanoscale-based agricultural intelligence with artificial intelligence and remote sensing systems in precision farming settings through facilitating autonomous sensing, real-time monitoring, as well as adaptive data collection.

4. ARTIFICIAL INTELLIGENCE ARCHITECTURES FOR INTELLIGENT AGRICULTURAL ANALYTICS

Artificial Intelligence (AI) has emerged as a prime facilitator of intelligent agricultural systems by converting raw agricultural data into useful insights to be used in making decisions. With the current growing dependence on modern farming environments on sensor-produced data, remote sensing platforms, and agricultural machinery, the necessity to have sophisticated computational models that could be used to interpret extensive and heterogeneous datasets has become urgent (Yao et al., 2020). The AI-based analytical systems facilitate predictive, prescriptive, and adaptive decision intelligence to help farmers and agricultural managers maximize the crop output, resource distribution, and environmental sustainability. Machine learning applications are used as a pillar in agricultural analytics as they can predict soil and crop analysis. Historical agricultural data can be processed by supervised and unsupervised learning algorithms to discover patterns related to crop development, deficiencies, and soil fertility changes and yield measures. These models are used to complete tasks like classifying crops, predicting soil nutrients and yield forecasting using environmental variables and their associations with agricultural results. Through the use of soil sensor and weather data, as well as historical farm activity data, machine learning systems can be used to implement evidence-based management practices to achieve better efficiency and waste less.

Recent progress in the field of deep learning has increased the analysis opportunities of AI in agriculture, especially when it comes to the interpretation of remote sensing data. Spectral information in multispectral and hyperspectral imagery produced by satellites and unmanned aerial vehicles is rich and is able to detect the subtle changes in crop physiology and vegetation health. Convolutional neural networks are deep neural networks that have the potential to detect crop stress, disease symptoms, and manage vegetation indices with high precision using high-dimensional spectral data. These models can resolve complicated spatial trends of agricultural images, and they can monitor the status of crops of big farmland areas at once. Reinforcement learning provides an adaptive solution to the agricultural decision-making process as it allows AI technology to learn the most appropriate management strategy by constantly interacting with the changing farm environments. An intelligent agent in this paradigm considers alternatives to different agricultural activities like scheduling of irrigation, application of fertilizer or control of pests and gets feedback on the outcome in form of crop performance or resource efficiency. The system eventually learns policies that are optimal to long-term agricultural yield and keep the impact on the environment low. The technique is especially effective in adaptive farm management where there is uncertainty and variable climatic conditions.

The transparency and interpretability of AI-driven recommendations have been necessary as AI systems continue to have an effect on farming decision-making. Explainable AI methods seek to give information about how AI models are used to make predictions or recommendations, which will boost the confidence of farmers, agronomists and policymakers (Chavula & Kayusi, 2025). Explainable AI can help make sure that automated decisions are understandable and responsible by determining the main environmental conditions that have an impact on model results, e.g. soil moisture, nutrient contents, or vegetation indices. The development of edge AI technologies is another important development in the field of agricultural AI as it allows real-time data processing directly within the field. The sensor data can be processed on-farm or at a local gateway with edge-based computational systems that can process them without necessarily using remote cloud infrastructures. Such functionality minimizes the latency, enhances responsiveness, and time-sensitive agricultural responses, including automated irrigation alterations or timely detection of diseases. To integrate these different AI functions into a single set of analytical functions, this chapter proposes Adaptive Agro-Intelligence Learning Framework (AAILF). The framework AAILF is a combination of machine learning, deep learning, reinforcement learning, and explainable AI into a single agricultural intelligence architecture that can process multisource agricultural data. Using a predictive model, adaptive learning, and real-time analytics, AAILF allows the optimization of the agricultural processes continuously and assists intelligent decision-making in next-generation precision farming systems.

5. REMOTE SENSING AND EARTH OBSERVATION FOR PRECISION AGRICULTURE

The technology of remote sensing and Earth observation has become one of the main elements of the modern precision agriculture, as these technologies allow monitoring the condition of crops, the processes in the environment, and the productivity of agriculture on a large scale (Thakur & Yadav, 2025). These technologies offer round-the-clock and spatially spread measurements whereby farmers and agricultural managers can gauge the condition of their crops and identify environmental stressors and optimize their agricultural activities. Knowing the Earth observation data and artificial intelligence-based analytics, the precision agriculture system will be able to switch to predictive and adaptive agricultural intelligence, rather than reactive field management. One of the most popular Earth observation methods in the agricultural industry is the use of satellite-based crop monitoring systems. Big data on vegetation conditions, soil moisture patterns, surface temperature of the land, and climatic variables are recorded with the help of satellite platforms that have sophisticated imaging sensors. The observations would enable the agricultural stakeholders to monitor the progress of crops over time and gauge the agricultural performance of the regions. Satellite-based vegetation indices can give useful information regarding the health of the plants and help in detecting stress in crops due to drought, deficiency of nutrients, or infestations of pests. Due to the fact that the satellites are able to observe large areas of farms at the same time, they are especially useful in national and regional agricultural planning, food security observation, and climate resilience evaluation.

As a supplement to satellite technologies, unmanned aerial vehicles (UAVs) have become a significant high-resolution agricultural imaging resource. The spatial resolution of UAVs is able to provide detailed images of agricultural fields compared to most platforms of satellites. It can be used to identify the localized crop stress patterns, disease outbreaks and nutrient deficiencies down to the field or plant level. Precision farming processes, in which the interventions to be applied are targeted (e.g., variable-rate fertilization, localized pesticide dispensing, or irrigation control) are another area of use of UAV-based imaging systems. The UAV deployment is also flexible, which means that it can collect data quickly in response to the emergent agricultural processes. One of the main advantages of advanced remote sensing systems is that these systems are capable of collecting multispectral and hyperspectral data, which gives a lot of information regarding plant physiology and biochemical features. Multispectral sensors observe reflectance in multiple unique spectral bands, which can be used to compute vegetation indices that reflect plant activity and chlorophyll content. Hyperspectral sensors go a step further to record hundreds of narrow spectral bands, which allow analyzing small deviations in the biochemical characteristics of plants, including the concentration of pigments, water content, and stress indicators. These spectral data sets are very rich and can be studied with high level of computation that can detect anomalies in the health of crops at early stages.

The artificial intelligence will be instrumental in converting the raw remote sensing information into practical agricultural intelligence. Vegetation health modeling methods based on AI can be used to assess spectral data changes over time and identify crop stress conditions, predict potential yield, and identify emerging pest or disease threats. Temperature, rainfall, and soil conditions are also environment parameters that can be incorporated in the machine learning models in conjunction with remote sensing to provide detailed information about the agriculture. Besides crop surveillance, remote sensing technologies facilitate climate-wise agricultural management through assessing the environmental change on agricultural productivity contributes on a continuous basis. Earth-observation platforms offer quality data on drought situation, severe weather and variability of soil moisture that would be crucial in the development of climate resistant agricultural strategies. Agricultural systems are able to respond better to climate changes and resource scarcity by merging remote sensor data and predictive analytics. In an attempt to improve the analytical qualities of remote sensing technology in the agricultural sector, the concept of Dynamic Crop Health Intelligence Mapping (DCHIM) is presented in this chapter. DCHIM is a framework where multi-source remote sensing data are integrated with AI-based analytics in order to produce constantly updated crop health intelligence maps. These maps dynamically visualize the changes in crop conditions, nutrient status and environmental stress in farmlands. By enabling real-time visualization of crop health patterns and predictive monitoring of agricultural risks, DCHIM supports data-driven decision-making and strengthens the role of Earth observation technologies within intelligent precision farming ecosystems.

6. INTEGRATED NANO-AI-REMOTE SENSING ARCHITECTURE FOR PRECISION FARMING

The synergies between nanotechnology, artificial intelligence, and remote sensing are a game-changing opportunity to realize intelligent agricultural ecosystems that can monitor, track, predict and manage farms in real time. When current precision farming systems are used, the current systems utilize isolated sensing or analysis technologies, which has limited the performance of the systems since an integrated architecture is missing. The solution to this drawback is the implementation of a single framework that has the potential to integrate nanoscale sensing, large scale Earth observation, and intelligent decision making mechanisms in one system of agricultural intelligence. This chapter offers a new Nano-AI Precision Farming Framework (NAPF) combining nano- capabilities of sensing technologies, remote sensing facilities, and AI-based models of analysis to assist in sustainability and responsive agricultural activities as shown in Figure 1. The suggested architecture can be described as a multi-layered intelligence system in which various technological elements can be used to create actionable information that can be used in precision farming. The framework enables the agricultural intelligence between data acquisition and automated farm actuation to flow continuously by integrating micro-level soil sensing and macro-level environment observation with sophisticated AI analytics.

Figure 1. Integrated Nano-AI Precision Farming Framework (NAPF)

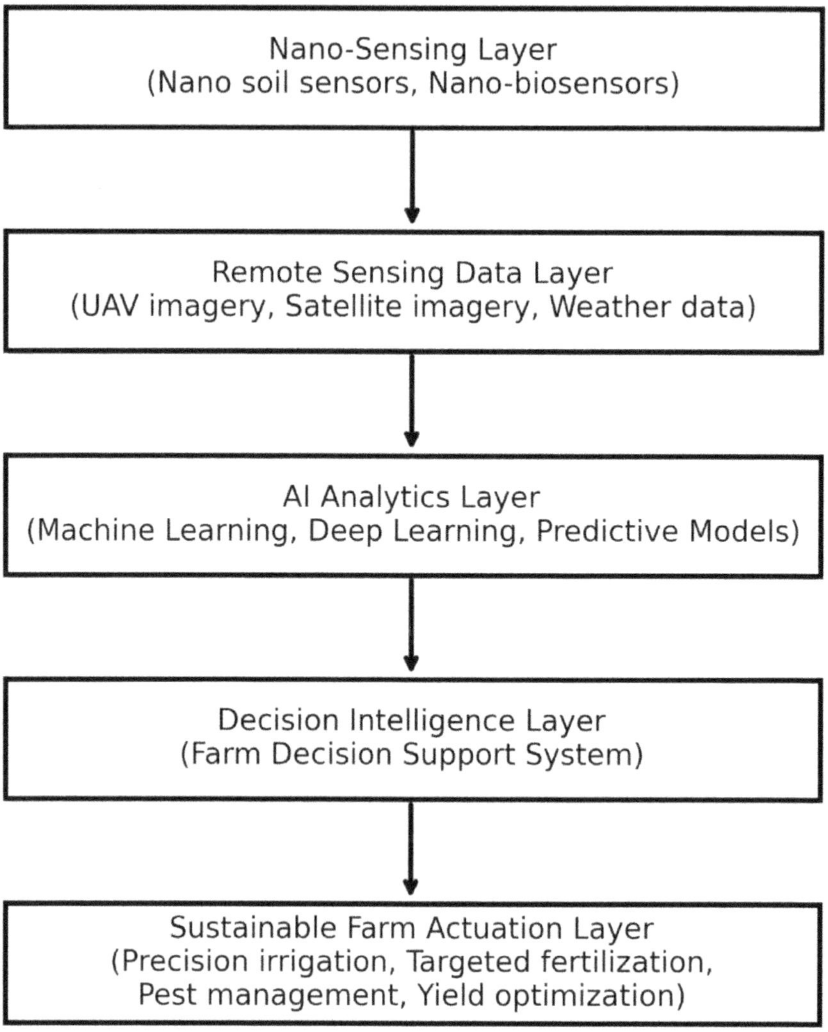

The NAPF architecture is organized into five interconnected layers, each responsible for a specific stage of agricultural data processing and decision execution.

Nano-Sensing Layer

The Nano-Sensing Layer is the bottom layer of the architecture and it is also concerned with the real-time monitoring of soil and crop biochemical conditions. It is a layer that has distributed nanosensors installed in the agricultural environment to detect important parameters like the level of nutrients in the soil, pH, and moisture content in soil, microbial activity, and plant physiological indicators. These sensors are sensitive and are engineered in nanoscale size and thus are able to record minute changes in soil chemistry and plant health that might be unnoticed by the conventional agricultural sensors. The real-time nature of nanosensor monitoring enables farmers to have access to current information about the dynamics of nutrients in the soil and the stress level of the crops (Takaloo et al., 2024). In contrast to periodic soil testing, which only give a static picture on the current state of the soil, nanosensors produce continuous flow of biochemical data, which is dynamic in nature of agricultural ecosystems. These data streams are the much needed input of smart agricultural analytics and intelligent nutrient management.

Remote Sensing Data Layer

Although nanosensors can be used to obtain localised data on the soil and plants, large-scale monitoring of the environment is also necessary in precision agriculture so that bigger pictures of the agricultural conditions are can be obtained. The Remote Sensing Data Layer takes into consideration this need by integrating the data that is obtained by satellite platforms, UAVs and other Earth observation systems. These systems record multispectral and hyperspectral images that depict changes in the health of vegetation, soil moisture distribution and environmental stress in agricultural landscapes. The framework allows monitoring the agriculture ecosystems comprehensively with a combination of macro-scale environmental studies and micro-scale nanosensor measurements. Areas of crop stress, drought, or pest infestations which can cover a complete field or area are given spatial context by remote sensing platforms via identifying patterns of nanosensor readings. Such a multi-scale data acquisition is a major boost to the accuracy and reliability of the agricultural intelligence systems.

AI Analytics Layer

The central processing unit of the NAPF architecture is the AI Analytics Layer. Machine learning and deep learning models will be used in this layer to process heterogeneous datasets obtained using nanosensors, remote sensing platforms, and environmental monitoring systems (Dhinakaran et al., 2024). The goal of this layer is to convert raw agricultural information to meaningful insights that can help in supporting predictive and adaptive farm management. There are AI models in this layer that carry out a number of important analytical operations. Predictive models are used to develop forecasts on how crops are going to grow, what their nutritional needs are, and what yields they would yield on a historical and real-time basis. Extensive imagery (multispectral and hyperspectral) Deep learning models learn to detect stress in vegetation, disease manifestations, nutrient deficiencies. Besides, different farm management strategies can be assessed by the reinforcement learning algorithms and optimal actions are discovered in consideration of environmental feedback and resource efficiency goals. The AI Analytics Layer, which combines various analytical strategies, allows exploring the complex world of agricultural data in a holistic way, thus, aiding the smart decision-making process in precision farming systems.

Decision Intelligence Layer

The Decision Intelligence Layer translates the analytical outputs of AI Analytics Layer into actionable recommendations to manage the farm. This layer acts as a smart decision-supported system that is able to analyze the state of the farm and choose the right management approach. Using the predictive information obtained through the use of AI models, the system can suggest the use of optimal schedules of irrigation, fertilization, pest control, and crop protection. A significant aspect of this layer is that it can be used to facilitate adaptive decision-making. As more and more data is constantly fed into the nanosensors and remote sensing platforms, the decision engine is constantly updated on its recommendations as the agricultural conditions change. This adaptive potential enables the farming system to react to any changes in the environment fast like sudden shortages in the supply of nutrients, water scarcity, or the occurrence of an outbreak. Additionally, this layer may include the explainable AI processes that would offer transparency on the process of particular recommendations creating. Such transparency is necessary to develop trust among farmers and other agricultural stakeholders who are dependent on the AI-assisted decision systems.

Sustainable Farm Actuation Layer

The last NAPF architecture layer is the Sustainable Farm Actuation Layer that provides a way of translating the decision intelligence into physical agricultural actions. The layer includes automated farm management systems like precision irrigation systems, variable-rate fertilizer applicators, autonomous agricultural systems, and intelligent pest control systems (Keerthana et al., 2023). These technologies are used to accomplish management procedures in accordance with the recommendations produced by the decision intelligence layer. To give an example, in case the AI analytics system detects localized nutrient deficiencies in particular areas of a field, the actuation layer may be triggered to open variable-rate fertilizer delivery systems such that nutrients are only applied where they are required. On the same note, the accuracy irrigation systems may also control water delivery based on the moisture content of soil detected by nanosensors. The actuation layer is highly efficient in terms of resource usage with minimal environmental damage by allowing targeted agricultural interventions.

Novel Contribution of the NAPF Framework

The novelty of the Nano-AI Precision Farming Framework (NAPF) is that the framework can bring nanoscale sensing, large scale environmental monitoring and smart decision systems together into a single agricultural intelligence architecture. Conventional precision agriculture systems in many cases use single sensors or single model analytical systems, which restrict their capacity to measure the complete complexity of the agricultural ecosystem. Conversely, the NAPF architecture creates a multi-spectral intelligence architecture, one that links soil-level measurements of biochemical to satellite-level measurements of the environment and decision intelligence that is enabled by AI. By enabling continuous data flow across nanosensors, remote sensing platforms, and AI analytics systems, the framework supports real-time monitoring, predictive nutrient optimization, and automated agricultural actuation. This integrated architecture provides a robust foundation for developing next-generation precision farming systems capable of improving crop productivity, enhancing sustainability, and supporting resilient agricultural ecosystems in the face of climate and resource challenges.

7. AI-DRIVEN PREDICTIVE MODELS FOR SUSTAINABLE CROP MANAGEMENT

The integration of artificial intelligence into agricultural systems has significantly enhanced the capability to predict crop behavior, optimize resource utilization, and improve farm-level decision-making. Predictive modeling plays a crucial role in precision agriculture by enabling the anticipation of environmental changes, crop responses, and potential risks before they significantly impact agricultural productivity as shown in Figure 2. By analyzing historical agricultural data, real-time sensor inputs, and environmental observations, AI-driven models can generate accurate forecasts that support sustainable crop management strategies.

Figure 2. Multi-Source Agricultural Data Fusion for AI Analytics

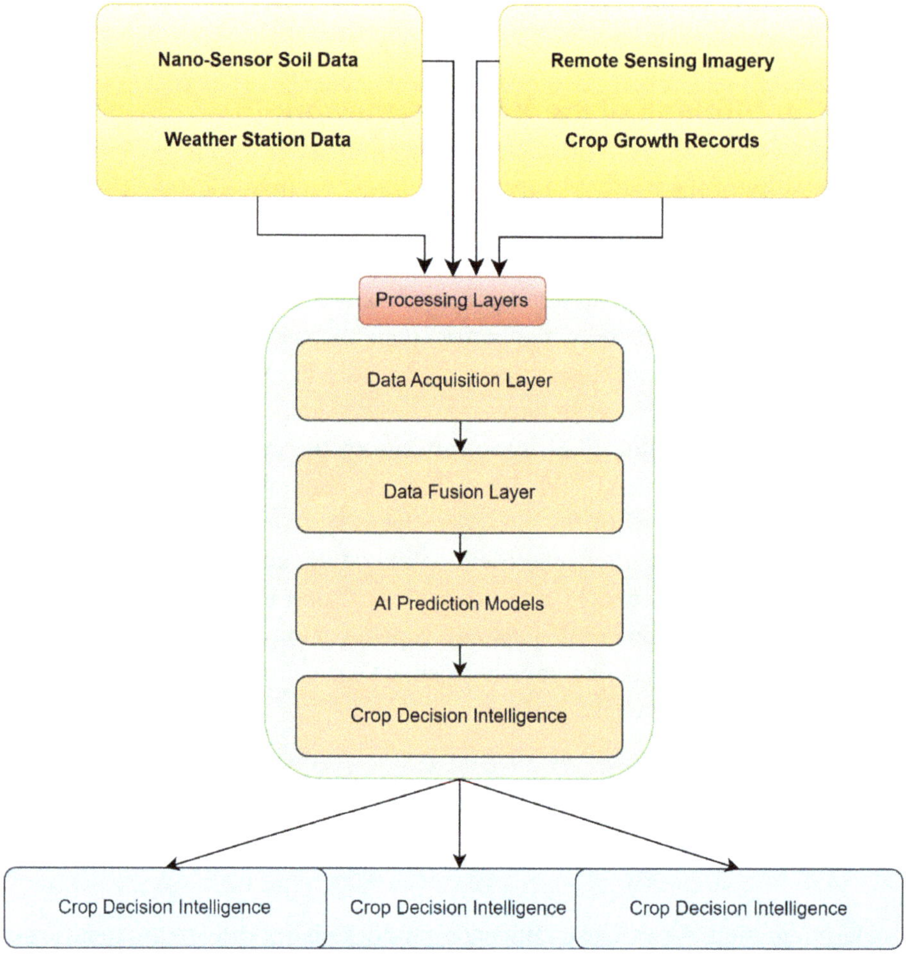

The estimation of the dynamics of the soil nutrients is one of the most significant uses of predictive analytics in agriculture. The soil fertility is very unpredictable within the agricultural fields and the nutrient levels tend to vary with time because of the uptake, or microbial processes, and the environment. The AI-based prediction models of predicting nutrients in the soil consider a dataset obtained through soil sensors, nanosensors, remote sensing platforms, and past agricultural records to assess the patterns of nutrient distribution (Sudharson et al., 2023). The algorithms of machine learning are capable of detecting interaction between intricate relations among the soil characteristics, weather conditions, and growing conditions of crops so that farmers anticipate loss of nutrients and schedule the fertilizer application. These prediction tools aid in effective management of nutrients and minimize the environmental hazards of overuse of fertilizers. Crop yield prediction is another highly important field where AI-based predictive models can be used to achieve sustainable agriculture. Conventional methods of yield prediction are usually based on a historical average or simplified models of agronomic processes, and these conditions may not reflect the dynamic nature of the interactions between the environmental conditions and crop development. Higher precision in predicting yields can be offered by hybrid AI models that integrate machine learning algorithms with environmental data analytics, as they can add more variables of soil conditions, weather patterns, irrigation schedules, and vegetation health indicators based on remote sensing imagery to predict yields more accurately. Such models enable farmers and related stakeholders to predict the likely crop yields, to enable them plan well their harvesting date, supply chain operations, and marketing.

The predictive modeling is also important in early detection and prediction of crop pests and diseases. Plant pathogens and pest infestations are usually progressive with minimal symptoms occurring on crops, before they are visible. With the use of AI, based analysis methods, and nano-biosensors that detect biochemical indicators of stress, soon the agricultural systems will be able to identify the initial stages of infecting plants or pests. Machine learning models process the signals of biosensors and environmental factors (humidity, temperature, conditions of crop canopy, and others) in order to predict possible outbreaks of disease. Early forecasting can help farmers to initiate specific measures, which will not result in significant losses of crops, or rather dependence on extensive pesticides use (Kishore et al., 2023). Water management is another important aspect of sustainable agriculture, especially in the areas where the water shortage and climate changes are becoming more frequent. Smart irrigation systems that run on AI can assess soil moisture and weather prognoses, evapotranspiration levels, and crop water needs to develop the most effective irrigation programs. Such predictive models are used to ensure that water is discharged at the right time and location to avoid over-irrigation and water stress. Effective irrigation management will also enhance crop productivity besides the conservation of precious water resources.

The success of the AI-based predictive systems in the agricultural industry heavily relies on the capability of joining a wide range of datasets created by various sources of information. The contemporary agricultural settings produce vast amounts of data in the form of soil sensors, nanosensors, satellite surveillance, UAV surveillance, and weather surveillance. The high-end data fusion methods help to fuse these heterogeneous datasets into integrated analytical models that allow AI models to learn the complicated interaction between the soil conditions, crop health, and environmental variables. In order to enhance predictive abilities of intelligent farming systems, the Nano-Integrated Predictive Farming Model (NIPFM) is presented in this chapter. NIPFM framework combines nanosensor-based soil monitoring, remote sensing data, and AI-based predictive analytics within an entire modeling platform of sustainable crop management (Dhinakaran et al., 2023). Through real-time biochemical monitoring and massive environmental surveillance and deep machine learning algorithms, NIPFM makes it possible to continuously predict nutrient levels, crop vitality, water needs, and the possibility of a pest attack. This combined predictive solution helps to make proactive decisions in agriculture and create a solid base on the development of sustainable and data-driven precision farming ecosystems.

8. DIGITAL TWIN AGRICULTURE: NANO-AI ENABLED FARM SIMULATION

Digital twins have become a revolutionary technological paradigm in a range of industries that allows creating dynamic virtual copies of real ones. In the agricultural landscape, the technology of digital twins has the opportunity to recreate and monitor the real-time conditions in the agricultural areas to enable farmers and researchers to analyze the crop performance, environmental conditions, and manage the agricultural operations using computational models. Digital twins have the potential to mimic the dynamics of agricultural ecosystems and enable evidence-based decision-making regarding sustainable crop production by solving problems through a combination of real-time data streams and predictive analytics as shown in Figure 3.

Figure 3. Nano-Agricultural Digital Twin (NADT) Framework

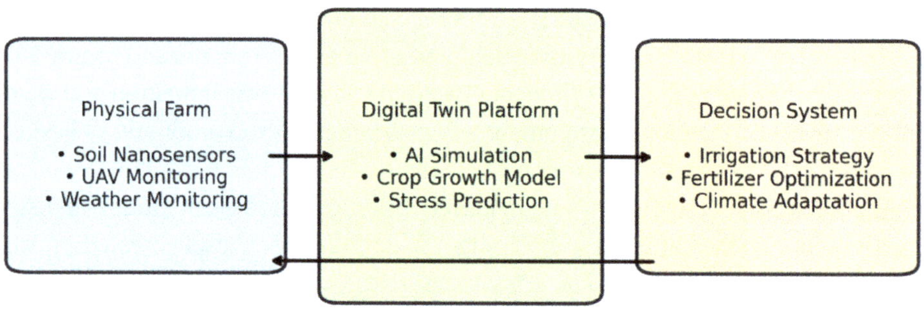

A digital twin is an agricultural setup, which is a virtual representation of an agricultural landscape that continuously receives field-based measurements, environmental monitoring tools and remote sensing infrastructures. This computerized system is a replica of real farm environment like soil characteristics, crop development, weather changes, and use of resources. Having physical farm environments aligned with digital counterparts, the agricultural stakeholders would observe crop health, identify anomalies, and simulate the possible management measures without having to interfere with the actual farming work. The artificial intelligence is instrumental in augmenting the functions of the digital twin in the agricultural sector. Farm simulation models based on artificial intelligence analyze the incoming data on agriculture and replicate multifaceted interactions between the condition of the soil, the process of plant growth, the weather, and agricultural interventions. Machine learning and deep learning algorithms allow the digital twin to make use of historical agricultural data to learn and constantly update its predictive models with real-time observations (Harini et al., 2023). By these means, the digital twin environment can be used to model the way crops can react to various environmental factors, irrigation methods, or fertilization policies.

Digital twins agriculture highly relies on the accessibility of precise and up-to-date data of the physical farm setting. The data source is very important in this respect and this is offered by nano- sensor technologies. Nanosensors integrated into the soil and plant systems would be able to monitor biochemical indicators that include nutrient content, moisture content, pH change and stress indicators. These micro-resolution data sets give detailed information on the micro level dynamics of agricultural ecosystems. Nanosensor data when incorporated in digital twin settings can enable the virtual farm model to reflect true real-time soil and plant conditions and therefore the accuracy of the simulation and prediction process. It is one of the most powerful features of digital twins systems, which allows predicting the scenario of crop growth and farm management initiatives. Farmers are capable of testing the consequences of various agricultural practices by simulating them in the digital environment and understanding the results of the application to the field to make their evaluation. As an illustration, digital twin models would be able to model the impact of various irrigation patterns, fertilizer application levels, or pest management plans on crop yields and resource use. This predictive capability helps identify optimal farming practices while minimizing risks associated with uncertain environmental conditions.

The digital twin technologies can also be very useful in climate-resilient agriculture. Extreme weather events and climate variations are becoming a greater threat to agricultural systems in the world. By monitoring the environment and simulating it in advance, the digital twins can also assess the sensitivity of crops to the life of stress influences caused by the climate i.e. drought, changes in temperature, or irregular rainfall patterns. The knowledge would allow farmers to adopt adaptive mechanisms that they can use to enhance the resilience of crops and ensure productivity even in the face of changing climatic conditions. In order to promote the usage of the digital twin in the field of precision agriculture, in this chapter the author presents the idea of the Nano-Agricultural Digital Twin (NADT). The NADT architecture combines networks of nanosensors, artificial intelligence-based analytics, and remote sensors to form an ever-changing digital model of agricultural locations. In that context, nanosensors can offer high-resolution biochemical measurements of soil and plant systems whereas remote sensing platforms will offer large-scale environmental measurements. These multi-source datasets are processed by artificial intelligence models to produce predictive simulation and adaptive management strategies. The Nano-Agricultural Digital Twin is an advanced agricultural intelligence platform that is able to bridge the gap between the real-world farm environment and intelligent digital ecosystems. NADT will deliver a potent technological base of sustainable, resilient, and data-driven precision farming in the digital age of agricultural technologies by ensuring real-time tracking, predictive simulation, and adaptive decision-making.

9. SUSTAINABILITY, ENVIRONMENTAL IMPACT, AND CLIMATE RESILIENCE

The increasing demand on food production coupled with the increasing environmental pressures has heightened the pressure especially on agricultural systems that are both productive and sustainable. Recent innovations in modern precision farming technologies (especially those that combine the fields of nanotechnology, artificial intelligence, and remote sensing) provide substantial prospects to enhance the efficiency of agriculture without adversely affecting the ecological situation. These technologies can help facilitate the environmentally friendly production processes that are in line with the long-term sustainability objectives by allowing the management of the soil nutrients, water resources, and crop health on a data basis. The application of nano-fertilizers in the effective management of nutrients through efficient nano-fertilizers is one of the greatest contributions of nano-enabled technologies to sustainable agriculture. The traditional fertilizers are characterized by inefficient nutrient utilisation whereby much of the nutrients applied are washed away as a result of leaching, volatilising or even run off. Not only do such losses inhibit the effect of fertilizers but they also lead to environmental issues like water contamination and soil erosion. These challenges are solved through the nano-fertilizer where desirable nutrients are transported through nanoscale carriers that can be released gradually depending on the demand of the plant or environmental stimulus (Islam & Islam, 2025). The mechanism is a controlled-release whereby crops obtain nutrients in a more efficient way and little fertilizer is overused. This leads to the fact that nano-fertilizer systems are able to increase the productivity of crops as well as minimizing environmental pollution and wastage of resources.

The other dimension that is significant in sustainable precision agriculture is the establishment of carbon-conscious agricultural management practices. Agricultural practices are contributing to the greenhouse gases due to the use of fertilizers, soil disturbance, and farming practices that require a lot of energy. The high-tech agricultural systems can also have carbon monitoring and optimization processes that assess the environmental imprint of the farming processes. Combining remote sensing with AI-based analytical models allows farmers to check the change of soil carbon processes, vegetation growth rates, and the processes which are connected to emissions during agriculture. These knowledge can be used to implement carbon efficient farming such as optimized fertilizers use, reduced tillage activities and better crop rotation measures that improve soil carbon sequestration. Climate variability is another challenge to agricultural sustainability because erratic and unpredictable weather patterns and extreme weather can greatly interfere with the productivity of crops. Upcoming technologies will provide the opportunity to create climate adaptive crop intelligences, which will monitor the environment conditions constantly and modify the production methods accordingly. Analytical frameworks based on AI can be used to analyze weather predictions, soil water levels, and crop development signals to suggest adaptive management techniques. These systems enable the farmers to predict the stress factors of climate like drought or heat waves and carry out prevention measures in good time to ensure that crops remain healthy and yield does not decrease.

In addition to productivity and climate adoptions, sustainable agriculture also has to take into consideration the maintenance of ecological balance and biodiversity. Precision technologies in farming may help to preserve biodiversity by minimizing the use of large amounts of chemicals and allow making specific interventions in agriculture. In theory, with AI-based pest detection systems, the localized infestations of pests can be detected and the pesticides applied only to those areas required, and no longer to all fields. On the same note, nano-pesticides in a controlled-release form can reduce the exposure of chemicals to non-target organisms, hence decreasing the adverse effects of the technology on beneficial insects, soil microorganisms and the surrounding ecosystems. The smart farming technologies contribute to preserving the ecological diversity in agricultural landscapes by encouraging more selective and efficient agricultural interventions. Although nanotechnology has many potential applications in the sustainability of agriculture, it also poses some significant concerns about the safety of the environment and the ecological effects in the long run. When engineered nanoparticles are introduced into soil ecosystems, there is a need to exercise caution and make sure that these materials are not accumulated in toxic levels or that they do not alter the normal processes of soils. It is therefore necessary that environmental risk assessment frameworks be adopted to assess the behaviour, mobility, and biodegradability of agricultural nanomaterials. Efforts in research are now being put towards creating nanomaterials that are environmentally friendly, which can be safely degraded once they have served their purpose in agriculture. Linear combination of nanotechnology, AI, and remote sensing into sustainable agricultural systems eventually facilitates a shift towards climate resistant and environmentally friendly farming systems. With help of these technologies, it is possible to create a new generation of smarts agricultural ecosystems, which are able to balance between productivity and environmental sustainability as the technologies will allow managing all resources more accurately, reducing agricultural waste, and helping to adapt to changing environmental conditions.

10. PERFORMANCE EVALUATION AND SUSTAINABILITY IMPACT ANALYSIS OF NANO-AI PRECISION FARMING SYSTEMS

In order to show the possible benefits of the nanotechnology, artificial intelligence, and remote sensing integration into the precision agriculture, in this section, the author provides a conceptual performance analysis of the suggested intelligent farming ecosystem. Though the provided models are framework-based but not necessarily a strictly experimental implementation, the analytical comparison can be used to show how the combination of nanoscale sensing, AI-based analytics and Earth observation technologies can help to achieve agricultural performance, sustainability, and decision intelligence significantly. Figure 4 5, and 6 are all demonstrating the beneficial impact on agricultural efficiency, crop monitoring capacity, and environmental sustainability as brought about by the proposed Nano-AI precision farming paradigm.

Figure 4. Performance Comparison of Smart Agriculture Systems

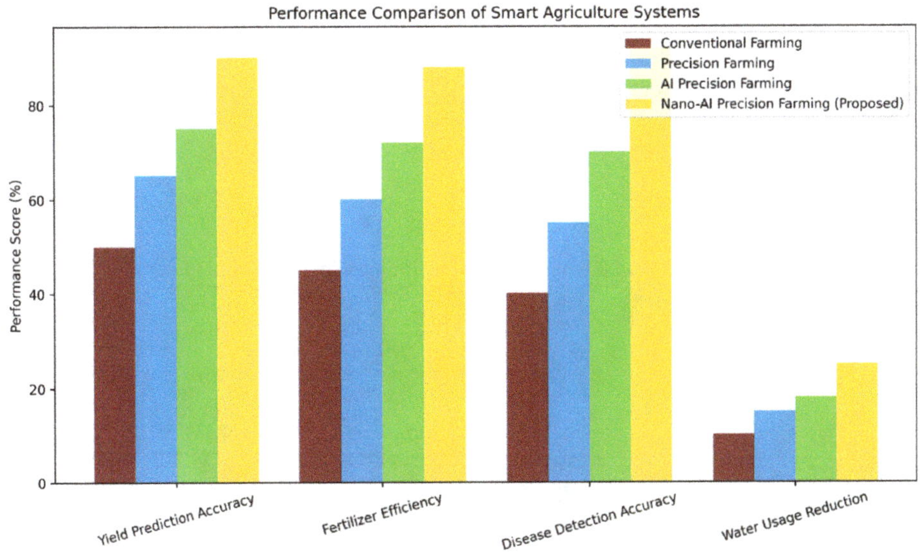

As shown in figure 4, there is a comparative performance analysis of four agricultural paradigms, conventional farming, precision farming, AI-enabled precision farming, and the proposed Nano-AI precision farming framework. The comparison is based on the main performance measures of agriculture, such as predictability and performance of crop yield, fertilizer efficiency, disease prediction and water saving. The findings have shown that there is gradual increase in the agricultural performance due to the change in farming systems between the traditional farming system to the modern technologically developed intelligent farming environment. Traditional farming systems have a generalized method of agriculture, and intermittent observation of the field, which can lead to inaccurate prediction and a waste of available resources. Precision farming enhances performance because it introduces new technologies like GPS-based field mapping, sensor networks, and variable-rate input management. Artificial intelligence also adds to the further development of agricultural decision-making, facilitating predictive analytics and the possibility to monitor crops automatically. Through AI-based models, remote sensing images and historical farming records may be analyzed to enhance crop yields and detection of the presence of diseases. Nevertheless, the offered Nano-AI precision farming system presents the greatest performance gains, mainly because of the usage of nanoscale sensing technologies that offer real-time biochemical analysis of the soil and plant systems. The framework enhances the accuracy of agricultural intelligence systems significantly by merging the data of nanosensors with AI-analytical predictive analysis and remote sensing. Such a combined solution will allow a more accurate delivery of nutrients, timely identification of stress in crops, and water management strategies.

Figure 5. Real-Time Crop Intelligence Map using Dynamic Crop Health Intelligence Mapping (DCHIM).

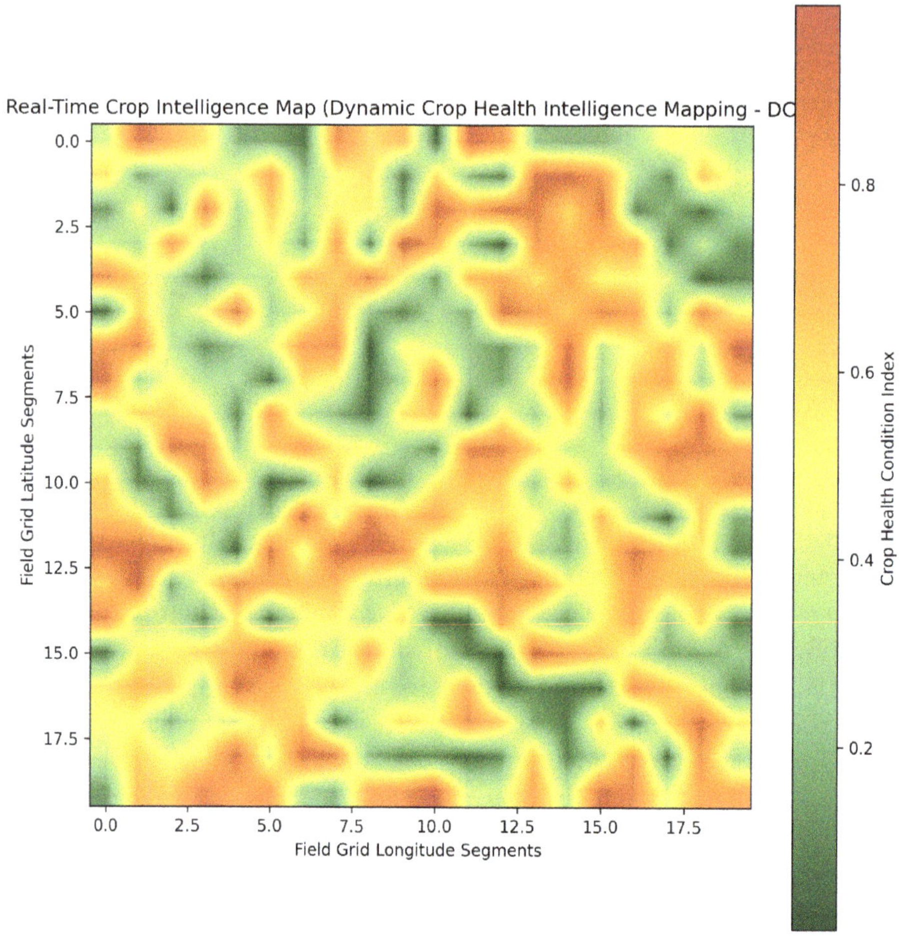

Figure 5 is a concept of Dynamic Crop Health Intelligence Mapping (DCHIM) that is an example of a spatial visualization of crop health conditions at agricultural fields. The map shows that AI-driven analytics can combine data of remote sensing platforms and nanosensor networks to create high-resolution crop health intelligence maps. Various colors used in the visualization indicate the various health conditions of the crops in the field. Green areas denote healthy agricultural areas that contain the best nutrient and water levels. Yellow is a medium stress condition, which is usually linked to early nutrient deficiency or slight environmental stress. Orange areas show that there is a state of water stress and irrigation changes might be necessary. Red zones are regions where it is possible to have pest infestation or outbreak of plant diseases. DCHIM allows real-time tracking of spatial variability of crops, therefore, facilitating its spread in a bigger agricultural area than it can be detected by farmers. The intelligence of space is essential to such spatial interventions as localized fertilization, optimization of irrigation, and the control of pests. The system allows identifying crop anomalies early enough and therefore enhanced farm-level decision-making without wasting many resources.

Figure 6. Sustainability Impact Metrics in Smart Agriculture Systems

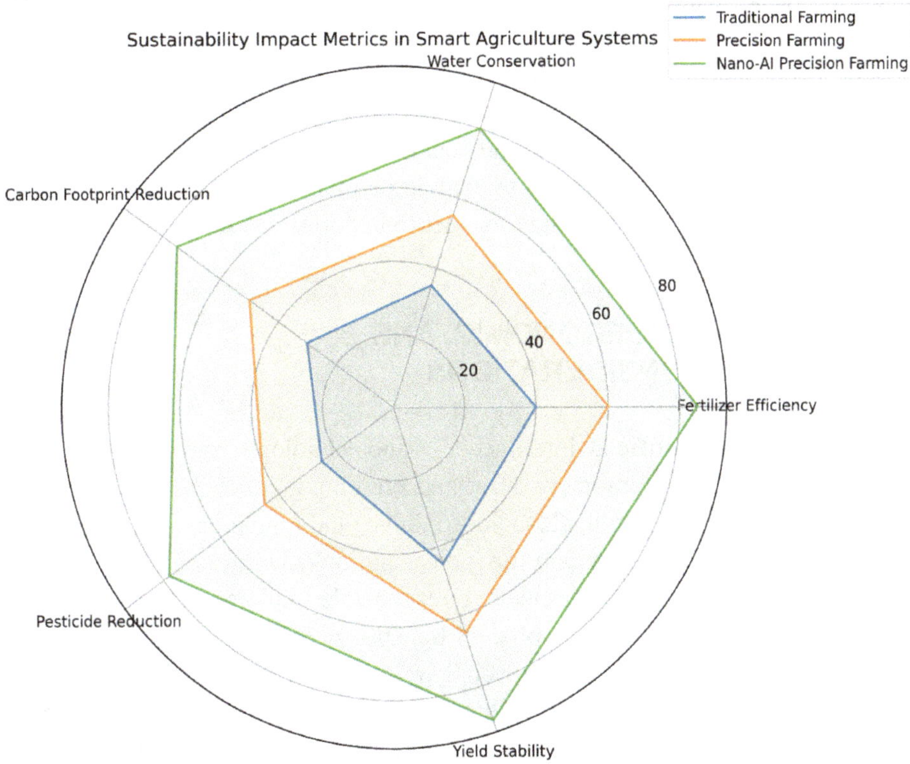

Figure 6 introduces a sustainability impact analysis of the traditional farming systems, the technologies of precision farming and the suggested Nano-AI precision agriculture framework. The radar chart determines performance in terms of five crucial sustainability indicators, which are fertilizer efficiency, water conservation, reduction of carbon footprint, reduction of pesticides and stability of yields. The analysis shows that the traditional way of farming has a lower sustainability performance because of the inefficiency in the use of inputs and the ability to monitor them. Precision farming technologies enhance sustainability through variable rate application of farm inputs and also enhanced resource management. Nevertheless, the application of nanotechnology and artificial intelligence to the framework offered can substantially increase these sustainability advantages. Nano-fertilizers and nanosensors can allow specific regulation of the kinetics of nutrients in the soil, which will lead to a decrease in the number of fertilizers applied to the soil and a decrease in the level of pollution of the environment. The AI-based irrigation models will help optimise water allocation with regards to real-time soil moisture status and climatic information, which will result in enhanced water savings. Also, AI-powered crop surveillance technology and nano-enabling pesticide detection enables specific pesticide spraying and minimize pollution to the nearby environment. The other significant sustainability outcome demonstrated in the figure is the yield stability which is the capacity of the system to offer stable agricultural productivity in different environmental conditions. Through the combination of predictive analytics with real-time environmental scanning, the Nano-AI precision farming systems will be able to be adjusted to evolving climate conditions and reduce the agricultural risks in a more efficient manner than traditional methods of farming.

11. ETHICAL, SOCIOECONOMIC, AND REGULATORY CONSIDERATIONS

The blending of artificial intelligence, nanotechnology, and remote sensing into the agricultural system is very fast and presents great technological values, yet, it poses some serious ethical, socioeconomic, and regulatory issues. With the rise of data-driven agriculture and the increasing intensive use of technologies, it is important to take measures to ensure a responsible implementation of the technologies to continue to make the process transparent and fair and to preserve the environmental sustainability. It is especially relevant that these issues be addressed in order to enhance a balanced agricultural innovation that is beneficial to farmers, ecosystems, and the society as a whole. A significant ethical issue can be associated with the increasing use of artificial intelligence when making farm-related decisions. Increasingly, AI-based systems are applied to control such key farming processes as irrigation, pest management, crop choice, and nutrient optimization. Although these technologies enhance efficiency and prediction ability, they come with the problem of transparency and accountability. The farmers and other stakeholders in the agricultural industry should have the capacity to know how the AI applications come up with the recommendations and the sources of the data used in making the decisions. The lack of transparency can also result in unintended biases or inappropriate suggestions that are made by automated decision systems without appropriate transparency and have adverse effects on crop productivity and the economics of farms. Thus, intelligent agricultural systems must continue to be trusted because explainable and interpretable AI models are developed.

The other most important concern is the question of data governance and agricultural data sovereignty. Current precision farming technologies create large volumes of agricultural data in sensor, drones, satellite platform, and farm management software. Such datasets have quality data on soil conditions, crop performance as well as farm operations. Nevertheless, the question comes up on whether this data belongs to anyone, who is entitled to access this information, and how this data can be exploited by providers of agricultural technology. The use of digital platforms to collect and analyze farm data may bring about imbalance in the control of agricultural information where farmers get addicted to these platforms. Creating equitable data governance systems that uphold the rights of farmers and provide clear data usage guidelines is critical to the creation of just digital agriculture systems. The issue of regulations is also becoming more complicated with the advent of nanotechnology in the farming systems. Nano-enabled fertilizers, pesticides and biosensors have a high potential to be more efficient and more precise but the environmental and health effects of nanomaterials should be properly considered. The regulatory bodies are on a dilemma to come up with principles of testing, authorization, and supervision of the agricultural nanomaterials to make certain that the technologies do not represent a danger on ecosystems and human well-being. The absence of uniform regulatory regimes in various nations can also pose obstacles in the popularization of nano-enabled technologies used in agriculture. This means there should be interdisciplinary cooperation between scientists, policymakers and regulators in order to come up with all-inclusive safety and regulatory standards.

When the effects of advanced agricultural technologies are considered, socio-economic aspects should be taken into account as well. Precision farming systems are usually costly in terms of sensor, analytical platform, and automated machine investments that might exclude access to these systems by the smallholder farmers as well as the resource-deprived agricultural societies. While the smallholder farms comprise a large percentage of agricultural output in the world, the technological advancements should be fashioned in such a manner that they are affordable. The development of scalable solutions, collaborative data-sharing portals and government-subsidized digital agriculture programs that facilitate inclusive agricultural development may be the solutions to ensure equitable access to digital agricultural technologies. Lastly, the notion of responsible AI is necessary to inform the future of intelligent agricultural systems. The responsible AI frameworks focus on fairness, transparency, accountability, and sustainability of AI-powered technologies. In farming, the responsible use of AI systems must be defined by the need to make the tools used in the decision-support systems work in a manner that allows benefiting farmers, safeguarding the natural ecosystems, and enhancing the food security in the long-term. The principles of ethical design must also promote the use of collaborative human-AI interaction in which technological systems must supplement the knowledge of farmers and not substitute it. With these considerations, ethical, regulatory, and socioeconomic, the combination of nanotechnology, artificial intelligence, and remote sensing will develop in a responsible and sustainable way. Clear governance, equal access to technology, and innovation that is safe to the environment will become a very important factor in forming the future of intelligent precision agriculture.

12. FUTURE RESEARCH DIRECTIONS: TOWARDS AUTONOMOUS INTELLIGENT AGRICULTURE

The emergence of the new generation of the agricultural systems marked by autonomy, adaptability, and connectivity to the global markets are being made possible by the fast development of the artificial intelligence, nanotechnology, and remote sensing tools. Although existing technologies in the field of precision farming mostly assist in making decisions that can be formed based on the data, more advanced agricultural ecosystems in the future will tend to shift towards fully autonomous intelligent agricultural systems that can learn, adapt, and react dynamically to environmental and operational conditions. Further studies in the new areas of technology will be of a significant contribution in developing these sophisticated agricultural systems.

12.1 Self-Learning Autonomous Farms

The future agricultural systems are determined to be shifting toward the system of self-learning autonomous farms where the intensive systems will constantly monitor and check the environmental conditions and crop performance and the use of resources and allow the systems to alter the work of the farms on their own. The machine learning and reinforcement learning algorithms can examine both historical and real-time agricultural data in order to optimize decision-making processes in such systems with time. Self-driving tracts, automated planting, and intelligent irrigation systems would work together with AI-based analytics to carry out farm management roles with minimal human supervision. The evolution of this sort of systems will need progress in the integration of sensors, adaptive learning algorithms and dependable autonomous farming equipment.

12.2 Swarm Robotics and AI-Assisted Crop Monitoring

Swarm robotics is a potential field of study in large scale agricultural observation and control. Swarm-based systems are systems in which several small autonomous robots are used to carry out some tasks like crop monitoring, pest detection and field inspection in agriculture. Such robotized agents will be able to communicate between themselves and manage their operations through distributed AI algorithms, which will allow them to effectively cover extensive farming lands. Swarm robotics can also be very beneficial in crop monitoring through offering local and continuous observation of the fields. Combined with AI-based analytics, the systems are capable of quickly identifying abnormalities in the health of crops, pest infestations, and aiding with specific agricultural solutions.

12.3 Quantum AI for Agricultural Optimization

Quantum computing and artificial intelligence can potentially revolutionize complicated optimization in agriculture. Agricultural systems are associated with a large number of variables, including the state of soil and weather conditions, dynamics of crop growth, and resource allocation policy. Conventional computational techniques might not be able to effectively solve this high-dimensional problem. The quantum AI technology has the potential to provide more effective and quicker optimization to farm activities, such as crop planting, irrigation planning, and nutrient control. Even though quantum-enhanced agricultural analytics is still in its infancy, the concept can be described as a promising area of research that can potentially greatly increase the computational power of future agricultural intelligence systems.

12.4 Nano-Robotic Soil Remediation Technologies

The existence of soil degradation and contamination poses significant threats to the use of sustainable agriculture. Future studies can be done in the development of nano-robotic systems to do specific soil remediation and nutrient balancing. Such nanoscale systems have a potential of detecting the presence of harmful pollutants, controlling the chemistry of soil and providing useful nutrients directly in the soil environment. Nano-robotic technologies have the potential to heal the soil by working on a microscopic level, which is more effective than conventional remediation processes. Such innovations can as well facilitate the accuracy of strategies of treating soil which will lessen the disturbances on the environment and enhance the fertility of soil and also the yield.

12.5 Global Smart Agriculture Networks

The future of digital agriculture can also entail the development of the world-widely networked agricultural intelligences. These networks would combine the information of farms, research centers, satellite systems, and climate monitoring systems in various geographical areas. These networks might improve early warning mechanisms on the spread of crop diseases, pests, and agricultural risks associated with climatic conditions by exchange of agricultural knowledge at an international level. Global agricultural platforms based on AI might be able to assess and analyze pooled data to create predictive methods that assist in food security preparation and climate-adaptable agriculture tactics in every part of the world. Taken together, these new research focuses point to the move towards autonomous intelligent agriculture, where high-tech technologies collaborate to establish the world of adaptive, resilient and highly efficient farming ecosystems. Further interdisciplinary studies that incorporate artificial intelligence, nanotechnology, robotics, and enhanced computing will be critical in determining the future of sustainable world agriculture.

13. CONCLUSION

Modern agriculture is changing as nanotechnology, artificial intelligence, and remote sensing converge, making it possible to establish intelligent and sustainable processes of precision farming. Nano-sensors give real-time updates in soil nutritional content, moisture, and plant health whereas remote sensing facilities can give environmental data at a large scale. These datasets are combined in AI-based analytics to anticipate crop performance, identify stress conditions, and make decisions in agriculture. Nano-fertilizers and predictive models are some of the technologies that can be used to facilitate adaptive nutrient management and effective resource use. Systems such as NAPF, NIPFM, and NADT establish comprehensive digital agricultural systems, which are more productive, resilient to climatic changes, and sustainable. All these developments can help humanity achieve data-driven types of farming that would be able to sustain food security in the world and environmental conservation in the future.

REFERENCES

Ashique, S., Raikar, A., Jamil, S., Lakshminarayana, L., Gajbhiye, S. A., De, S., & Kumar, S. (2025). Artificial intelligence integration with nanotechnology: A new frontier for sustainable and precision agriculture. *Current Nanoscience*, *21*(2), 242–273. DOI: 10.2174/0115734137275111231206072049

Chavula, P., & Kayusi, F. (2025). Systematic Review on the Application of Nanotechnology and Artificial Intelligence in Agricultural Economics. *LatIA*, *3*, 322. DOI: 10.62486/latia2025322

D, C. S., & Devarajan, Y. (2025). D, C.S., Devarajan, Y. Investigation of Emerging Technologies in Agriculture: An In-depth Look at Smart Farming, Nano-agriculture, AI, and Big Data. *Journal of Biosystems Engineering*, *50*(2), 170–192. DOI: 10.1007/s42853-025-00258-z

Dhinakaran, D., Edwin Raja, S., Ramathilagam, A., Vennila, G., & Alagulakshmi, A. (2025). Ethical and legal challenges with IoT in home digital twins. *MethodsX*, *14*, 103409. DOI: 10.1016/j.mex.2025.103409 PMID: 40852565

Dhinakaran, D., Jagadish Kumar, N., Raja Brundha, A., Subiksha, N., & Karpagam, T. (2024). Precision Maintenance with PARM and Augmented Reality for Asset Optimization. *Navigating the Augmented and Virtual Frontiers in Engineering, Pages*, *28*. Advance online publication. DOI: 10.4018/979-8-3693-5613-5.ch002

Dhinakaran, D., Udhaya Sankar, S. M., Latha, B. C., Anns, A. E. J., & Sri, V. K. "Dam Management and Disaster Monitoring System using IoT," *2023 International Conference on Sustainable Computing and Data Communication Systems (ICSCDS)*, Erode, India, 2023, pp. 1197-1201, DOI: 10.1109/ICSCDS56580.2023.10105132

Fathi Qarachal, J., & Alizadeh, M. (2025). Harnessing precision agriculture and nanotechnology for sustainable farming: A review. *Nanotechnol. Environ. Eng.*, *10*(2), 37. DOI: 10.1007/s41204-025-00429-5

Goyal, V., Rani, D., Mehrotra, S., Deng, C., & Wang, Y. (2023). Unlocking the Potential of Nano-Enabled Precision Agriculture for Efficient and Sustainable Farming. *Plants*, *12*(21), 3744. Advance online publication. DOI: 10.3390/plants12213744 PMID: 37960100

Harini, M., Dhinakaran, D., Prabhu, D., Udhaya Sankar, S. M., Pooja, V., & Kokila Sruthi, P. (2023). "Levarging Blockchain for Transparency in Agriculture Supply Chain Management Using IoT and Machine Learning," 2023 World Conference on Communication & Computing (WCONF), RAIPUR, India, pp. 1-6, DOI: 10.1109/WCONF58270.2023.10235156

Islam, F. A. S., & Islam, M. A. N. (2025). AI-Driven Integration of Nanotechnology and Green Nanotechnology for Sustainable Energy and Environmental Remediation. *Journal of Engineering Research and Reports*, *27*(7), 260–311. DOI: 10.9734/jerr/2025/v27i71574

Keerthana, M., Dhinakaran, D., Ananthi, M., Harish, R., Udhaya Sankar, S. M., & Sree, M. S. (2023). "IoT Based Automated Irrigation System for Agricultural Activities," 2023 12th International Conference on Advanced Computing (ICoAC), Chennai, India, pp. 1-6, DOI: 10.1109/ICoAC59537.2023.10249426

Kishore, K. V., Dhinakaran, D., Sankar, S. U., Chandu, K., & Visuwasam, L. M. M. (2023, November). Fish farm monitoring system using IoT. In *2023 International Conference on System, Computation, Automation and Networking (ICSCAN)* (pp. 1-6). IEEE. doi:DOI: 10.1109/ICSCAN58655.2023.10395013

Singh, S., Jakhar, S. K., & Kulhar, K. S. (2024). Harnessing nanotechnology and artificial intelligence for precision agriculture in smart cities. In *E3S Web of Conferences* (Vol. 540, p. 13002). EDP Sciences., DOI: 10.1051/e3sconf/202454013002

Sudharson, K., Alekhya, B., Abinaya, G., Rohini, C., Arthi, S., & Dhinakaran, D. (2023, February). Efficient soil condition monitoring with IoT enabled intelligent farming solution. In 2023 IEEE International Students' Conference on Electrical, Electronics and Computer Science (SCEECS) (pp. 1-6). IEEE. doi:DOI: 10.1109/SCEECS57921.2023.10063050

Takaloo, S., Xu, A. H., Zaidan, L., Irannejad, M., & Yavuz, M. (2024). Towards point-of-care single biomolecule detection using next generation portable nanoplasmonic biosensors: A review. *Biosensors (Basel)*, *14*(12), 593. DOI: 10.3390/bios14120593 PMID: 39727858

Thakur, N., & Yadav, A. N. (2025). Nanotechnology in agriculture: A review on precision farming and sustainable crop production. *BioNanoScience*, *15*(2), 243. DOI: 10.1007/s12668-025-01861-2

Yadav, A., & Yadav, K. (2025). Harnessing Nanotechnology and Artificial Intelligence to Revolutionize Sustainable Smart Agriculture. In Yadav, S. K., & Kumari, A. (Eds.), *Nanobiotechnology in Agri-food Sector. Sustainability Sciences in Asia and Africa*. Springer., DOI: 10.1007/978-981-96-7756-6_7

Yao, S., Yuan, X., Jiang, L., Xiong, T., & Zhang, J. (2020). Recent progress on fullerene-based materials: Synthesis, properties, modifications, and photocatalytic applications. *Materials (Basel)*, *13*(13), 2924. DOI: 10.3390/ma13132924 PMID: 32629789

Zain, M., Ma, H., Nuruzzaman, M., Chaudhary, S., Nadeem, M., Shakoor, N., & Ahamad, T. (2023). Nanotechnology based precision agriculture for alleviating biotic and abiotic stress in plants. *Plant Stress*, *10*, 100239. DOI: 10.1016/j.stress.2023.100239

Chapter 7

AI Applications in Agricultural Monitoring, Crop Yield Estimation, and Food Security

Nilotpal Das
http://orcid.org/0009-0005-0083-4093
Uttaranchal University, India

Rohit Kumar
http://orcid.org/0009-0003-4713-2625
GLA University, India

Atin Kumar
http://orcid.org/0000-0002-1653-2146
Uttaranchal University, India

Sharad Sachan
Lovely Professional University, India

ABSTRACT

Artificial intelligence (AI) is transforming the agriculture industry by facilitating the proper estimation of crop yields, real-time data acquisition, and evidence-based decision-making to enhance food security. Combining AI with remote sensing, IoT sensors, UAVs, and geospatial technologies can be used to monitor the health of crops, soil, variability of the weather, and dynamics of pests and diseases continuously. Machine learning and deep learning models are used to analyze the multi-source data and predict the yields, optimize the use of resources, and facilitate the timely interventions. Climate changes also enhance AI systems to give early warning of

DOI: 10.4018/979-8-2600-0778-5.ch007

droughts, floods, nutrient deficiencies and biotic stresses. Although there are issues regarding data access, infrastructure, and ethics, in particular, in developing regions, AI can fulfill its potential to establish sustainable agriculture and global food security with favorable policies and increased usage.

1. INTRODUCTION

The accelerated rates of population increase, climate change unpredictability, land and water resources pressure have rendered the agricultural monitoring system and crop yield assessment to be essential resources in protecting the food security since they allow detecting crop stress early enough, allocating resources efficiently, and planning to ensure that production and market crashes do not occur (Sishodia et al., 2020; Benos et al., 2021; Çelik, 2026). Real-time data on crop status via field sensors, remote sensors and farm records can be used to make decisions on irrigation, fertilization, pests and diseases, enhancing inputs, reducing loss of yield and improving efficiency of supply chains through shocks and this information can support national and global evaluations of food availability as well as informing the policies of storage and trade; when combined with effective yield prediction models, this information can also be used to make national and global estimates of food availability, guide storage and trade policies and resilience to shocks. However, conventional methods of empirical and statistical analysis cannot manage the complexity of the interactions between weather, soil, genotype, and management and the flood of information that is currently produced by satellites, UAVs, IoT sensors, and farm machinery (Barbedo, 2025; Soussi et al., 2024). It is on this backdrop that, as a key enabler of contemporary, information-intensive agriculture, artificial intelligence (AI) has quickly become a fundamental component of machine learning, deep learning, and similar data-driven methodologies. According to systematic reviews, today, AI applications are used across soil and crop sensors, predictive analytics, and agricultural robotics, and farmers are increasingly accessing sensors, imaging, and AI-based decision tools in fields like crop protection, variable rate input application, and yield forecasting (Elbasi et al., 2023; Alahmad et al., 2023). Most models of machine learning and deep learning, frequently trained on multispectral and hyperspectral satellite or drone-observed images in combination with weather and management data have proven highly accurate about yield predictions of large crop species and identifying factors that could drive yield changes during atypical climatic conditions, thus enabling precision management at the field level but strategic management in the region level (Choi et al., 2025; Wang et al., 2022; Benos et al., 2021). Simultaneously, AI-powered mapping and surveillance technologies lead to better food security, through the better representation of production in space, infection and pest detection, and efficient transportation of inputs and harvesting logistics, which in combination allow for compensating yield differences and decreasing wastes (Espinel et al., 2024; Alahmad et al., 2023). In that context, the current chapter places AI applications in agricultural monitoring and crop yield estimation very squarely as applied to the goals of food security. Its area includes three closely interconnected themes, namely: (i) the significance of real time agricultural surveillance to diagnose and treat farm stresses, to regulate inputs and to stabilize yields; (ii) the creation and application of AI based crop yield prediction and forecasting models, which take advantage of multi-source data to aid on farm decision making and upper level policy; and (iii) how these AI enabled capabilities contribute to the four dimensions of food security, which are availability, access, utilization, and stability under climate and market uncertainty (Dhal and Kar, 2024; Sishodia et al., 2020). The aims are to review the existing academic findings regarding the role of AI and smart sensing in revolutionizing crop surveillance and yield forecasting; to explain how the technologies may be used to help farmers achieve higher productivity, resource efficiency, and risks management and to critically comment on the bottlenecks of the data quality, model generalization, interpretability and adoption barriers; and to outline the research and implementation priorities that the adoption of AI driven agricultural information systems can be scaled to in technologically advanced and data poor areas. By so doing, the chapter is meant to offer a conceptual and methodological underpinning to researchers, practitioners, and policymakers who wish to leverage AI not as some technological new phenomenon, but as a part of the overall food systems that are sustainable and resilient and able to support the demand of future societies within the environmental constraints.

2. OVERVIEW OF ARTIFICIAL INTELLIGENCE IN AGRICULTURE

2.1 Definition and Evolution of AI in Agricultural Systems

The concept of artificial intelligence (AI) in agriculture may be interpreted as a collection of computational approaches and intelligent systems capable of sensing, interpreting, learning, and performing actions on agricultural data to assist or automate previously performed decisions by farmers, agronomists, and managers. Initially, applications in the 1980s and 1990s were predominantly expert systems that coded expert knowledge in the form of if-then rules to provide advice on pest control, fertilization, and choice of crop, but were limited by inflexible rule bases and limited availability of digital information (Elbasi et al., 2023). As precision agriculture and information and communication technologies emerged, farms started to bring forth a continuous volume of data from field sensors, weather stations, satellites, unmanned aerial vehicles (UAVs), as well as machinery. This data tsunami generated the necessity and chance of more adaptable, data-driven approaches to AI, which can automatically find patterns and connections in the overwhelming and high-dimensional data (Benos et al., 2021). Within the past ten years or so, AI has become an essential enabler of sustainable intensification, resource use efficiency, and climate resilient production as cloud and edge computing, wireless sensor networks, robotics, and big data analytics have advanced to the point where they are viewed as integrated and operational components of digital farm management and automation (Elbasi et al., 2023; Subeesh and Mehta, 2021).

2.2 Key AI Techniques: Machine Learning, Deep Learning, Computer Vision, and Expert Systems

In this development, machine learning (ML) is now the most prevalent branch of AI in agriculture. ML algorithms are trained on historical and real-time data to do tasks like crop and variety suggestions, soil type and soil fertility classification, plant disease and pests, and irrigation schedules, yield prediction, and livestock behaviour (Benos et al., 2021; Akkem et al., 2023). Classification and regression Supervised ML techniques (e.g., support vector machines, random forests, artificial neural networks) are widely used in crop, water, soil, and livestock management, and unsupervised techniques are used to cluster management zones and identify anomalies. Use of deep learning (DL) multi-layer neural networks that learn hierarchical features has further enhanced the performance, particularly on complex data, including image and time series, to provide high-accuracy solutions in plant disease, weed-crop, fruit detection and counting, land cover classification, and yield estimation (Dhanya et al., 2022; Saleem et al., 2021). Other advanced models, such as the fully convolutional networks, ResNet, and the U Net, have demonstrated recurrently better performance compared to traditional ML in most vision-based agricultural tasks, yet consume more data and computation (Dhanya et al., 2022; Saleem et al., 2021). These DL approaches are closely related to computer vision, which offers non-contact, image-based sensing throughout the production cycle, such as seed and soil surface evaluation to canopy health sensing, weed detection, livestock monitoring, and automated harvesting using RGB, multispectral, hyperspectral, thermal, and 3D images of satellites, UAVs and ground robots (Dhanya et al., 2022). Meanwhile, expert systems are still valuable where information is sparse and agronomic principles and other heuristics governing crop protection, nutrient management, greenhouse climate control, and advisory services are considered well-structured and are becoming more and more integrated with ML and DL in hybrid decision support systems that integrate human knowledge with data-driven prediction (Elbasi et al., 2023).

2.3 Integration of AI With Digital Agriculture and Smart Farming

Modern digital agriculture and smart farming represent the combination of AI with facilitating technologies like the Internet of Things (IoT), smart sensors, wireless sensor networks, robotics, cloud/edge computing, big data platforms, and high-speed communications (4G/5G). Smart farms are outfitted with ubiquitous devices of soil, climate, and plant sensors, and UAVs and autonomous or semi-autonomous equipment to collect multimodal data continuously, which is transferred over IoT infrastructures to centralised or edge platforms where real-time monitoring, prediction, and control are performed by AI models (Elbasi et al., 2023; Subeesh and Mehta, 2021) (Figure 1).

Figure 1. Integration of AI with digital agriculture and smart farming

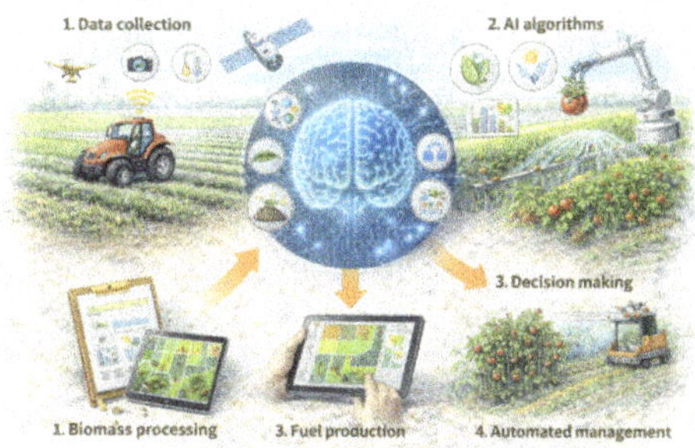

ML-powered DSSes optimize irrigation, fertilization, and crop protection in precision agriculture, and DL based computer vision modules navigate weeding and spraying robots, assist with precision harvesting, as well as automated grading and sorting during post-harvest tasks (Dhanya et al., 2022; Akkem et al., 2023). On a larger spatial level, AI, infrastructure based on remote sensing and geographic information systems, is used in agricultural mapping, in crop classification, drought and disease risks, and yield forecasting in a region, to aid in resource management and policy decisions (Espinel et al., 2024). Smart farming and IoT based agriculture reviews highlight that such convergence of AI, IoT, robotics and 5G is turning traditional farming into data intensive, predictive and partially autonomous systems, with the potential to have benefits in productivity, input efficiencies, environmental sustainability, and labour savings and raises issues around cost, interoperability, data ownership and skills, particularly in the environments of smallholders and developing countries (Elbasi et al., 2023; Subeesh and Mehta, 2021).

3. DATA SOURCES FOR AI-BASED AGRICULTURAL APPLICATIONS

3.1 Remote Sensing Data (Satellite, UAV, Aerial Imagery)

Remote sensing is the basis of the AI-powered agricultural solutions as it offers a repeated and spatially explicit measurement of crops, soils, and farm landscapes. Satellite sensors (e.g., multispectral and hyperspectral) can be used to monitor crop vigor, water stress, nutrient status, and land use on a large area with vegetation indices and biophysical variable extraction and apply AI models to crop classification, disease and pest risk mapping, and yield forecasting on the scale of the field to regions (Sishodia et al., 2020; Weiss et al., 2020). Unmanned aerial vehicles (UAVs) and other aerial platforms are used to complement satellites with centimetre-scale resolution that can measure fine variability in the canopy structure, weed patches, lodging, or localised damage, which is vital in training and verifying deep learning and computer-vision models on precision agriculture (Sishodia et al., 2020; Wang et al., 2022). The combination of these multiscale datasets (satellite and UAV, and manned aircraft) can be used to constitute powerful spatiotemporal data for machine learning pipelines predicting yield, processing variable-rate inputs, and assisting decision-making with site-specific management.

3.2 Internet of Things (IoT) and In-Field Sensor Networks

The sensor networks of IoT create continuous, high-frequency in situ sensors of environmental and crop conditions, the basis of real-time, data-driven agriculture. After measuring the soil moisture, temperature, pH, nutrient (NPK), and salinity (distributed soil moisture sensors), plant stress, microclimate, and livestock sensors send the data over wireless networks (LoRa, Wi Fi, cellular, 5G) to edge or cloud operating systems where AI algorithms analyse trends, identify anomalies, and can cause automated events (smart irrigation or greenhouse climate control), among others (Morchid et al., 2023; Alahmad et al., 2023). The resulting volume and range of smart data on these smart sensor systems is enormous compared to what is achievable through remote sensing, and these systems enable AI models to combine proximal sensing with remote sensing and historical data to more accurately predict water demand, fertilizer requirements, and disease outbreaks (Soussi et al., 2024; Alahmad et al., 2023).

3.3 Weather, Climate, and Soil Databases

The crucial contextual layers of AI models in agriculture would be gridded weather and climate data (historical data, forecasts, seasonal predictions) and soil information on its texture, depth, organic matter, and hydraulic characteristics. Combining station data, reanalysis and forecast products with field and remote-sensing observations can enable the machine learning models to reveal the mechanism of growth, phenology and yield variability of crops, which rely on temperature, rainfall, radiation, and extreme events, which can be used in applications like drought early warning, irrigation scheduling, and climate smart planning (Alahmad et al., 2023; Awais et al., 2023; Çelik et al., 2026). Likewise, the soil information systems of the region and the world, in combination with AI and geostatistics, assist in high-resolution mapping of soil characteristics, soil water content, and soil fertility limitations, which are inputted into the yield prediction models, crop suitability analysis models, and input optimization models (Awais et al., 2023).

3.4 Farm Management and Historical Yield Data

Raw sensing and environmental information are converted into a learning resource in the form of farm-level management records and past yield data on the effect of particular decisions. Varied management approaches, weather, soil states, and performance when observed in conjunction with detailed records of varieties, planting periods, fertilizer and pesticides, irrigation instances, tillage activities, and equipment settings can help AI models to acquire cause-effect connections among management approaches, weather, and soil conditions, as well as outcomes (Alahmad et al., 2023; Mohyuddin et al., 2024). Decision-support tools based on these datasets suggest the best rate of input, areas to use, and the best way to cultivate crops and assist in quantifying trade-offs between economics and the environment. With the spread of precision technologies, the combination of remote sensing, IoT sensors, weather-soil databases, and long-term farm records to integrated platforms based on big-data is making AI-based applications more sustainable, climate-resilient agricultural practices increasingly more accurate and powerful (Soussi et al., 2024; Alahmad et al., 2023).

4. AI APPLICATIONS IN AGRICULTURAL MONITORING

4.1 Crop Health and Stress Detection (Biotic and Abiotic Stresses)

Machine learning and deep learning algorithms are progressively used in AI-based monitoring systems to identify both biotic and abiotic stresses (diseases, pests, water stress, nutrient stress, heat stress, salinity stress, etc.) (Figure 2) based on imaging and sensor data. Early detection of stress patterns is possible in multiscale sensing, such as RGB, multispectral, hyperspectral, thermal, chlorophyll fluorescence and LiDAR, along with ML, which help to protect against the loss of vitality and offer proactive approaches to managing the problem before it begins to be visible (Muhammad et al., 2025; Wang et al., 2022) (Table 1). The applications of deep learning models (e.g., convolutional neural networks) have been extensively utilized in the classification of stressed plants vs. healthy plants, stress severity, and the mapping of stress spatial patterns, which are often more accurate, scalable, and runnable in a real-time scenario (Muhammad et al., 2025; Gao et al., 2020).

Figure 2. Crop health and stress detection

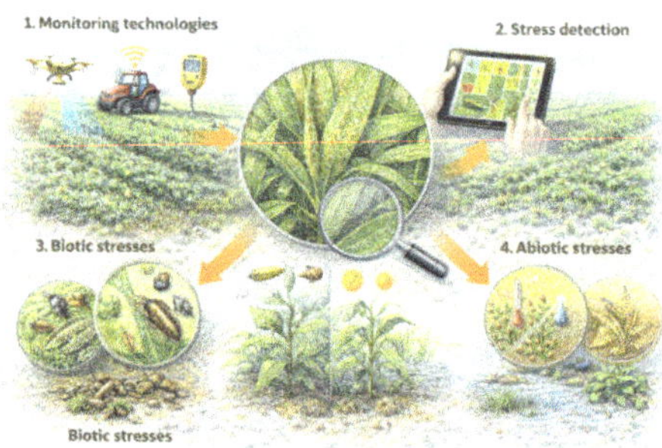

4.2 Soil Moisture, Nutrient Status, and Fertility Monitoring

The techniques used in AI are also key in the monitoring of soil moisture and fertility. Machine and deep learning algorithms that are trained using remote-sensing data (optical, SAR, hyperspectral), and in situ sensor measurements enhance large-scale soil moisture prediction, root zone moisture, and temporal prediction, which are vital in smart irrigation and drought management (P et al., 2025; Ramaiah et al., 2025). Equally, AI-processed hyperspectral and multispectral data enable mapping of soil organic carbon, texture, and essential nutrients to address gaps in yields and provide location-specific management of nutrients, especially where traditional sampling may be both sparse and costly (Bouanani et al., 2025; Awais et al., 2023). The multi-sensor data is also used in these AI-based frameworks to improve accuracy and spatial coverage over the conventional empirical or purely physical models (P et al., 2025).

4.3 Pest, Disease, and Weed Detection Using Computer Vision

Deep learning-based computer vision has become a leading trend in the detection of plant diseases, pests, and weeds. Image-based pipelines are characterized by the following processes: image acquisition, preprocessing, segmentation, feature extraction, and classification based on such models as CNNs, object detectors, and semantic segmentation networks (Jafar et al., 2024; Upadhyay et al., 2025). It has been observed that RGB, multispectral, and hyperspectral smartphone imaging, ground cameras, UAVs, or satellites have high recognition accuracies of leaf diseases, pests, and weed weed/crop discrimination, making them superior to diagnose and apply control on a disease, pest, and weed/crop discrimination rapidly and without destruction (Jafar et al., 2024; Upadhyay et al., 2025; Gao et al., 2020). Recent efforts have also focused on the combination of IoT and multi-sensor data with vision models and the utilization of transfer learning and large pre-trained networks to enhance robustness in real-world field conditions.

4.4 Phenological Stage Identification and Growth Tracking

Phenological stage identification and tracking growth are utilized more and more with the help of AI models and are necessary to schedule management activities, optimise inputs, and enhance yield prediction. Growth stages, canopy development, and biomass estimation at plot to regional scales are classified with the help of deep-learning and other ML algorithms using the time series imagery of satellites, UAVs, and proxim sensors (Sishodia et al., 2020; Wang et al., 2022). Continuous monitoring of plant development, nutrient status, and yield components, which replaces the labour-intensive visual scoring of crop development in the field and controlled conditions, has been achieved in the field and controlled environments by CNN-based methods (Wang et al., 2022; Ojo and Zahid, 2022).

Table 1. AI Applications in Agricultural Monitoring

Monitoring task	Main data & AI approach	Citations
Crop stress (biotic/abiotic) detection	Multi-sensor imaging + ML/DL classification/forecast	(Muhammad et al., 2025; Wang et al., 2022)
Soil moisture and fertility monitoring	Remote sensing + in-situ data + ML/DL regression	(Bouanani et al., 2025; P et al., 2025)
Pest, disease, and weed detection (computer vision)	RGB/multispectral images + CNNs/object detection	(Jafar et al., 2024; Upadhyay et al., 2025)
Phenology and growth stage tracking	Time-series imagery + DL/ML stage classification	(Sishodia et al., 2020; Wang et al., 2022)

5. AI-BASED CROP YIELD ESTIMATION AND FORECASTING

Crop yield estimation and crop forecasting AI-based applications combine various data and modelling approaches to make predictions at the field to regional levels to assist in food security planning and precise management.

5.1 Machine Learning Models for Yield Prediction

The classical and more modern machine learning (ML) methods that are popular include linear regression, Random Forest, Support Vector Machines, and Artificial Neural Networks, and they are used to model nonlinear relationships between yield and predictors such as weather, soil properties, management, and spectral indices. Crops and region review. Studies document prediction accuracies of between 50 and over 90 percent with simpler models frequently applied to small datasets and more intricate ensemble or tuned models common in instances where large and heterogeneous datasets exist (Trentin et al., 2024; Piekutowska and Niedbała, 2025). Random Forest and other tree-based ensembles can also make a formidable baseline because they can process mixed data types, are able to analyze variable importance, and perform well even under abnormal climatic conditions (Piekutowska et al., 2025). Nonetheless, investigations on thousands of fields of farmers around the globe indicate that even advanced ML can only explain only a fraction of the variability of on farm yields even in very heterogeneous smallholder systems, and that this points to limitations of entirely data driven prediction and the necessity of an improvement of explanatory variables and cautious validation both across space and time (Silva et al., 2023).

5.2 Deep Learning Approaches Using Multi-Temporal Remote Sensing Data

Deep learning (DL) is now a key to yield estimation, based on multi-temporal remote sensing, utilizing rich spatial and temporal occurrences in satellite (e.g., Sentinel 2, MODIS) and UAV images. Convolutional Neural Network and recurrent models (e.g., LSTM) can take image sequences or vegetation index time series as input to learn growth patterns and predict yields weeks before harvest and are often more effective than traditional ML when there is enough labelled data (Saha et al., 2025; Yuan et al., 2024; Maestrini et al., 2022). Recent studies on Sentinel 2 and UAV data demonstrate that CNN based models trained on multi date images can identify the canopy structure, phenology and stress dynamics and can accurately map yields of major grains, fruits and tree crops, although the effectiveness depends heavily on the quality of data, extent of the phenology and the choice of features (Trentin et al., 2024; Aslan et al., 2024; Yuan et al., 2024).

5.3 Integration of Weather, Soil, and Management Practices in Yield Models

Higher performing yield models are utilizing more and more multi-source data, a combination of remote sensing and meteorological, soil, and management data, to obtain both crop status and driving factors. Reviews of potato, tree crops and general crop yield prediction point to the fact that additions of weather time series, soil fertility variables and management variables, including sowing date, fertilizer and irrigation, have a substantial positive effect on robustness and generalizability, especially in the case of climate variability (Piekutowska et al., 2025; Luo et al., 2023; Maestrini et al., 2022). Spatial and temporal continuity can also be further enhanced with data assimilation structures that interface process based crop models with remote sensing (e.g., assimilating leaf area index, soil moisture or vegetation indices), whereas hybrid methods apply crop models to synthesize biophysical features or use ML to simulate and accelerate process based simulations (Luo et al., 2023; Maestrini et al., 2022).

5.4 Spatial and Temporal Yield Variability Analysis

Spatial and temporal varying knowledge is essential to site-specific management and risk evaluation. Multi-year image time series and geostatistical or ML techniques are employed in remote sensing ML pipelines to investigate the relationship between patterns and soil heterogeneity, topography, climate extremes, and management intensity to analyse the relationship between yield variation in fields and across landscapes (Piekutowska & Niedbała, 2025; Torre et al., 2021; Amsalu et al., 2026). Empirical research using on-farm yield data in thousands of fields suggests that climate variables are more important to variability in some high-input systems and management, and socio-economic factors are more important in smallholder settings, and that predictive ability tends to reduce when models are applied to unobservable years or locations (Silva et al., 2023; Dutta et al., 2020). Multi-scale analyses that integrate UAV, satellite, and ground-based data assist in determining consistent high and low yield areas, aiding in the determination of variable rate input requirements, and give insights into where extra data (e.g., on management or socio-economic conditions) is necessary to clarify the remaining variability (Yuan et al., 2024; Silva et al., 2023). The achievements of these innovations make AI-based yield estimation a key element of precision and climate-smart agriculture and highlight the current issues with data quality, representativeness, and interpretability.

6. ROLE OF AI IN CLIMATE-SMART AGRICULTURE

The key role of AI in climate-smart agriculture is that it facilitates the superior identification of climate stress, the advancement of warning, and planning to improve adaptation and resilience through data-driven planning (Figure 3).

Figure 3. AI in Climate-Smart Agriculture

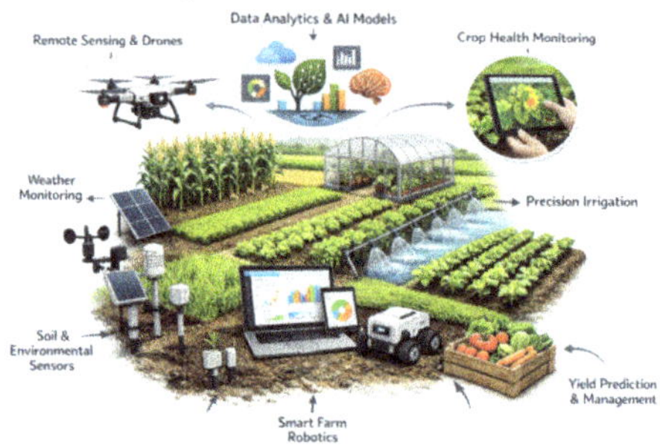

6.1 Drought, Flood, and Heat Stress Assessment

Drought, floods, and heat stress are more frequently evaluated using AI and machine learning methods using vast amounts of climate, hydrological, and crop data. Assessment reviews indicate that ML and deep learning models (e.g., artificial neural networks, support vector machines, recurrent networks) are more effective in monitoring and predicting the meteorological and agricultural drought indices with greater accuracy and lead time than most conventional statistical methods (Kikon & Deka, 2021; Prodhan et al., 2022). Hydrology AI models are applicable to predict streamflow, the groundwater level, and the likelihood of floods, which aid in the more comprehensive assessment of flood risks and early preparedness in water-scarce and prone to flood areas of agriculture (Biazar et al., 2025). On a field scale, AI in conjunction with remote sensing and thermal imaging can be used to conduct non-destructive assessment of crop water stress by identifying water deficits associated with drought and heat prior to actual damage and supporting smarter irrigation and minimizing yield and quality loss to climate change (Cho et al., 2024). All of these methods enable a more accurate mapping of the agro ecosystems regarding climate-related hazards and their effects.

6.2 Early Warning Systems and Risk Prediction

The use of climate-smart agriculture is based on the use of early warning that can be used to translate the climate information into a risk alert that can be acted upon. It has been demonstrated that machine learning models and deep learning models that are trained on historical climate, hydrological, and impact data enhance the quality and promptness of early warnings of extreme events like floods, droughts, and heatwaves and disaster preparedness among farming populations (Mălina et al., 2025; Materia et al., 2024). It is noted that AI has the potential to improve drought early warning through more accurate prediction of onset, duration, and severity, and by combining different indicators into the working systems to reduce risks and contingency planning in agriculture (Kikon and Deka, 2021; Prodhan et al., 2022). In a broader context, climate impact modelling with AI relies on predictive analytics to discuss climate scenarios in the future, test the possibilities of adaptation, and facilitate fair governance by converting the complex climate information into risk information relevant to making decisions by policymakers and local stakeholders (Ukoba et al., 2025). These objective forecasts and danger prevention features assist in decreasing the losses, direct emergency techniques, and prioritize susceptible areas.

6.3 Adaptation and Resilience Building Under Climate Change

The direct benefits of AI include adaptation and resilience building through the provision of information on climate-resilient practices, the optimal utilization of resources, and long-term planning. AI, GIS, and remote sensing are approached in conjunction in climate-resilient agriculture, meaning that the risk and vulnerability to climate are mapped, the planning of climate-resilient land use and crop suitability under future climatic scenarios is planned, and the monitoring of the consequences of adaptation interventions over time is conducted (Mălina et al., 2025). Decision-support systems powered by AI have the potential to optimize irrigation, fertilization, and other management in case of climate uncertainty, e.g., with the help of reinforcement learning, jointly improving yield and greenhouse gas emissions and adapting to changing weather conditions (Mălina et al., 2025). At greater levels, forecasting of climate extremes and food system disturbances with the help of AI can be used to design climate-prudent policies, food security, and healthier supply chains (Dhal and Kar, 2024; Ukoba et al., 2025). Surveys of AI in agricultural production systems point to the overall idea that the combination of AI technologies: sensing, robotics, predictive models, and decision-support can lead to increased productivity, lower environmental pressures, and more resilience of agricultural systems to climate change and also note the problem of data gaps, model transferability, and context specific (inclusive) deployment (Mohyuddin et al., 2024; Usigbe et al., 2023). All these functions put AI at the forefront of climate-smart agriculture that will be able to both address existing threats and contribute to transformative adaptation.

7. AI CONTRIBUTIONS TO FOOD SECURITY

The use of AI enhances food security by enhancing the production, monitoring, transporting, and regulating food in the four pillars of availability, access, utilization, and stability.

7.1 Yield Forecasting for Food Supply Planning

Weather and soil data, combined with data on management and remote sensing, are used to predict crop yield and production with the help of AI-driven forecasting models (ML, deep learning, ARIMA/SARIMA). Governments and markets can understand when there will be surpluses or shortages and stabilize prices (Dhal and Kar, 2024; Alahmad et al., 2023). Neural networks, Random Forest, and hybrid models of machine learning can achieve high levels of forecast accuracy and are already applied at the regional level in key staples, to support proactive import-export policies, inventory control, and input planning (Dhal and Kar, 2024; Gavasso-Rita et al., 2023).

7.2 Regional and National Food Security Assessment

Several indicators, including availability, accessibility, affordability, and utilization, are used to evaluate the food security of countries and regions with the help of AI models (Sarku et al., 2023). Machine learning, optimization, and geospatial decision models are being used in studies to synthesize data on poverty and food consumption, prices, climate, and infrastructure into composite food security indices that can be used to demonstrate hotspots and causes of vulnerability and intervene (Sarku et al., 2023; Chukwuma et al., 2024). Narrative early warning reports are also being analyzed using text mining and ML on narrative reports to identify global determinants of food insecurity, including market shocks, flooding, rainfall, and price volatility, which can be used to enhance strategic risk-based analysis (Carneiro et al., 2025).

7.3 Decision Support Systems for Policymakers and Stakeholders

AI analytics utilize big data, web-based decision support systems (DSS) to combine climate information and supply chain data with food safety and food security decisions in the context of climate change (Talari et al., 2021). DSS help, made possible by AI, can be used to simulate scenarios, rank policy options, prioritize indicators (e.g., sustainability, groundwater, income, food waste), and negotiate on the food system trade-offs between multiple actors, and complex data is made usable by planners and regulators (Sarku et al., 2023; Birgani et al., 2022).

7.4 Reducing Post-Harvest Losses and Improving Supply Chain Efficiency

AI assists in reducing post-harvest losses and optimization of the supply chain by predictive and optimization models programmed in logistics, cold chain, and processing systems. Artificial intelligence-based predicting of disruptions, demand, and market can result in more informed inventory, routing, and storage choices, less spoilage, and enhanced resilience to shocks, including pandemics or climate extremes (Dhal and Kar, 2024; Tamasiga et al., 2023). AI application is used with IoT and blockchain in processing and distribution to increase traceability, real-time condition monitoring, and quality control, reduce waste, and optimize efficiency across the farm-to-consumer chain (Dhal and Kar, 2024; Talari et al., 2021).

Table 2. Example AI Roles Across the Food System

Food-system stage	Main AI contribution	Citations
Production	Crop yield and loss prediction for planning	(Dhal & Kar, 2024; Alahmad et al., 2023)
Assessment	Regional/national food security indices & drivers	(Sarku et al., 2023; Chukwuma et al., 2024; Carneiro et al., 2025)
Governance	Policy DSS, indicator prioritization, scenario analysis	(Sarku et al., 2023; Birgani et al., 2022; Talari et al., 2021)
Supply chain	Demand/disruption forecasting, logistics optimization, loss reduction	(Dhal & Kar, 2024; Talari et al., 2021; Tamasiga et al., 2023)

8. AI TOOLS AND PLATFORMS FOR AGRICULTURAL DECISION SUPPORT

Agricultural decision support is increasingly becoming the focus of AI tools and AI platforms, which transform various data streams into feasible, site-specific advice to farmers and connect field decisions to cloud analytics and precision technologies.

8.1 Farm-Level Decision Support Systems (DSS)

Farm-level DSS integrates sensor, weather, soil, and management data with AI/ML algorithms to give instructions on the irrigation, fertilization, pest management, and cropping strategies. Agroecology and smart farming replace DSS with modern versions that combine crop models, expert systems, and GIS to help in making decisions related to soil and water management, integrated pest management, and resource optimization, and are often designed with user-friendly interfaces that farmers can understand (Petraki et al., 2025; Gebresenbet et al., 2023). DSS in precision agriculture are being developed into automated decision-making systems that combine real-time sensing with AI-based analytics to provide farmers with dynamic control of irrigation or nutrient delivery and to guide farmers not with prescriptive advice but with continuous advice (Alahmad et al., 2023; Subeesh and Mehta, 2021).

8.2 Cloud-Based and Mobile AI Applications

Mobile applications and cloud computing bring AI-based decision support to the industry by storing heavy analytics in the cloud and sending recommendations to mobile devices. It has been reviewed that large remote-sensing and sensor datasets are processed on cloud platforms with the help of ML and deep learning and returned insights on crop health, disease risk, or yield predictions, which reduces on-farm hardware needs (Sishodia et al., 2020; Subeesh and Mehta, 2021). Disease diagnosis, species identification, and soil health assessment are now available on the phones of farmers through mobile expert systems, which are now capable of providing AI models and remote imagery as point-of-use advisory services, particularly valuable in the contexts of the smallholder and resource-poor (Marinello et al., 2023; Benos et al., 2021) (Figure 4).

Figure 4. Cloud-based and mobile AI applications

8.3 Integration With Precision Agriculture Technologies

AI decision support is closely combined with precision agriculture infrastructures (smart sensors, Internet of Things networks, UAVs, autonomous machines, and others). Sensor-IoT platforms send real-time data on the soil moisture, nutrients, and crop status to AI engines to drive variable-rate applications and targeted interventions to enhance efficiency and sustainability (Soussi et al., 2024; Mansoor et al., 2025). All-terrain robotic platforms with multispectral/thermal sensors and AI vision, and UAVs, assist in automated scouting, disease detection, and site-specific operations with further feedback to DSS that complete the observation, analysis, and act loop on the farm (Agrawal and Arafat, 2024; Padhiary et al., 2024).

9. CASE STUDIES AND PRACTICAL APPLICATIONS

9.1 AI-Based Monitoring Systems in Smallholder and Large-Scale Farms

Farm sensor technology is now starting to be applied with smart sensors, UAVs, and vision systems to offer AI-enabled monitoring to track the status of the crops, pests, and resource utilization in real-time. Integrated smart-farming architectures are described in the reviews, which enhance the efficiency of small and large farms in terms of productivity and resource consumption by refining information on soil, climate, and crop sensors to AI analytics to inform irrigation, fertilizing, and stress detection (Elbasi et al., 2023; Soussi et al., 2024). Autonomous Greenhouses Challenge case studies show improvements in operations on a large scale in greenhouse and controlled-environment agriculture, as they prove how the AI-based temperature management and growth monitoring can cut heating energy consumption by a factor of 2 and maintain yields at the same level (Hoseinzadeh and Garcia, 2024).

UAV-AI platforms with large maize and soybean farms have also demonstrated scalability to industrial farms, with deep learning models and drones-mounted multispectral sensors mapping crop health, directing pesticide application, and generating hundreds of dollars per hectare of more revenue, and these applications, too (Agrawal & Arafat, 2024). At the same time, the low-cost sensor-IoT systems with AI analytics are undergoing modification to the smallholders so that they can monitor the soil moisture, classify the nutrients in the soil, and recommend the rate of crop or fertilizer to be used in the areas with limited extension services (Marinello et al., 2023).

9.2 National and Regional Yield Estimation Programs

Other than the individual farms, AI methods are becoming the focus of national and regional-level yield estimation. Machine-learning and deep-learning frameworks trained on the data of satellites, UAVs, and in situ sensors assist in predicting precision under changing climate and management with reviews of high accuracy with models like Random forests, neural networks, and convolutional networks on key crops (Alahmad et al., 2023; Choi et al., 2025). Precision agriculture systems based on remote sensing demonstrate the application of vast amounts of high-resolution images with cloud computing and machine learning to track crop health, outline variability, and predict yields on regional scales to support the input supply, storage, and trade planning (Sishodia et al., 2020). These AI-powered solutions are used to fill information gaps in data-poor areas with the help of images and environmental data available on a global scale and provide solutions to monitor production shortages and create specific support to the vulnerable regions (Dhal and Kar, 2024; Alahmad et al., 2023).

9.3 Successful AI-Driven Food Security Initiatives

A number of AI projects demonstrate first-hand donations to the food system. The AI models have been used to run global early warning mechanisms like the Hunger Map by the World Food Program to integrate satellite imagery, economic indicators, and local reports to track food insecurity in close real time and predict new hotspots, so that more aggressive humanitarian responses can be made (Dhal & Kar, 2024) (Figure 5).

Figure 5. Successful AI-driven food security initiatives

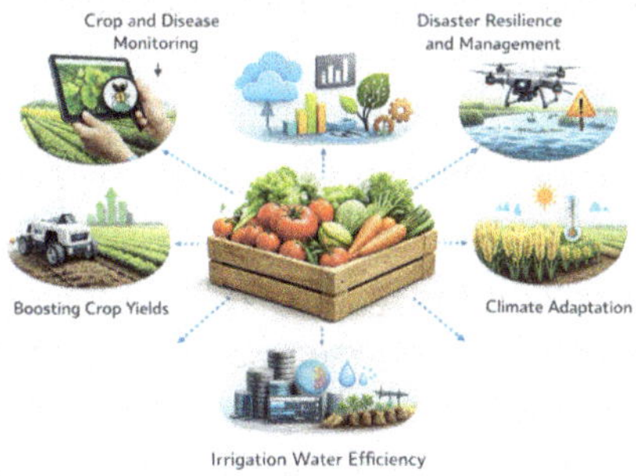

Real-time sensor data applies to AI controllers to control nutrients, water, and microclimate to generate better yields and lower resource use in instances of AI-enhanced hydroponic, aquaponic, and vertical systems in both urban and resource-limited environments (Dhal and Kar, 2024). AI-enabled field-level precision systems containing smart sensors and IoT have been demonstrated to decrease yield gaps, waste, and input-use efficiency, all of which have enhanced food availability and food sustainability (Alahmad et al., 2023; Mansoor et al., 2025). According to the reviews of smart-farming implementations, the implementation of AI systems alongside the knowledge of farmers, extension services, and supportive policy can help to increase resilience to climate shocks, decrease post-harvest losses, and create more stable and fair food systems (Usigbe et al., 2023; Gebresenbet et al., 2023).

10. CHALLENGES AND LIMITATIONS

The use of AI in agriculture has numerous constraints and difficulties, leading to its ineffectiveness and unfair use. One of the underlying problems is the idea of data quality, availability, and interoperability: agricultural data remain mostly scarce, noisy, or biased to well resourced areas and many smallholder and emerging systems (e.g., hydroponics, aquaponics) do not have consistent, labeled, and long term data, which leaves the model training and prediction effectiveness questionable (Dhal & Kar, 2024; Villa-Henriksen et al., 2020). Heterogeneous format, unstandardised schemas, and decentralised local databases make integration and reuse challenging, and interoperability across sensors, platforms, and organisations has remained a bottleneck that is addressed by shared standards, ontologies, and FAIR data principles (Villa-Henriksen et al., 2020; Šestak and Copot, 2023; Araújo et al., 2021). Such restrictions contribute to the issues of the model scalability and interregional transferability, as models trained on a certain crop, climate, or management regime are hardly generalizable to new agro ecosystems or socio economic settings without significant adaptation and retraining, which restricts the global use of AI forecasting and decision support systems (Dhal and Kar, 2024; Barbedo, 2025; Villa-Henriksen et al., 2020). Simultaneously, infrastructure, cost, and digital divide continue to be significant obstacles: initial investment in sensors, connectivity, cloud/edge computing and digital-twin infrastructures, and rural broadband, power, and unavailability of technical expertise all limit its adoption especially by smallholders and developing areas, and risks creation of a two tier agriculture, where the benefits only go to capital rich farms (Marinello et al., 2023; Mansoor et al., 2025; Alahmad et al., 2023; Zhang et al., 2025). Lastly, there is ethical, privacy, and data governance issues, which cross the ecosystem: lack of clarity about who owns data, proprietary platforms, and cybersecurity threats create mistrust and may lock farmers into unequal relations with technology companies; establishing the respectable agricultural data space with effective access guidelines, transparency, and accountability is the key to ensuring the right of farmers and ensuring that the digitalization process benefits and does not harm the sustainable and fair agri food systems (Šestak and Copot, 2023; MacPherson et al., 2022).

11. FUTURE PERSPECTIVES AND EMERGING TRENDS

The future of agricultural AI is shifting towards systems that are not only powerful but transparent as well as closely linked with automation and throughout the supportive institutional structures. One of the trends is the emergence of explainable AI (XAI) and interpretable models, in general, in complex tasks like crop yield prediction and plant phenotyping, where stakeholders need to recognize why a model encourages actions in a particular way. Crystalization Review The reviews on crop yield prediction indicate that XAI will be crucial to the interpretation of multidimensional models, enhance user trust, and reveal possible biases or spurious correlations, which might restrict transferability in abnormal climatic conditions (Choi et al., 2025). Equally, explainable AI research in the field of plant phenotyping demonstrates how model explanations can be useful in identifying the image properties that connect genotype, environment, and phenotype, and convert black box models into biological understanding, as opposed to merely predictive ones (Danilevicz et al., 2025). Simultaneously, AI and robotics, as well as automation, are developing in concept to more functional field systems. The AI-enhanced all-terrain vehicles and agricultural robots are reviewed and note gains in yield, input efficiency, and timeliness with an AI vision and machine learning, and offer future directions including more adaptable locomotion, robust sensing, and energy-efficient models to operate with true autonomy (Padhiary et al., 2024; Oliveira et al., 2021). Such robotic platforms are being more and more integrated with AI-based real-time monitoring and digital twins, where large sensor networks and IoT infrastructure are used to send real-time data to the simulation of crop, greenhouse, plant factory, or even machine fleets. Digital twins in agriculture are discussed in terms of bibliometric analyses that explain the transition to system-level, cyber-physical architectures with support of real-time simulation, optimization, and predictive maintenance, which allow conducting scenario analysis with climate resilience and resource utilization before interventions are applied in the field (Gund et al., 2025; Zhang et al., 2025). Smart environmental management of vertical farms and greenhouses is another example of how real-time AI models will be used to control light, temperature, humidity, and nutrients, and how novel technologies, such as edge computing and transfer learning, will enable more rapid local decision-making in a wide variety of crops and settings (Kaya, 2025). To realize these possibilities, the policy support and AI adoption capacity-building will be necessary: surveys of farmer acceptance indicate that the perceived behavioral control, evident economic benefits, and the belief in data rights have a strong influence on the willingness to use AI, which is why it is necessary to conduct training and extension programs and establish a robust governance of an agricultural data space (Mohr and Kuhl, 2021). The reviews of AI-based greenhouse and smart-farming systems underline that investment in data infrastructure, education, and skills, as well as regulation of privacy, interoperability, and ethical use, are the key factors to proceed with pilot projects to scalable, inclusive AI that facilitates sustainable and robust food systems instead of contributing to expanding the existing digital gaps (Hoseinzadeh and Garcia, 2024; Soussi et al., 2024).

12. CONCLUSION

Artificial intelligence has become an effective means of changing agricultural monitoring, crop yield prediction, and food security by being able to handle large and complex data sets to produce timely and actionable information. The chapter indicates that an AI-based combination of remote sensing, IoT sensors, geospatial information, and advanced machine learning algorithms can be used to provide precise measurements of crop health, yield predictions, and early warnings of biotic and abiotic stress determinants. The capabilities facilitate precision agriculture, resource optimization, and minimize production risks in the face of rising levels of climate variability. The future of AI and agricultural monitoring and food security is that it should provide real-time monitoring, dependable yield forecasting, and decision support on a farm, regional, and national scale, which will enhance food supply planning and policy development. Nevertheless, achieving these advantages on a large scale is only possible through a sustainable and inclusive strategy. The next course of action includes enhancing the level of data access and quality, building explainable and geo-specific AI models, reinforcing digital infrastructure, and building capacity between farmers and institutions. Other supportive policies, partnerships between the government and the business, and ethical data management are also necessary to have fair adoption. Through responsible applications, AI could be of great use in building resilient agricultural systems and global food security in the long-term perspective.

REFERENCES

Agrawal, J., & Arafat, M. (2024). Transforming Farming: A Review of AI-Powered UAV Technologies in Precision Agriculture. *Drones (Basel)*, *8*(11), 664. Advance online publication. DOI: 10.3390/drones8110664

Akkem, Y., Biswas, S., & Varanasi, A. (2023). Smart farming using artificial intelligence: A review. *Engineering Applications of Artificial Intelligence*, *120*, 105899. DOI: 10.1016/j.engappai.2023.105899

Alahmad, T., Neményi, M., & Nyéki, A. (2023). Applying IoT Sensors and Big Data to Improve Precision Crop Production: A Review. *Agronomy (Basel)*, *13*(10), 2603. Advance online publication. DOI: 10.3390/agronomy13102603

Amsalu, A., Garedew, W., Melka, G. A., Çelik, M. A., & Gemeda, D. O. (2026). Trends in climate extreme indices and implications for coffee farming in The Kafa Biosphere Reserve, southwest Ethiopia. *Climate Services*, *42*, 100637. DOI: 10.1016/j.cliser.2026.100637

Araújo, S., Peres, R., Barata, J., Lidon, F., & Ramalho, J. (2021). Characterising the Agriculture 4.0 Landscape—Emerging Trends, Challenges and Opportunities. *Agronomy (Basel)*, *11*(4), 667. DOI: 10.3390/agronomy11040667

Aslan, M., Sabancı, K., & Aslan, B. (2024). Artificial Intelligence Techniques in Crop Yield Estimation Based on Sentinel-2 Data: A Comprehensive Survey. *Sustainability (Basel)*, *16*(18), 8277. Advance online publication. DOI: 10.3390/su16188277

Awais, M., Zaigham, S., Naqvi, A., Zhang, H., Li, L., Zhang, W., Awwad, F., Ismail, E., Khan, M., Raghavan, V., & Hu, J. (2023). AI and machine learning for soil analysis: An assessment of sustainable agricultural practices. *Bioresources and Bioprocessing*, *10*(1), 90. Advance online publication. DOI: 10.1186/s40643-023-00710-y PMID: 38647622

Barbedo, J. (2025). A Review of Artificial Intelligence Techniques for Wheat Crop Monitoring and Management. *Agronomy (Basel)*, *15*(5), 1157. Advance online publication. DOI: 10.3390/agronomy15051157

Benos, L., Tagarakis, A., Dolias, G., Berruto, R., Kateris, D., & Bochtis, D. (2021). Machine Learning in Agriculture: A Comprehensive Updated Review. *Sensors (Basel)*, *21*(11), 3758. Advance online publication. DOI: 10.3390/s21113758 PMID: 34071553

Biazar, S., Golmohammadi, G., Nedhunuri, R., Shaghaghi, S., & Mohammadi, K. (2025). Artificial Intelligence in Hydrology: Advancements in Soil, Water Resource Management, and Sustainable Development. *Sustainability (Basel), 17*(5), 2250. Advance online publication. DOI: 10.3390/su17052250

Birgani, R., Takian, A., Djazayery, A., Kianirad, A., & Pouraram, H. (2022). Climate Change and Food Security Prioritizing Indices: Applying Analytical Hierarchy Process (AHP) and Social Network Analysis (SNA). *Sustainability (Basel), 14*(14), 8494. Advance online publication. DOI: 10.3390/su14148494

Bouanani, N., Laamrani, A., Hajji, H., Bourriz, M., Bourzeix, F., Abdelali, H., El-Battay, A., Amazirh, A., & Chehbouni, A. (2025). Estimating Soil Attributes for Yield Gap Reduction in Africa Using Hyperspectral Remote Sensing Data with Artificial Intelligence Methods: An Extensive Review and Synthesis. *Remote Sensing (Basel), 17*(9), 1597. Advance online publication. DOI: 10.3390/rs17091597

Carneiro, B., Resce, G., Caravaggio, N., Santangelo, A., Ruscica, G., Tucci, G., Pacillo, G., Eilerts, G., Läderach, P., & Coffey, K. (2025). Text mining and machine learning reveal global determinants of food insecurity. *Scientific Reports, 15*(1), 36709. Advance online publication. DOI: 10.1038/s41598-025-20670-x PMID: 41120495

Çelik, M. A. (2026). Agro-climatic classification of Türkiye's Agricultural Basins using long-term ERA5-Land temperature data. *Physics and Chemistry of the Earth Parts A/B/C, 142*, 104307. DOI: 10.1016/j.pce.2026.104307

Çelik, M. A., Kıraç, A., & Gouveia, C. M. (2026). North Atlantic Oscillation modulation of drought and vegetation activity in Türkiye. *Theoretical and Applied Climatology, 157*(2), 74. Advance online publication. DOI: 10.1007/s00704-025-05971-5

Cho, S., Soleh, H., Choi, J., Hwang, W., Lee, H., Cho, Y., Cho, B., Kim, M., Baek, I., & Kim, G. (2024). Recent Methods for Evaluating Crop Water Stress Using AI Techniques: A Review. *Sensors (Basel), 24*(19), 6313. Advance online publication. DOI: 10.3390/s24196313 PMID: 39409355

Choi, J., Hidayat, M., Cho, S., Hwang, W., Lee, H., Cho, B., Kim, M., Baek, I., & Kim, G. (2025). Recent Trends in Machine Learning, Deep Learning, Ensemble Learning, and Explainable Artificial Intelligence Techniques for Evaluating Crop Yields Under Abnormal Climate Conditions. *Plants, 14*(18), 2841. Advance online publication. DOI: 10.3390/plants14182841 PMID: 41011993

Chukwuma, E., Afolabi, O., Okonkwo, C., Olamigoke, O., & Okonkwo, C. (2024). Application of geospatial technology and decision model in the development of an improved food security index. *Scientific Reports, 14*(1), 30204. Advance online publication. DOI: 10.1038/s41598-024-77845-1 PMID: 39632976

Danilevicz, M., Upadhyaya, S., Batley, J., Bennamoun, M., Bayer, P., & Edwards, D. (2025). Understanding plant phenotypes in crop breeding through explainable AI. *Plant Biotechnology Journal, 23*(10), 4200–4213. DOI: 10.1111/pbi.70208 PMID: 40574309

Dhal, S., & Kar, D. (2024). Transforming Agricultural Productivity with AI-Driven Forecasting: Innovations in Food Security and Supply Chain Optimization. *Forecasting, 6*(4), 925–951. Advance online publication. DOI: 10.3390/forecast6040046

Dhanya, V., Subeesh, A., Kushwaha, N., Vishwakarma, D., Kumar, N., Ritika, G., & Singh, A. (2022). *Deep learning based computer vision approaches for smart agricultural applications.* Artificial Intelligence in Agriculture., DOI: 10.1016/j.aiia.2022.09.007

Dutta, S., Chakraborty, S., Goswami, R., Banerjee, H., Majumdar, K., Li, B., & Jat, M. (2020). Maize yield in smallholder agriculture system—An approach integrating socio-economic and crop management factors. *PLoS One, 15*(2), e0229100. Advance online publication. DOI: 10.1371/journal.pone.0229100 PMID: 32092077

Elbasi, E., Mostafa, N., AlArnaout, Z., Zreikat, A., Cina, E., Varghese, G., Shdefat, A., Topcu, A., Abdelbaki, W., Mathew, S., & Zaki, C. (2023). Artificial Intelligence Technology in the Agricultural Sector: A Systematic Literature Review. *IEEE Access : Practical Innovations, Open Solutions, 11*, 171–202. DOI: 10.1109/ACCESS.2022.3232485

Espinel, R., Herrera-Franco, G., García, J., & Escandón-Panchana, P. (2024). Artificial Intelligence in Agricultural Mapping. *Agriculture, 14*(7), 1071. Advance online publication. DOI: 10.3390/agriculture14071071

Gao, Z., Luo, Z., Zhang, W., Lv, Z., & Xu, Y. (2020). Deep Learning Application in Plant Stress Imaging: A Review. *AgriEngineering*. https://doi.org/DOI: 10.3390/agriengineering2030029

Gavasso-Rita, Y., Papalexiou, S., Li, Y., Elshorbagy, A., Li, Z., & Schuster-Wallace, C. (2023). Crop models and their use in assessing crop production and food security: A review. *Food and Energy Security*. Advance online publication. DOI: 10.1002/fes3.503

Gebresenbet, G., Bosona, T., Patterson, D., Persson, H., Fischer, B., Mandaluniz, N., Chirici, G., Zacepins, A., Komašilovs, V., Pitulac, T., & Nasirahmadi, A. (2023). A concept for the application of integrated digital technologies to enhance future smart agricultural systems. *Smart Agricultural Technology, 5*, 100255. Advance online publication. DOI: 10.1016/j.atech.2023.100255

Gund, R., Badgujar, C., Samiappan, S., & Jagadamma, S. (2025). Application of Digital Twin Technology in Smart Agriculture: A Bibliometric Review. *Agriculture*, *15*(17), 1799. Advance online publication. DOI: 10.3390/agriculture15171799

Hoseinzadeh, S., & Garcia, D. (2024). Ai-driven innovations in greenhouse agriculture: Reanalysis of sustainability and energy efficiency impacts. *Energy Conversion and Management*, *X*, 100701. Advance online publication. DOI: 10.1016/j.ecmx.2024.100701

Jafar, A., Bibi, N., Naqvi, R., Sadeghi-Niaraki, A., & Jeong, D. (2024). Revolutionizing agriculture with artificial intelligence: Plant disease detection methods, applications, and their limitations. *Frontiers in Plant Science*, *15*, 1356260. Advance online publication. DOI: 10.3389/fpls.2024.1356260 PMID: 38545388

Kaya, C. (2025). Intelligent Environmental Control in Plant Factories: Integrating Sensors, Automation, and AI for Optimal Crop Production. *Food and Energy Security*, *14*(1), e70026. Advance online publication. DOI: 10.1002/fes3.70026

Kikon, A., & Deka, P. (2021). Artificial intelligence application in drought assessment, monitoring, and forecasting: A review. *Stochastic Environmental Research and Risk Assessment*, *36*(5), 1197–1214. DOI: 10.1007/s00477-021-02129-3

Luo, L., Sun, S., Xue, J., Gao, Z., Zhao, J., Yin, Y., Gao, F., & Luan, X. (2023). Crop yield estimation based on assimilation of crop models and remote sensing data: A systematic evaluation. *Agricultural Systems*, *210*, 103711. Advance online publication. DOI: 10.1016/j.agsy.2023.103711

MacPherson, J., Voglhuber-Slavinsky, A., Olbrisch, M., Schöbel, P., Dönitz, E., Mouratiadou, I., & Helming, K. (2022). Future agricultural systems and the role of digitalization for achieving sustainability goals. A review. *Agronomy for Sustainable Development*, *42*(4), 70. Advance online publication. DOI: 10.1007/s13593-022-00792-6 PMID: 35818482

Maestrini, B., Mimić, G., Van Oort, P., Jindo, K., Brdar, S., Athanasiados, I., & Van Evert, F. (2022). Mixing process-based and data-driven approaches in yield prediction. *European Journal of Agronomy*, *139*, 126569. Advance online publication. DOI: 10.1016/j.eja.2022.126569

Mălina , C., Matei, F., Pop, I., Sălăgean, T., & Mălina , A. (2025). *Individual and Synergistic Contributions of GIS, Remote Sensing, and AI in Advancing Climate-Resilient Agriculture.* AgriEngineering., DOI: 10.3390/agriengineering7070230

Mansoor, S., Iqbal, S., Popescu, S., Kim, S., Chung, Y., & Baek, J. (2025). Integration of smart sensors and IOT in precision agriculture: Trends, challenges, and future prospects. *Frontiers in Plant Science*, *16*, 1587869. Advance online publication. DOI: 10.3389/fpls.2025.1587869 PMID: 40438737

Marinello, F., Zou, X., Liu, Z., Zhu, X., Zhang, W., Qian, Y., Li, Y., Karunathilake, E., Le, A., Heo, S., Chung, Y., & Mansoor, S. (2023). The Path to Smart Farming: Innovations and Opportunities in Precision Agriculture. *Agriculture*, *13*(8), 1593. Advance online publication. DOI: 10.3390/agriculture13081593

Materia, S., García, L., Van Straaten, C. O. S., Mamalakis, A., Cavicchia, L., Coumou, D., De Luca, P., Kretschmer, M., & Donat, M. (2024). Artificial intelligence for climate prediction of extremes: State of the art, challenges, and future perspectives. *Wiley Interdisciplinary Reviews: Climate Change*, *15*(6), e914. Advance online publication. DOI: 10.1002/wcc.914

Mohr, S., & Kühl, R. (2021). Acceptance of artificial intelligence in German agriculture: An application of the technology acceptance model and the theory of planned behavior. *Precision Agriculture*, *22*(6), 1816–1844. DOI: 10.1007/s11119-021-09814-x

Mohyuddin, G., Khan, M., Haseeb, A., Mahpara, S., Waseem, M., & Saleh, A. (2024). Evaluation of Machine Learning Approaches for Precision Farming in Smart Agriculture Systems: A Comprehensive Review. *IEEE Access : Practical Innovations, Open Solutions*, *12*, 60155–60184. DOI: 10.1109/ACCESS.2024.3390581

Morchid, A., Alami, R., Raezah, A., & Sabbar, Y. (2023). Applications of Internet of Things (IoT) and sensor technology to increase food security and agricultural Sustainability: Benefits and challenges. *Ain Shams Engineering Journal*. Advance online publication. DOI: 10.1016/j.asej.2023.102509

Muhammad, A., Khan, Z., Khan, J., Mashori, A., Ali, A., Jabeen, N., Han, Z., & Li, F. (2025). A comprehensive review of crop stress detection: Destructive, non-destructive, and ML-based approaches. *Frontiers in Plant Science*, *16*, 1638675. Advance online publication. DOI: 10.3389/fpls.2025.1638675 PMID: 40978773

Ojo, M., & Zahid, A. (2022). Deep Learning in Controlled Environment Agriculture: A Review of Recent Advancements, Challenges, and Prospects. *Sensors (Basel)*, *22*(20), 7965. Advance online publication. DOI: 10.3390/s22207965 PMID: 36298316

Oliveira, L., Moreira, A., & Silva, M. (2021). Advances in Agriculture Robotics: A State-of-the-Art Review and Challenges Ahead. *Robotics (Basel, Switzerland)*, *10*(2), 52. DOI: 10.3390/robotics10020052

Padhiary, M., Saha, D., Kumar, R., Sethi, L., & Kumar, A. (2024). Enhancing Precision Agriculture: A Comprehensive Review of Machine Learning and AI Vision Applications in All-Terrain Vehicles for Farm Automation. *Smart Agricultural Technology*, *8*, 100483. Advance online publication. DOI: 10.1016/j.atech.2024.100483

Petraki, D., Gazoulis, I., Kokkini, M., Danaskos, M., Kanatas, P., Rekkas, A., & Travlos, I. (2025). Digital Tools and Decision Support Systems in Agroecology: Benefits, Challenges, and Practical Implementations. *Agronomy (Basel)*, *15*(1), 236. Advance online publication. DOI: 10.3390/agronomy15010236

Piekutowska, M., & Niedbała, G. (2025). Review of Methods and Models for Potato Yield Prediction. *Agriculture*, *15*(4), 367. Advance online publication. DOI: 10.3390/agriculture15040367

Prodhan, F., Zhang, J., Hasan, S., Sharma, T., & Mohana, H. (2022). A review of machine learning methods for drought hazard monitoring and forecasting: Current research trends, challenges, and future research directions. *Environmental Modelling & Software*, *149*, 105327. DOI: 10.1016/j.envsoft.2022.105327

Ramaiah, M., Settu, P., & Ravi, V. (2025). Artificial Intelligence Techniques Enabled Soil Moisture Estimation Frameworks Using Remote Sensing Satellite Images: Challenges and Future Directions- Review. *Wiley Interdisciplinary Reviews. Data Mining and Knowledge Discovery*, *15*(3), e70032. Advance online publication. DOI: 10.1002/widm.70032

Saha, S., Kucher, O., Utkina, A., & Rebouh, N. (2025). Precision agriculture for improving crop yield predictions: A literature review. *Frontiers in Agronomy*, *7*, 1566201. Advance online publication. DOI: 10.3389/fagro.2025.1566201

Saleem, M., Potgieter, J., & Arif, K. (2021). Automation in Agriculture by Machine and Deep Learning Techniques: A Review of Recent Developments. *Precision Agriculture*, *22*(6), 2053–2091. DOI: 10.1007/s11119-021-09806-x

Sarku, R., Clemen, U., & Clemen, T. (2023). The Application of Artificial Intelligence Models for Food Security. *Agriculture*, *13*(10), 2037. Advance online publication. DOI: 10.3390/agriculture13102037

Šestak, M., & Copot, D. (2023). Towards Trusted Data Sharing and Exchange in Agro-Food Supply Chains: Design Principles for Agricultural Data Spaces. *Sustainability (Basel)*, *15*(18), 13746. Advance online publication. DOI: 10.3390/su151813746

Silva, J., Heerwaarden, J., Reidsma, P., Laborte, A., Tesfaye, K., & Ittersum, M. (2023). Big data, small explanatory and predictive power: Lessons from random forest modeling of on-farm yield variability and implications for data-driven agronomy. *Field Crops Research*, *302*, 109063. Advance online publication. DOI: 10.1016/j.fcr.2023.109063 PMID: 37840838

Sishodia, R., Ray, R., & Singh, S. (2020). Applications of Remote Sensing in Precision Agriculture: A Review. *Remote Sensing (Basel)*, *12*(19), 3136. DOI: 10.3390/rs12193136

Soussi, A., Zero, E., Sacile, R., Trinchero, D., & Fossa, M. (2024). Smart Sensors and Smart Data for Precision Agriculture: A Review. *Sensors (Basel)*, *24*(8), 2647. Advance online publication. DOI: 10.3390/s24082647 PMID: 38676264

Subeesh, A., & Mehta, C. (2021). *Automation and digitization of agriculture using artificial intelligence and the Internet of Things*. Artificial Intelligence in Agriculture., DOI: 10.1016/j.aiia.2021.11.004

Talari, G., Cummins, E., McNamara, C., & O'Brien, J. (2021). State-of-the-art review of Big Data and web-based Decision Support Systems (DSS) for food safety risk assessment with respect to climate change. *Trends in Food Science & Technology*. Advance online publication. DOI: 10.1016/j.tifs.2021.08.032

Tamasiga, P., Ouassou, E., Onyeaka, H., Bakwena, M., Happonen, A., & Molala, M. (2023). Forecasting disruptions in global food value chains to tackle food insecurity: The role of AI and big data analytics – A bibliometric and scientometric analysis. *Journal of Agriculture and Food Research*, *14*, 100819. Advance online publication. DOI: 10.1016/j.jafr.2023.100819

Torre, D., Gao, J., & Macinnis-Ng, C. (2021). Remote sensing-based estimation of rice yields using various models: A critical review. *Geo-Spatial Information Science*, *24*(4), 580–603. DOI: 10.1080/10095020.2021.1936656

Trentin, C., Ampatzidis, Y., Lacerda, C., & Shiratsuchi, L. (2024). Tree Crop Yield Estimation and Prediction Using Remote Sensing and Machine Learning: A Systematic Review. *Smart Agricultural Technology*, *9*, 100556. Advance online publication. DOI: 10.1016/j.atech.2024.100556

Ukoba, K., Onisuru, O., Jen, T., Madyira, D., & Olatunji, K. (2025). Predictive modeling of climate change impacts using Artificial Intelligence: A review for equitable governance and sustainable outcomes. *Environmental Science and Pollution Research International*, *32*(17), 10705–10724. DOI: 10.1007/s11356-025-36356-w PMID: 40183866

Upadhyay, A., Chandel, N., Singh, K., Chakraborty, S., Nandede, B., Kumar, M., Subeesh, A., Upendar, K., Salem, A., & Elbeltagi, A. (2025). Deep learning and computer vision in plant disease detection: A comprehensive review of techniques, models, and trends in precision agriculture. *Artificial Intelligence Review*, *58*(3), 92. Advance online publication. DOI: 10.1007/s10462-024-11100-x

Usigbe, M., Asem-Hiablie, S., Uyeh, D., Iyiola, O., Park, T., & Mallipeddi, R. (2023). Enhancing resilience in agricultural production systems with AI-based technologies. *Environment, Development and Sustainability*, *26*(9), 1–29. DOI: 10.1007/s10668-023-03588-0

Villa-Henriksen, A., Edwards, G., Pesonen, L., Green, O., & Sørensen, C. (2020). Internet of Things in arable farming: Implementation, applications, challenges, and potential. *Biosystems Engineering*, *191*, 60–84. DOI: 10.1016/j.biosystemseng.2019.12.013

Wang, D., Cao, W., Zhang, F., Li, Z., Xu, S., & Wu, X. (2022). A Review of Deep Learning in Multiscale Agricultural Sensing. *Remote Sensing (Basel)*, *14*(3), 559. DOI: 10.3390/rs14030559

Weiss, M., Jacob, F., & Duveiller, G. (2020). Remote sensing for agricultural applications: A meta-review. *Remote Sensing of Environment*, *236*, 111402. Advance online publication. DOI: 10.1016/j.rse.2019.111402

Yuan, J., Zhang, Y., Zheng, Z., Yao, W., Wang, W., & Guo, L. (2024). Grain Crop Yield Prediction Using Machine Learning Based on UAV Remote Sensing: A Systematic Literature Review. *Drones (Basel)*, *8*(10), 559. Advance online publication. DOI: 10.3390/drones8100559

Zhang, C., Song, J., Yin, X., Cai, J., Liang, Y., & Lu, J. (2025). Digital twin-based approaches for agricultural machinery damage prediction and maintenance: A review. *Journal of Computational Design and Engineering*, *12*(10), 87–117. DOI: 10.1093/jcde/qwaf097

Zhang, R., Zhu, H., Chang, Q., & Mao, Q. (2025). A Comprehensive Review of Digital Twins Technology in Agriculture. *Agriculture*, *15*(9), 903. Advance online publication. DOI: 10.3390/agriculture15090903

Chapter 8
A Framework for Robust and Interpretable Yield-Risk Classification Using Country-Year-Crop Data (X-AMLF)

Sk Muktadir Hossain
http://orcid.org/0009-0006-9073-9264
American International University-Bangladesh, Bangladesh

Md Sadiqur Rahman
http://orcid.org/0009-0002-4476-7564
American International University-Bangladesh, Bangladesh

Sharif Eime Akhter
http://orcid.org/0009-0005-2310-8430
American International University-Bangladesh, Bangladesh

Dipta Gomes
http://orcid.org/0000-0003-3019-6051
American International University-Bangladesh, Bangladesh

Ayswarjo Sarkar
http://orcid.org/0009-0000-2621-2130
American International University-Bangladesh, Bangladesh

Mahfujur Rahman
http://orcid.org/0000-0002-3245-8767
American International University-Bangladesh, Bangladesh

Kazi Samia Mostofa
http://orcid.org/0009-0007-9275-4991
American International University-Bangladesh, Bangladesh

Kazi Tanvir
http://orcid.org/0009-0001-0612-6101
American International University-Bangladesh, Bangladesh

DOI: 10.4018/979-8-2600-0778-5.ch008

ABSTRACT

This study introduces the Explainable Agro-Intelligence Machine Learning Framework (X-AMLF), which is a structured evaluation framework for yield risk classification in agriculture using aggregated country-year-crop data. A harmonised dataset was created by combining data on crop yield, annual average temperature, rainfall, pesticide use, crop identity, and country indicators to create 28,242 complete observations on the yield of 101 countries (1990–2016) for 10 major crops. Multiple classification models were evaluated, such as linear baselines and tree-based ensembles, and hyperparameters were optimised with Optuna to investigate variations in prediction and error behaviour. Under random splitting, ensemble models were able to get very high scores (e.g. Random forest F1 ≈ 0.976), but under a stricter evaluation, performance dropped. With country-holdout validation, the best-performing models had F1 ≈ 0.843, attenuation of performance when tested on unseen regions.

1. INTRODUCTION

The precision agriculture revolution represents a transformative shift in modern farming systems, driven by advances in data availability, computational methods, and artificial intelligence (Rane et al., 2024; Nasir et al., 2025). This transformation is increasingly viewed as essential for achieving global sustainability and food security goals, particularly in rural and climate-vulnerable regions (Çelik, 2026). Despite these technological advances, food insecurity remains a critical challenge: approximately 828 million people experience hunger worldwide, and more than 700 million live in extreme poverty (Akintuyi, 2024; Nasir et al., 2025). Climate change further exacerbates this challenge (Amsalu et al., 2026; Çelik et al., 2026). The IPCC Sixth Assessment Report (AR6) indicates that global production of major staple crops wheat, rice, maize, and soybeans has already declined by approximately 9.5%. Under a 1.5 °C global temperature increase, projected yield reductions include 5.7% for maize, 4.5% for rice, 6.3% for wheat, and 6.7% for soybeans (Liu et al., 2025; Mansoor et al., 2025; Razak et al., 2024). These trends highlight the urgent need for more robust, data-driven approaches to agricultural decision support.

Conventional farming systems typically rely on manual observation, experiential farmer knowledge, and localized agronomic heuristics. While effective in stable environments, these approaches often struggle under increasing climatic variability, resulting in inconsistent production, inefficient pesticide usage, and unstable yields. Such inefficiencies contribute to agroecosystem degradation and biodiversity loss (Logeshwaran et al., 2024; Rane et al., 2024; Liu et al., 2025; Turgut et al., 2024; Mallik et al., 2025). In contrast, data-driven precision agriculture has been widely explored as a means of improving yield prediction and resource-use efficiency by leveraging historical records and sensor-derived data through machine learning and deep learning techniques. Prior studies have reported that integrating these approaches into operational farming systems can contribute to higher yields, improved water-use efficiency, and reduced chemical inputs (Akintuyi, 2024; Rane et al., 2024; Cambrin et al., 2024; Mansoor et al., 2025; Mallik et al., 2025; Nasir et al., 2025).

Deep learning models including convolutional neural networks, recurrent neural networks, and ensemble-based architectures have been extensively studied for agricultural tasks such as yield forecasting, crop monitoring, and resource optimization (Logeshwaran et al., 2024; Mallik et al., 2025; Mansoor et al., 2025; Nasir et al., 2025; Tariq et al., 2024; Mohan et al., 2025). Despite their strong predictive performance, these models are often characterized as black-box systems because of the difficulty in interpreting and validating their internal decision-making processes (Ryo, 2022; Chen et al., 2023; Mohan et al., 2025). This lack of transparency limits trust, interpretability, and adoption among farmers and practitioners, consequently constraining real-world deployment and reducing the practical impact of otherwise accurate predictive models (Razak et al., 2024; Tariq et al., 2024; Rane et al., 2024; Turgut et al., 2024).

Explainable Artificial Intelligence (XAI) has emerged as a critical approach for enhancing the transparency and reliability of machine learning systems by providing human-interpretable explanations for model predictions. XAI methods enable stakeholders to understand how input features influence predictions, identify potential biases in training data, and interpret model behavior at both global and local levels (Chen et al., 2023; Mohan et al., 2025; Nasir et al., 2025). Broadly, XAI operates through two paradigms: (i) inherently interpretable models, such as decision trees or rule-based systems, that provide transparent decision logic; and (ii) post-hoc explanation methods, such as SHAP and LIME, which explain predictions from high-performance, yet opaque, models after training (Chen et al., 2023; Laktionov et al., 2023).

Although explainable AI has gained traction in agricultural research, significant gaps remain. Many studies prioritize predictive accuracy using random train–test splits, which can overestimate real-world performance when data exhibit geographic or crop-specific dependencies (Logeshwaran et al., 2024; Mallik et al., 2025; Mohan et al., 2025). Furthermore, explanations are frequently treated as secondary visual aids rather than tools for diagnosing generalization failures or error behavior. Much of the XAI literature also focuses on domain-agnostic benchmarks, with limited adaptation to the unique characteristics of agriculture, including heterogeneous agro-climatic features, nonlinear crop–climate interactions, and the need for explanations comprehensible to non-technical stakeholders in data-sparse regions (Cambrin et al., 2024; Cartolano et al., 2024; Mansoor et al., 2025).

To address these challenges, this study introduces the Explainable Agro-Intelligence Machine Learning Framework (X-AMLF), an evaluation framework designed for agricultural machine learning utilizing aggregated, multisource agro-climatic data rather than real-time sensor or control systems. X-AMLF integrates country–year–crop-level information, including historical crop yield, average temperature, rainfall, pesticide usage, crop identity, and geographic indicators, to support yield-risk classification and systematic model assessment. Through structured data preprocessing and comparative evaluation, X-AMLF enables rigorous testing of multiple machine learning models under realistic validation scenarios, incorporating geographic, crop-based, and temporal holdout strategies.

To enhance transparency, X-AMLF incorporates model-agnostic explainability techniques, such as SHAP and LIME, providing both global and local interpretations of model predictions. These explanations reveal how climatic variables and crop characteristics influence yield-risk outcomes and aid in diagnosing model performance–specifically, why models succeed or fail when transferred across regions, crops, or time periods. By combining explainability with deployment-oriented validation, X-AMLF addresses a critical gap in agricultural machine learning research and offers a more realistic basis for assessing the suitability of yield prediction models for decision-support applications under climate variability. The conceptual workflow of X-AMLF is illustrated in Figure 1. The primary contributions of this study are as follows:

• Explainable Evaluation Framework for Agricultural ML

This study introduces the Explainable Agro-Intelligence Machine Learning Framework (X-AMLF), a systematic evaluation framework that combines machine learning, explainable AI techniques, and deployment-oriented validation protocols to support transparent and reliable agricultural yield prediction using aggregated agro-climatic data.

• Yield-Based Risk Prediction with Human-Interpretable Explanations

This study provides an accurate classification of yield outcomes with model-agnostic explanations generated using SHAP and LIME, enabling stakeholders to understand how climate variables, crop type, and management factors influence predictions.

• Multi-Feature Agro-Climatic Data Integration

We construct and analyze a harmonized country–year–crop dataset that integrates crop yield, average temperature, rainfall, and pesticide usage across multiple crops and regions, demonstrating how heterogeneous agro-climatic features can be effectively modeled using ensemble learning methods.

• Rigorous Generalization Analysis under Realistic Deployment Scenarios

Beyond standard random train–test splitting, we evaluate model robustness using country-holdout, crop-holdout, and temporal validation strategies. This analysis reveals the limitations of overly optimistic performance estimates and provides a realistic assessment of model transferability across regions, crops, and time.

• Error-Sensitive Performance Assessment for Agricultural Decision Support

The framework emphasizes false-negative and false-positive behavior alongside traditional accuracy metrics, highlighting the importance of error profiles in yield-risk prediction tasks, wherein missed low-yield events can have serious agronomic and economic consequences.

• Explainability-Driven Insight into Model Limitations and Transferability

Through global and local explanation analysis, we show that crop identity is the dominant driver of model predictions, explaining both strong performance under random splits and severe degradation under crop-holdout validation. This insight demonstrates how explainable AI can be used not only for transparency but also to diagnose generalization failure in agricultural machine learning systems.

The remainder of this paper is organized as follows. Section 2 "Literature Review" examines prior research on AI and precision agriculture. Section 3 "Methodology" describes the diverse roles of the proposed framework and the machine learning algorithms. The "Results" Section 4 displays the comparative performance outcomes. Section 5 "Discussion" analyzes the implications and practical applications of the findings. In Section 6, discussed "Future Work" that outlines directions for framework enhancement and broader implementation. The Section 7 "Conclusion" encompasses the fusion of findings and recommendations.

Figure 1. Conceptual framework of the X-AMLF

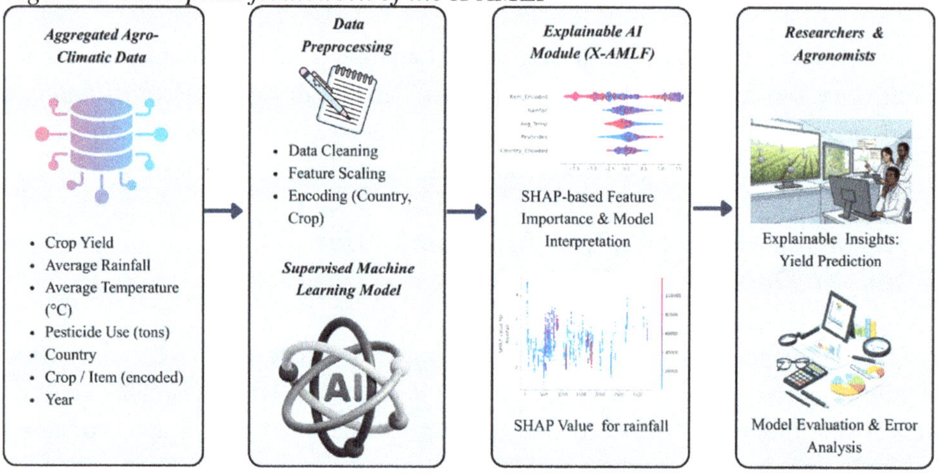

2. LITERATURE REVIEW

The use of artificial intelligence (AI) and machine learning (ML) in agriculture has led to advances in predictive and optimization solutions. However, AI models remain to be interpreted, as many models act as "black boxes" that require trust and acceptance among farmers. Efforts to increase transparency through explainable AI frameworks aim to provide predictive models with interpretable properties. This literature reviews the evolution of AI use in agriculture, focusing on the description of specific interpretive machine learning models that will bridge the gap between advanced predictions and practical agricultural adoption.

(Tahir et al., 2025) proposed an Explainable AI Framework (EPAF) for smart agriculture are analyzed using Learning and Light XAI technologies (SHAP, LIME, Grad-CAM). The framework achieved an accuracy of 85.2%. However, the limited coverage of crops and the complexity of the environment with limited resources are some of the limitations of this approach. (Martin et al., 2024) proposed an XAI-based smart agriculture framework, which combines XLNet-based feature extraction and optimization with various machine learning models for crop yield prediction. The XLNet+SVM model provided high accuracy with interpretable results using SHAP and LIME. However, this framework faces problems of high computational cost, reliance on datasets, and limited applicability in different climatic conditions. (Cartolano et al., 2024) developed an (XAI) model to enhance the transparency and reliability of AI-powered systems in smart agricultural environments. This framework used LIME and SHAP XAI techniques. These explainable models remarkably increased trust and transparency in AI systems while maintaining predictive accuracy. However, this approach faces challenges such as poor data quality, interpretation complexity, and the existence of different agricultural environments. (Shams et al., 2024) proposed an explainable AI-based crop recommendation system (XAI-CROP), where Decision Tree, Random Forest, Gradient Boosting, and Gaussian multi-feature Naive Bayes models are integrated with LIME. This framework showed high accuracy. However, it primarily relied on historical agricultural data. (Mohan et al., 2025) introduced an AI and XAI-based framework that aims for accurate and interpretable crop yield prediction. Researchers used decision tree regressor, random forest regressor, lightGBM regressor, SHAP, and LIME. The Random Forest Regressor achieved the best performance, with an $R^2 \approx 0.80$. But this study faces several limitations, such as limited data quality, lack of socioeconomic factors, and generalizability. Based on the Bayesian ensemble model (BM), (Hu et al., 2023) created an interpretable machine learning framework to examine the risks associated with the black-box model and the impact of climate change on crop production. The proposed framework was systematically compared with support vector machines, random forests, and XGBoost. The results showed that the BM achieved equivalent or greater predictive accuracy and significantly greater interpretability. However, the Bayesian method has additional costs and uncertainty in predicting crop yields in new climates.

In this research (Logeshwaran et al., 2024), the authors proposed an Agro Framework to enhance the effectiveness of precision agriculture. (El-Kenawy et al., 2025) proposed a machine learning approach for forecasting potato crops. This system combines KNN, Gradient Boosting, XGBoost. Despite the challenges, it provided higher accuracy compared to traditional methods and is helpful for resource management and food security for farmers.

(Sharma et al., 2024) proposed an AI-powered precision agriculture system that combines machine learning to monitor crop and maximize crop yield. Random forest regression model was used to predict crop yield. Mean absolute error (MAE) = 50.27 and root mean squared error (RMSE) = 59.35 were used to evaluate yield. Limitations here include data mismatch, high cost, lack of technical expertise, and poor rural connectivity. In (Qin et al., 2025), the authors proposed a two-step machine learning framework to predict the effects of mechanical soil compaction on soybean crop yields. In this method, MLR, XGBoost, and ANN algorithms were used to change production. The hybrid XGBoost-ANN model showed the best performance, with R2 = 0.889, MAE = 1.47, and RMSE = 1.964. However, this study is limited to soybean crops and field conditions in specific regions. (Zou et al., 2025) proposed a model that uses XGBoost and quantile regression to accurately predict crop yield and also takes into account the aspect of ignorance. When tested using data from a northern city, the model showed good accuracy. However, this model is dependent on local data, which limits its widespread use. In the study (Kumar et al., 2025), analyzed how machine learning models can be helpful in sustainable agriculture and help predict crop yields. However, the model is based on historical data.

(Singh et al., 2025) introduced the Transformative Crop Recommendation Model (TCRM) to enhance precision agriculture using machine learning and cloud computing. TCRM used logistic regression, K-nearest neighbors (KNN), and random forests algorithms. The model achieved 94% accuracy, 94.46% precision, 94% recall, and an F1-score of 93.97%. However, it faces challenges in rural availability. (Begum et al., 2024) proposed a machine learning-based crop recommendation system for the development of precision agriculture. In this study, two machine learning algorithms were used: KNN and RF. The results showed that Random Forest performed best, with an accuracy of approximately 96%, whereas KNN had an accuracy of 78%. However, this system primarily depends on historical data and does not use real-time sensor data. Therefore, its effectiveness may be limited in changing the field conditions.

(Lapajne et al., 2025) proposed a framework for potato crop analysis using XGBoost and explainable AI, with explainability ensured through SHAP. This framework achieved 97.10% accuracy in potato variety identification, an R^2 value of 0.57 for physiological trait prediction. However, the performance slightly decreased for spectrally similar varieties.

(Alsanoosy et al., 2025) developed a framework that integrates SAM-L (Self-Adaptive Meta-Learning) and XAI (Explainable Artificial Intelligence) to predict plant yield-risk. A three-layer LSTM network that can analyze sequential data and adjust to changing environmental conditions is part of the model design. The proposed model achieved 89.17% accuracy, 88.39% macro precision, 88.56% macro recall and 88.31% macro F1-score, demonstrating outstanding ability to moderate, and high yield-risk. In a study (Syari et al., 2025), proposed a smart farming with machine learning. Here used Random Forest and Linear Regression models. In a 30-day trial, water consumption and harvest time was reduced, and yield increased from 720 g to 850 g. However, there are some limitations to large-scale applications, as the research is limited to short-term trials, single crops, and specific-field environments.

(Ponnurajan et al., 2025) used the Federated Learning (FL) method to predict crop yield while maintaining the privacy of agricultural data. This study compared the performance of multiple machine learning and ensemble models, including Random Forest, SVM, KNN, Gradient Boosting, XGBoost, LightGBM, AdaBoost, and CatBoost. The results showed that XGBoost performed the best, with an accuracy of 93% and an F1-score of 0.92. However, this study was completed on a comparatively small dataset and in a simulated environment. (Petropoulos et al., 2025) developed an interpretive machine learning approach for the application of yield prediction. Here, different models such as Decision Tree, Random Forest, KNN, SVR, MPLR and XGBoost are trained. These models were tested using different features and spectral bands, including NDVI and EVI. Among the results, GBBoost achieved the best performance ($R^2 \approx 0.88$). However, its generalizability may be limited because it is restricted to a specific geographical area and does not include soil-based entities. (Tan et al., 2025) developed an optimized and interpretable random forest framework for predicting rice yield using various features, environmental, and phenological variables. However, this model depends on experimental trial data, and its relevance in actual agricultural fields and future climate scenarios may be limited.

(Fatimah et al., 2024) discussed how Agriculture 5.0 and Explainable AI (xAI) can work together to improve transparency and sustainability in agriculture. They showed how xAI makes complex AI models easier to understand and proposed a framework based on a policy and clear communication. However, the study also highlighted some challenges, such as limited discussion of xAI, difficulties in implementing the technology in resource-poor areas, and lack of sufficient data for different climates and crops. Mohammed et al. (2025) conducted a comprehensive review on recent advancements in precision agriculture, emphasizing that yield prediction remains one of the most extensively studied application areas. Similar findings have been reported in several studies focusing on crop yield prediction using machine learning techniques (Patil et al., 2025; Shawon et al., 2025; Ramesh & Kumaresan, 2025; Shastri et al., 2025). Despite the encouraging performance of these computational models, multiple challenges continue to limit their widespread adoption, including limited research coverage, inadequate infrastructure in developing regions, high implementation costs, data privacy concerns, and insufficient technical training for farmers. (Botero-Valencia et al., 2025) conducted a PRISMA-based systematic literature review on the use of machine learning in sustainable agriculture. The results showed that ensemble models, such as Random Forest and XGBoost, achieved 7-12% higher accuracy in crop yield prediction achieved 90-95% accuracy. Nevertheless, the lack of model interpretability, high computational costs, data quality issues, and limitations in adapting to local agricultural environments.

Table 1. Summary of recent studies on Precision Agriculture Analysis

Reference	Approach	Models	Metrics	Limitations
Kumar et al. (2025)	ML sustainable yield	MLP-ANN, DT	Accuracy ≈ 91%	Historical only, no real-time
Lapajne et al. (2025)	ML for potato	XGBoost; SHAP/ UMAP	Accuracy: 97.10%, F1: 0.826	Performance drop on similar varieties
Petropoulos et al. (2025)	ML for yield	XGBoost etc.	$R^2 \approx 0.88$	Geographic and soil limits
Begum et al. (2024)	ML crop recommendation	KNN, RF	Accuracy: 96% (RF)	Historical data, no real-time sensors
El-Kenawy et al. (2025)	Hybrid ML for potato	KNN, XGBoost, MLP, GNN	Higher than traditional	Data complexity, scalability issues
Botero-Valencia et al. (2025)	ML sustainable review	RF, XGBoost	Yield +7-12%	Interpretability, compute, data issues
Sharma et al. (2024)	AI crop yield	RF Regressor	MAE: 50.27, RMSE: 59.35	Data inconsistencies, high cost, expertise gaps, rural connectivity
Tahir et al. (2025)	Federated XAI	SHAP, LIME, Grad-CAM	Accuracy: 85.2%	Tropical crops limited, low-resource complexity
Martin et al. (2024)	XAI for yield	XLNet+SVM, RF, NN, DT; SHAP/ LIME	High accuracy (XLNet+SVM)	High compute cost, dataset and climate dependency
Cartolano et al. (2024)	XAI on multi-feature data	LIME, SHAP	Accuracy maintained + trust boost	Poor data quality, interpretation complexity, scalability issues
Shams et al. (2024)	XAI-CROP recommendation	DT, RF, GB, Naive Bayes; LIME	$R^2 \approx 0.94$	Historical data reliance, overfitting risk
Mohan et al. (2025)	AI+XAI yield prediction	DT/RF/LightGBM; SHAP/LIME	$R^2 \approx 0.80$	Limited data quality, no socioeconomic factors, generalizability
Alsanoosy et al. (2025)	SAM-L XAI yield-risk prediction	LSTM	Accuracy: 89.17%	Limited dataset
Fatimah et al. (2024)	Agriculture 5.0 + XAI review	Conceptual	--	Limited XAI lit, resource-poor implementation, data gaps
Mohammed et al. (2025)	AI precision ag review	Various	--	Infrastructure, cost, privacy gaps
Zou et al. (2025)	XGBoost-quantile yield uncertainty	XGBoost + Quantile	MAE ≈ 630 kg/ha	Location dependency

continued on following page

Table 1. Continued

Reference	Approach	Models	Metrics	Limitations
Tan et al. (2025)	Interpretable RF rice	RF	$R^2 = 0.77$	Trial data only, limited real-farm relevance
Ponnurajan et al. (2025)	FL yield privacy	XGBoost etc.	Accuracy: 93%	Small and simulated data
Qin et al. (2025)	Two-step soybean compaction	XGBoost-ANN hybrid	R^2: 0.889	Crop and region limited
Hu et al. (2023)	Bayesian ensemble climate risks	Bayesian vs others	Superior interpretability	Extra compute costs, new climate uncertainty
Logeshwaran et al. (2024)	Precision agriculture	Inception	Accuracy: 85.41%, F1-score: 88.91%	Black-box nature, data dependency

2.1. Research Questions

Recent work in precision agriculture reports strong performance for yield and yield-risk prediction, yet important gaps remain regarding generalization, error behavior, and interpretability. Many studies rely on random data splits that may inflate performance, provide feature-importance results without assessing their consistency, and study hyperparameter tuning gains without examining their impact on agriculturally critical errors such as missed yield-risk events. Addressing these limitations, this study investigates the following research questions:

RQ1. To what extent do tree-based ensemble models outperform linear baselines for yield and yield-risk classification using multisource country–year–crop data?
RQ2. Which modeling approach provides the most reliable trade-off between false negatives and false positives under yield-risk-sensitive agricultural decision requirements?
RQ3. How does hyperparameter optimization affect classification performance and error profiles across different models?
RQ4. How robust are model performance and explanatory patterns when evaluated across unseen countries and crops?
RQ5. Do model performance and explanatory attributions remain stable under temporal evaluation using earlier years for training and later years for testing?

These questions link predictive accuracy with interpretability, error sensitivity, and generalization, which are essential for transparent and reliable agricultural decision support.

3. METHODOLOGY

This section provides a comprehensive explanation of the study techniques and their associated workflows. We analyzed the dataset utilized in constructing the applied methods through evaluation, employing various hyperparameters to assess their performance. To offer a thorough understanding of the functionality of the feature engineering approach, a description of the architecture of the proposed approach is also included. Figure 2 illustrates the analytical approach in this study. We used a crop yield prediction dataset to conduct the recommended experiments. Using this dataset, we trained and tested the models through feature selection. The dataset was divided into two parts: 70% for training and 30% for testing. Optuna was used to tune multiple machine learning models to systematically assess how hyperparameter optimization affects classification performance and error behavior across different model families. The performance results of the proposed machine learning model were evaluated and utilized for precision agriculture.

Figure 2. Workflow of the X-AMLF evaluation pipeline

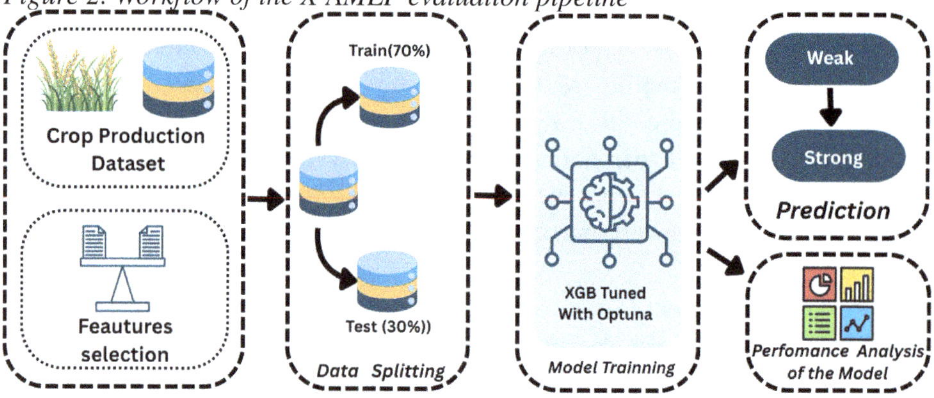

3.1 Datasets Description

Crop production (hg/ha) data for the ten major crops (maize, potato, rice, sorghum, soybean, cassava, sweet potato, wheat, yam, and banana) were collected from the yield.csv dataset file, which contains approximately 280,000 records for 282 countries/regions from 1961 to 2016 described in Table 2. Country-specific annual average temperature (°C) was obtained from Berkeley Earth and the Climate Research Unit temp.csv dataset file, providing approximately 35,000 records from 1843 to 2013. Long-term annual average rainfall (millimeters/year) was collected from the World Bank Climate Change Knowledge Portal and the FAO's climate archive rainfall.csv dataset file; a single climatological average value for each country (1985–2017) was used. Data on pesticide use (total active ingredient tons) were collected from the pesticides.csv dataset file, with approximately 4,500 records for 168 countries from 1990 to 2016.

Table 2. Summary of Dataset Analysis

Dataset	Time Period	Area covered	Features/Variable	Records	Link
yield_df.csv	1990 - 2016	101 countries	-Crop yield (hg/ha) for 10 crops - Average rainfall (mm/year) -Pesticides use (tones) -Average temperature (°C)	28,242	
yield.csv	1961-2016	282 countries	Crop yield (hg/ha) for 10 major crops (Maize, Potatoes, Rice paddy, Sorghum, Soybeans, Cassava, Sweet potatoes, Wheat, Yams, Plantains)	280,000	
rainfall.csv	1985-2017	282 countries	Average annual rainfall (mm/year)	9000	Dataset
temp.csv	1843-2013	200+ countries	Annual average temperature (°C)	35,000	
pesticides.csv	1990-2016	168 countries	Total pesticides use (tones of active ingredients)	4,500	

These four datasets were merged at the country-year-crop level to create the main modeling dataset file, yield_df.csv, which contains 28,242 complete observations for 101 countries between 1990 and 2016. Owing to the limited availability of pesticide information (especially before 1990 and in some regions), the timeframe was restricted to 1990–2016. Incomplete or missing pesticide records were excluded from the study to ensure data quality. Consequently, we obtained a clean, balanced panel dataset that is highly suitable for multi-feature machine learning and explainable artificial intelligence (AI) analyses to forecast crop yields and identify crop vulnerability.

3.2 Features Correlation

As shown in Figure 3, the feature correlation matrix indicates a strong positive correlation between the actual yield and the binary target Yield_High (0.67), which validates the target for the classification task. There was a slight negative correlation with the year (– 0.09) and a slight positive correlation with Yield_High (0.06), and the linear correlations of yield with average rainfall, temperature, and pesticide usage were very weak (maximum = 0.11). This demonstrates that the impact of rainfall and temperature varies nonlinearly and is context-dependent across different climates and crops. Independence is maintained among the one-hot-encoded country and crop variables without any information leakage, and there is no significant multicollinearity among the numerical features (< 0.32). Overall, the weak linear correlations of individual features with yield underscore the necessity of machine learning models to capture complex nonlinear relationships in crop production.

Figure 3. Correlation Heatmap of Datasets

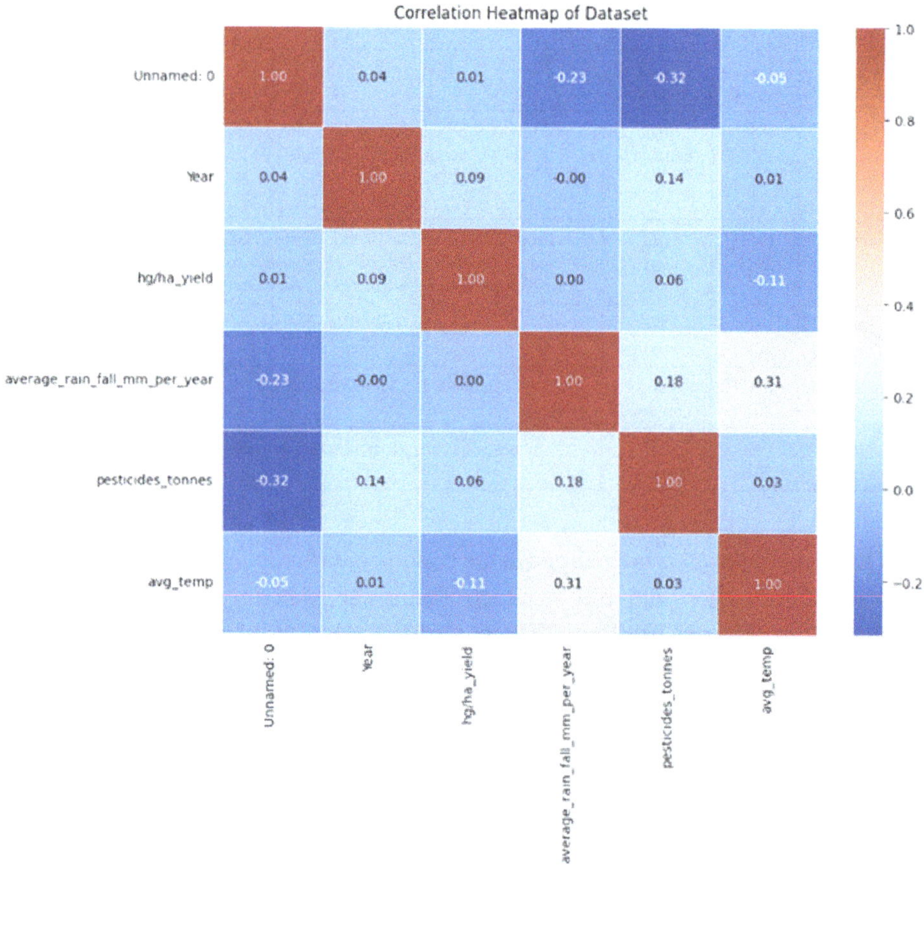

3.3 Data Analysis

The preprocessed dataset (yield_df.csv) used in this study contained 28,242 records covering information on 10 major crops in 101 countries from 1990 to 2013.

This dataset consolidates information on crop production (hg/ha), annual average temperature, country-specific average rainfall, and total pesticide use (tons).

Figure 4 shows the distribution of samples across countries. India had the highest number of records (4000 +), followed by Brazil, Mexico, Pakistan, and Japan. Data from Australia, Indonesia, South Africa, Turkey, and Ecuador were also included; however, there were fewer records. As shown in Figure 5, potatoes had the highest number of records (4,000+), followed by maize, wheat, rice, and soybeans. Yams and plantains had very few records. This disparity reflects the actual production and reporting.

Figure 4. Distribution of 10 Top Countries

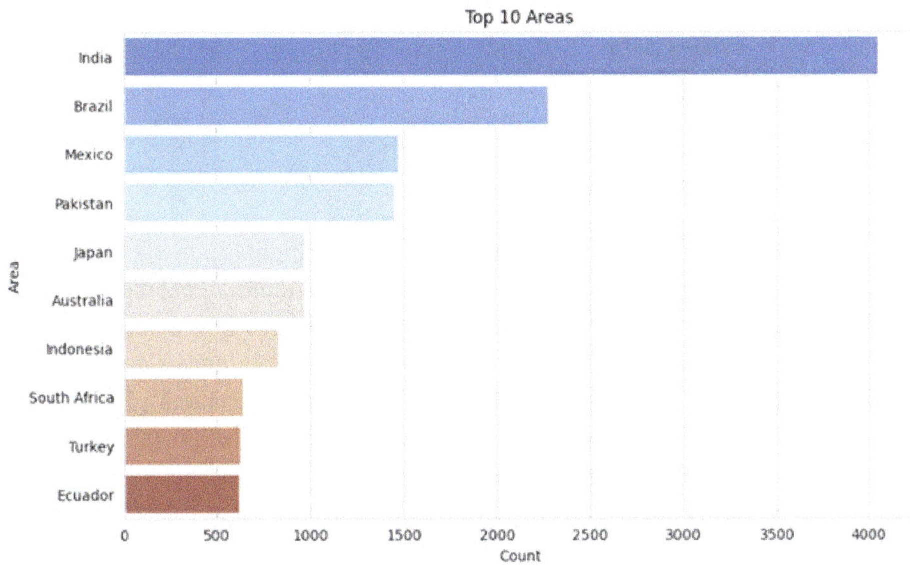

Figure 5. Showing Item Count

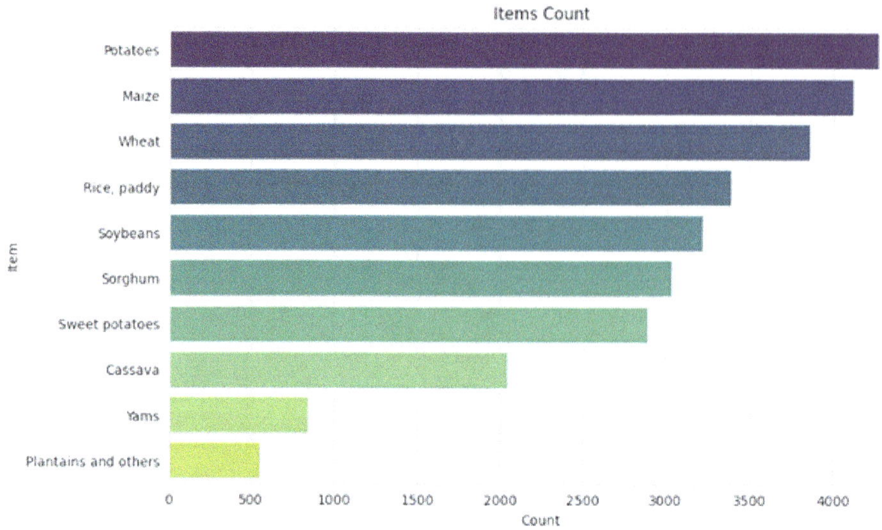

The distribution of the target variable hg/ha_yield is highly right-skewed (Figure 6). Most yields are below 100,000 hg/ha, but there is a long tail extending beyond 500,000 hg/ha, primarily driven by crops such as potatoes in temperate regions. As shown in Figure 7, the distribution of average annual rainfall clearly shows a multifaceted pattern, with approximately 600 mm (dry subtropical), 1,100–1,200 mm (monsoon), 1,600–1,800 mm (wet subtropical), and 2,500+ mm (tropical). This indicates that all types of climate zones are represented in the dataset.

Figure 6. Distribution of the target variable hg/ha_yield

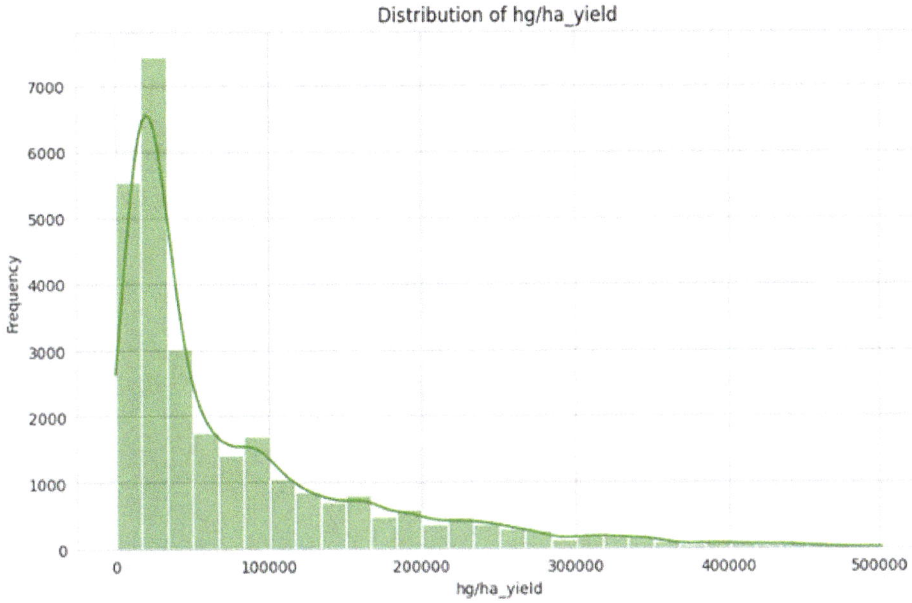

Figure 7. Distribution of average annual rainfall

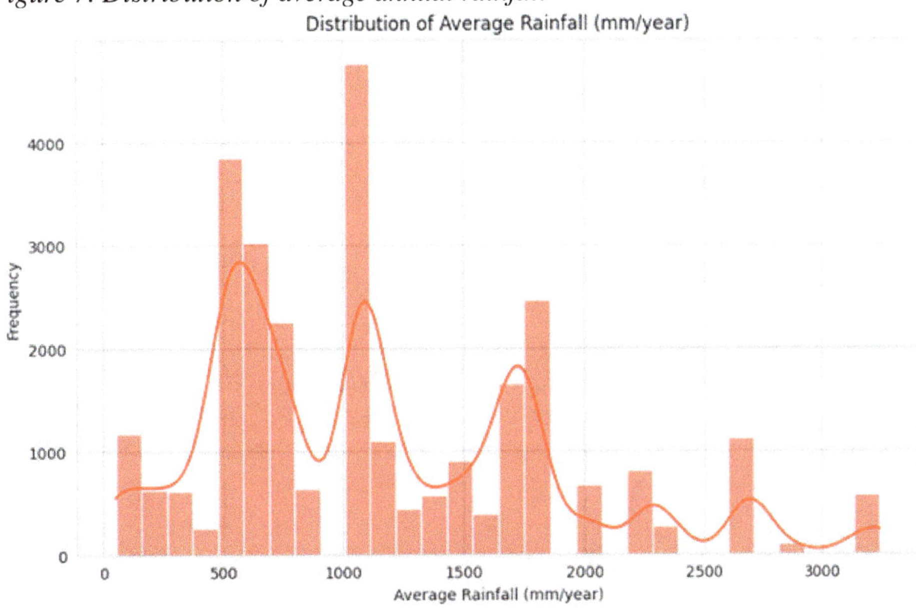

Figure 8 shows a normal distribution of the average temperature, peaking at 24–27°C, which is typical for tropical and warm-temperate regions. Figure 9 illustrates the distribution of pesticide use, which is most striking: more than 12,000 country-year instances use less than 10,000 metric tons, but some countries, such as Brazil, the United States, China, and India, use up to 200,000–367,000 metric tons. Consequently, its feature correlation coefficient with yield is +0.64, indicating that it is the strongest feature.

Figure 8. Distribution of the average temperature

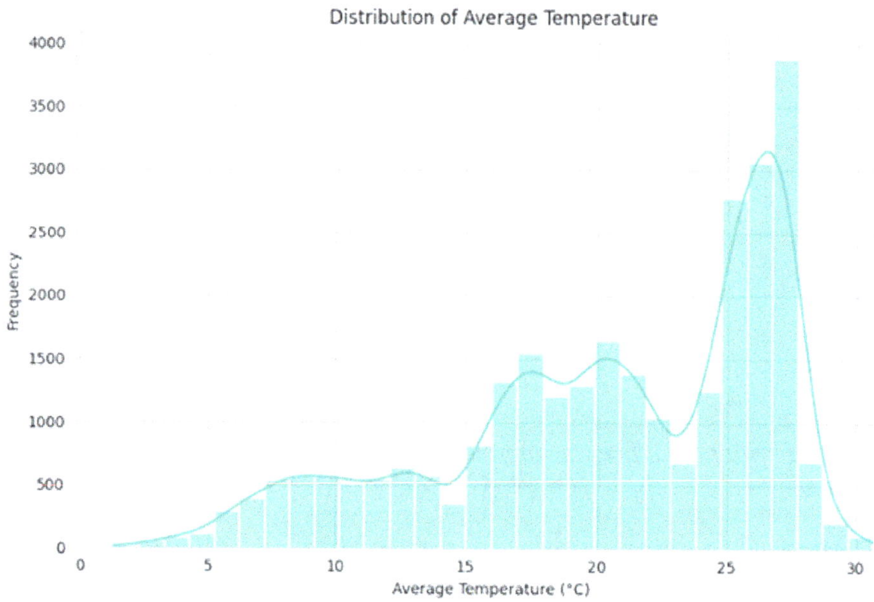

Figure 9. Distribution of pesticide use

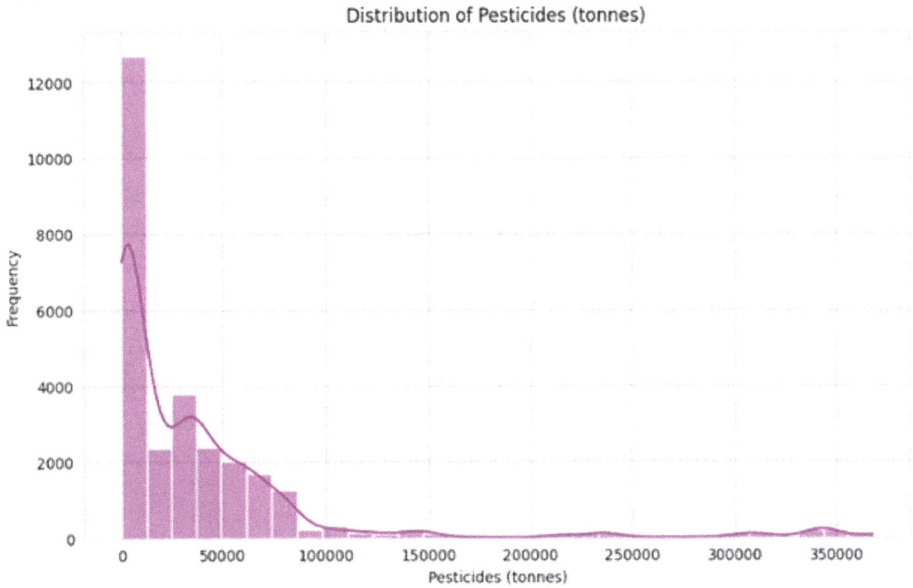

Figure 10. Number of entries over years

As shown in Figure 10, the number of entries over the years was the lowest in 1990 (approximately 1,170) and gradually increased to reach a peak (almost 1,240) after 2010, indicating an improvement in global information reporting.

3.4 Applied Artificial Intelligence Methods

To analyze crop yield outcomes within a precision agriculture context, this study employs a range of supervised machine learning algorithms applied to aggregated country-year-crop tabular data. The modeling dataset integrates historical crop yield records with corresponding annual average temperature, average rainfall, pesticide usage, crop identity, and country information, as described in the Dataset Description section. Prior to model training, the data underwent a structured preprocessing pipeline that included data cleaning, handling of missing values, feature scaling, and categorical encoding of crop and country variables.

3.4.1 Random Forest (RF)

Random Forest is an ensemble method that combines many Decision Tree classifiers (DT) trained using the bagging technique. It enhances the stability and accuracy of the model by utilizing an average model strategy. The RF classifier is essentially a group of DT classifiers, each of which has been built with a set of random vectors and can vote for the most preferred class for forecasting. RF is an ensemble learning method that constructs multiple DTs and merges their predictions to obtain more accurate and robust results. RF algorithm can be expressed as:

$$\hat{y} = \mathbf{mode}\left(\{T(x;\ \theta i)\}_{i=1}^{N}\right) \tag{1}$$

where,

- $\hat{y}$ refers to predicted output.
- T(x; θi) represents the prediction of the *i*-th DT with parameters θi for the input x. DT
- The mode function calculates the most common prediction among all the DT in the forest.

This Eq.1 shows how the ensemble of DT in an RF collectively determines the final prediction $\hat{y}$ for a given input x by considering the most frequent output among the individual tree predictions.

3.4.2 Decision Tree (DT)

The DT classifier is a widely used machine learning algorithm known for its interpretability and versatility. The algorithm creates a tree-like structure in which each internal node represents a decision based on a particular feature, and each leaf node corresponds to the predicted class label.

A splitting criterion is an essential aspect of Decision Trees, which determines the optimal feature and threshold for partitioning the data at each decision node. The common splitting criteria include Gini impurity and entropy. Entropy, denoted by S, is a measure of disorder or impurity in a set of label classes. The entropy formula is as follows:

$$Entropy(S) = -\sum_{i=1}^{c} pi\ log2(pi) \tag{2}$$

Here, c represents the number of classes, and pi is the proportion of samples belonging to class i in set S in Eq.2. A lower entropy value indicates a more homogeneous class distribution.

Information Gain is another key concept used in Decision Trees, representing the reduction in entropy after a split. It is calculated as the difference between the entropy of the parent set and the weighted sum of the entropies of the child sets:

$$Information\ Gain = Entropy(S) - \sum_{v \in values(A)} \frac{|Sv|}{|S|} Entropy(Sv) \tag{3}$$

Here, A is the feature being considered for the split, values(A) are its possible values, and Sv is the subset of samples for which feature A takes value v in Eq.3. Maximizing information gain leads to more effective feature selection.

Decision Trees may suffer from overfitting, especially with deep trees capturing noise in the training data. Pruning is a technique used to address this issue by removing unnecessary branches. Random Forests, an ensemble of Decision Trees, further enhance performance and robustness.

In conclusion, Decision Trees offer simplicity, interpretability, and versatility in handling various data types. However, careful consideration of hyperparameters, such as tree depth, is necessary to balance model complexity and generalization.

3.4.3 K-Nearest Neighbors Classifier (KNN)

The oldest and most fundamental supervised machine learning technique is KNN. It is a flexible method in statistical learning because it can solve both classification and regression problems. Separate data are categorized with each other according to the distance between data locations, which is the basis for its operation. The Euclidean or Manhattan distance between two points, where one is a new data point A(x1,y1) and the other is a previously accessible data point B(x2,y2), is used to calculate KNN. The basic formula for KNN is represented by the Eq.4:

$$d(x, y) = \sqrt{(x1 - x2)^2 + (y1 - y2)^2} \quad (4)$$

This method classifies a test sample by identifying the 'k' nearest neighbors from the training data and assigning the test sample to the class that is the most frequent among these neighbors (Khan et al., 2023).

3.4.4 GRADIENT BOOSTING (GB)

Gradient Boosting classifiers (GB) are an ensemble learning method and a group of ML algorithms that combine many weak algorithms to establish a robust predictive model and work with loss functions, weak learners, and adaptive models. The loss function is the primary difference between GB regression and GB classification. A GB classifier can be applied when the target attribute is binary. By mitigating the overfitting problem and executing regularization, the performance of the GB classifier improves. The algorithm can be defined as follows:

Initialize the model:

$$F_0(x) = mean(y_i) \quad (5)$$

For $t = 1$to T (number of iterations):

1) Compute the pseudo-residuals:

$$r_i^{(t)} = \frac{-\partial L(y_i, F(xi))}{\partial F(xi)}\Big|_{F(x)=Ft-1(x)} \quad (6)$$

2) Fit a weak learner (e.g., DT) to predict the pseudo-residuals $r_i^{(t)}$.
3) Update the model:

$$F_t(x) = F_{t-1}(x) + \nu \cdot h_t(x) \quad (7)$$

where ν is the learning rate and $h_t(x)$ is the weak learner at iteration t.

The final prediction:

$$F(x) = F_0(x) + \sum_{t=1}^{T} \nu \cdot h_t(x) \tag{8}$$

3.4.5 Gaussian Naive Bayes (GNB)

GNB is a classification algorithm based on the Naive Bayes theorem. It is specifically designed to handle continuous data, such as real-valued features, and assumes that the data follow a Gaussian (normal) distribution. This algorithm is a variation of the Naive Bayes classifier, which is a probabilistic ML algorithm used for classification tasks. The GNB formula is as follows:

The probability of class C_k given the features x_1, x_2,...,x_n can be calculated as follows:

$$P(C_k|x_1,x_2,...,x_n) = \frac{1}{Z} \cdot P(C_k) \cdot \prod_{i=1}^{n} P(x_i|C_k) \tag{9}$$

where:

- $P(C_k|x_1, x_2,...,x_n)$ is the posterior probability of class $C_k$given the features.
- $P(C_k)$ is the prior probability of class C_k.
- $P(C_k)$is the probability of observing feature x_i given class C_k, typically modeled as a Gaussian distribution.
- Z is known as a normalization constant. The normalization constant Z is often not explicitly calculated because it is the same for all classes, and it is sufficient to compare the abnormalized probabilities to make a classification decision.

3.4.6 Logistic Regression

Logistic regression is a statistical technique used to model the probability of a binary outcome, such as success or failure. It uses a logistic function to describe the relationship between independent variables (x) and the probability of the dependent variable (y) taking a specific value, as shown below:

$$p(y = 1|x) = \frac{1}{1 + e^{-z}} \tag{10}$$

where,

$$z = \beta_x = \beta_0 + \beta_1 x_1 + \cdots + \beta_p x_p \tag{11}$$

β is the vector of coefficients, and x is the vector of explanatory variables. The estimated coefficients β in a regression model indicate the influence of independent variables on the dependent variable. The coefficients β can be estimated using the maximum likelihood estimation (MLE) method. Mathematically, for a binary classification problem with a training set of N datapoints, where yi is the actual class label and is the vector of input features for the data point, then the likelihood function can be formulated as: The coefficients β can be estimated using the Maximum Likelihood Estimation (MLE) method. Mathematically, for a binary classification problem with a training set of N datapoints, where is the actual class label and is the vector of input features for the data point, then the likelihood function can be formulated as:

$$L(\beta) = \prod_{i=1}^{N} p(y_i|x_i,\beta)^{y_i}(1 - p(y_i|x_i,\beta))^{1-y_i} \tag{12}$$

Parameter estimation is performed by maximizing the natural logarithm of this equation, commonly known as the log-likelihood. Although logistic regression is preferred because of its ease of implementation, superior interpretability, and robustness to noise, it assumes a linear relationship between independent variables and the probability of the dependent variable on the logit scale, which may not always hold true (Sengupta ET AL., 2023).

3.4.7 Extra Trees Classifier

Extra trees classifier is a tree-based ensemble method that introduces strong randomization in both the attribute and cut-point selections when splitting nodes, resulting in the construction of highly randomized trees (Geurts et al., 2006). This approach aims to reduce variance and prevent overfitting by averaging the predictions from numerous unpruned decision trees (Beckmann et al., 2023; Talukder et al., 2024). Unlike other ensemble methods, the structures of these completely randomized trees are independent of the output values of the learning sample (Geurts et al., 2006).

For each row type r in R, train an Extra Trees Classifier model M_r using the preprocessed data X_r and y_r:

$$M_r = ExtraTreesClassifierTrain(X_r, y_r) \tag{13}$$

The Extra Trees Classifier prediction algorithm combines the predictions of multiple decision trees to make a final prediction for a given new row x_{new}:

$$\hat{y}_{new} = ExtraTreesClassifierPredict(x_{new}, M_r) \tag{14}$$

Here, $\hat{y}_{new}$ represents the predicted class label for the new samplex_{new}. M_r denotes the trained Extra Trees Classifier model specific to the row type *r*. (Rath & Date, 2023).

D is the dataset consisting of *N* rows and *M* columns.

R is set of distinct row types in *D* (where |*R*| is the total number of row types).

For each row type *r* in *R*,

D_r is subset of *D* containing rows of type *r*.

X_r is the feature matrix for rows of type *r*.

y_r is the target variable vector for rows of type *r*.

3.4.8 Linear Support Vector Regression

Linear Support Vector Regression (Linear SVR) is a machine learning model that was first used for regression tasks, distinctly when there is a linear relationship in the data. It is a type of Support Vector Machines (SVM), which is normally used for classification problems, but can also be applied to regression problems. Linear SVR creates a linear function that best fits a given dataset.

The goal of Linear SVR is to find a function $f(x) = w\ x + b$ that fits the data well and allows prediction of new data using this function.

Given a set of training instance-target pairs $\{(x_i , y_i)\}$, $x_i \quad R\ n$, $y_i \quad R$, $i = 1,...,l$, linear_SVR finds a model w such that $w^T x_i$ is close to the target value y_i. It solves the following regularized optimization problem. Where,

$$f(w) \equiv \frac{1}{2} w^T w + C \sum_{i=1}^{l} \xi_\varepsilon (w; x_i, y_i) \tag{15}$$

The objective is to minimize the model complexity while ensuring that prediction errors fall within a specified tolerance margin, thereby balancing performance and generalizability (Awad & Khanna, 2015).

3.4.9 Support Vector Classification With Radial Basis Function

Support Vector Classification with Radial Basis Function (SVC_RBF) is a classification method that is a special type of support vector machine (SVM). It uses the RBF kernel to effectively classify data with linear or nonlinear features. The Radial Basis Function kernel (RBF) is a kernel function that helps to identify linear relationships in the data. It maps the data into a higher-dimensional space, where they can be linearly separable. SVC_RBF is generally used to solve problems in which there are no linear constraints in the data, and we want to accurately separate the patterns. The decision function of the SVC_RBF classifier is,

$$f(x) = sign\left(\sum_{i=1}^{l} \alpha_i y_i K(x_i, x) + b\right) \quad (16)$$

Here, α is the Lagrange multiplier for the support vector, y is the class label (1 or -1), K(x , x) is the kernel function that determines the relationship between x and x and b is the bias.

3.4.10 Extreme Gradient Boosting Classifier

The Extreme Gradient Boosting (XGBoost) algorithm is an advanced ensemble method based on decision trees. It employs a boosting sampling technique to enhance the predictive accuracy and mitigate overfitting by regularizing the loss function (Huang et al., 2023). This advanced approach builds an ensemble of weak learners, typically decision trees, to minimize a differentiable loss function. XGBoost includes a regularization term to control the complexity of each tree and prevent overfitting:

$$\Omega(f_k) = \gamma T + \frac{1}{2}\lambda \sum_{j=1}^{T} w_j^2 \quad (17)$$

Where, $T$is the number of leaves in the tree,$\gamma$is the complexity regularization parameter, $\lambda$is the L2 regularization parameter for the leaf weights and $w_j$are the leaf weights.

4. RESULTS

A comparison and analysis of the outcomes achieved using ML strategies are presented in this section. This section provides a full overview of the approaches by offering a detailed discussion of the performance metrics of each model.

4.1 Setup of Experiment

This section discusses the experimental setup used to develop the applied ML techniques. Table 3 lists the libraries commonly used in the code of this crop yield prediction project: pandas, NumPy, matplotlib.pyplot, seaborn, scikit-learn, SciPy, and statsmodels. The pandas library is used for data manipulation and analysis, including loading crop data (e.g., yield, soil properties, and weather) from CSV files, cleaning, filtering, and handling missing values. NumPy is used for numerical computations, such as creating data arrays, calculating average yield, and feature scaling. Matplotlib. pyplot is used for data visualization, such as showing the relationship between yield and rainfall through histograms or scatter plots. Seaborn provides advanced visualizations, such as correlation heatmaps and boxplots between features. scikit-learn is crucial for machine learning modeling, including training models with Linear Regression or Random Forests, splitting data, and calculating metrics such as MSE. SciPy is used for scientific computing and statistical testing, such as correlation analysis and normality testing of yields. In contrast, the Statsmodels library supports statistical modeling, such as testing feature significance and p-values using OLS regression. Altogether, these libraries complete the entire process of data loading, EDA, feature engineering, modeling, and visualization.

Table 3. Developing the applied ML techniques by using different ML Libraries.

Library	Primary Function	Specific Applications
pandas	Data manipulation and analysis	Loading crop data from CSV files, cleaning, filtering, handling missing values
NumPy	Numerical computations	Creating data arrays, calculating average yield, feature scaling
matplotlib. pyplot	Data visualization	Showing relationships between yield and rainfall through histograms or scatter plots
seaborn	Advanced visualizations	Generating correlation heatmaps and boxplots between features
scikit-learn	Machine learning modeling	Training models, splitting data, calculating metrics
SciPy	Scientific computing and statistical testing	Correlation analysis, normality testing of yields
statsmodels	Statistical modeling	Testing feature significance and p-values using OLS regression

4.2 Evaluation Matrices

To assess the efficacy of these methodologies, we closely examined metrics such as accuracy, precision, recall, and F1 score. These pivotal metrics enable us to effectively gauge the performance of the chosen models. In classification, True Positive (TP) represents instances correctly identified as positive, while False Positive (FP) signifies incorrect positive predictions. True Negative (TN) denotes instances correctly identified as negative, and False Negative (FN) indicates instances incorrectly predicted as negative. These metrics play a crucial role in assessing a model's performance, aiding in the computation of accuracy, precision, recall, f1 score, and other evaluation measures.

$$Accuracy = \frac{\text{Number of Predictions}}{\text{Total number of Predictions}}$$

In this equation, the ''Number of Correct Predictions'' represents the count of correctly classified instances, and the ''Total Number of Predictions'' is the total number of instances in the dataset that were classified. Accuracy is a measure of the model's ability to correctly classify instances and is typically expressed as a percentage. To calculate the precision metric, the total number of classified positive samples is divided by the number of correctly classified positive instances. The formula for computing this metric is as follows:

$$Precision = \frac{TP}{TP + FP}$$

Recall is a metric in classification that measures the true positives correctly identified from all actual positives. This indicates the model's ability to detect all positive instances.

$$Recall = \frac{TP}{TP + FN}$$

The F1 score is a combination of precision and recall that measures a model's overall performance. It is calculated using the harmonic mean, which gives more weight to low values than to high values.

$$F1 - Score = \frac{2 \times Recall\ times\ Precision}{Recall + Precision}$$

4.3 Results With ML Models

Table 4 presents a comprehensive assessment of the ML models using the original dataset features. The evaluation centers on key performance metrics: F1 score, recall, precision, and accuracy, which are crucial for assessing the effectiveness of ML models. A detailed analysis indicated diverse performance levels among the models, including RF, ET, XGB, DT, KNN, GB, SVC_RBF, Logistic Regression, LinearSVC, and GNB. In this analysis, the RF model stood out, attaining a commendable accuracy score of 0.975. It also achieved 0.975 precision, 0.974 recall, and 0.974 F1-score. The most important point is that its false-negative rate is only 0.026, indicating that among areas where low-yield crop conditions actually existed, the model failed to detect them in only 2.6% of cases. Such misses are the most harmful in agriculture. Simultaneously, because the false-positive rate is only 0.0248, unnecessary alerts and associated costs are also minimized. As a result, the Random Forest showed an almost perfectly balanced performance on this multi-feature dataset. The Extra

Trees classifier achieved accuracy, 0.972 precision of 0.976, a recall, and a 0.974 F1-score. The most important aspect is that its FNR is very low at 0.024, which is extremely valuable in agriculture because missing low-yield crop conditions can lead to significant yield loss. The FPR is also very low (0.0276), and it has shown excellent robust performance on a multi-feature dataset. DT and KNN also showed very good results, with accuracies above 0.96. GB yielded moderate results (0.919). However, the SVC model with the RBF kernel performed much worse (accuracy of only 0.755). Linear models performed the worst logistic regression (0.614), linear SVC (0.613), and GaussianNB (0.609). The reason tree-based ensemble models perform better is that they are capable of effectively capturing the complex nonlinear relationships and interactions of the multi-source features used in the multi-feature agro-climatic dataset used in this study.

Table 4. Evaluating the performance of different ML approaches across diverse metrics.

Model	Accuracy	Precision	Recall	F1-Score	FNR	FPR
RandomForest	0.975	0.975	0.974	0.974	0.026	0.0248
ExtraTrees	0.974	0.972	0.976	0.974	0.024	0.0276
XGBClassifier	0.972	0.981	0.964	0.972	0.036	0.0188
DecisionTree	0.970	0.969	0.971	0.970	0.029	0.0312
KNN	0.964	0.966	0.962	0.964	0.037	0.0340
GradientBoosting	0.919	0.927	0.909	0.918	0.090	0.0719
SVC_RBF	0.755	0.776	0.719	0.746	0.281	0.2079
LogisticRegression	0.614	0.616	0.606	0.611	0.394	0.3775
LinearSVC	0.613	0.616	0.603	0.609	0.397	0.3764
GaussianNB	0.609	0.646	0.482	0.552	0.517	0.2638

Table 5 presents the values in the confusion matrix, which clearly show how effectively different machine learning models performed in identifying positive and negative classes. Here, True Negative (TN), True Positive (TP), False Positive (FP), and False Negative (FN) indicate not only the accuracy but also the actual decision-making behavior of each model.

First, tree-based ensemble models, such as random forest, extra trees, and XG-Boost, consistently provided higher-quality predictions than all other methods. In the case of Random Forest, TN and TP values indicated that the model could identify both classes with almost equal efficiency. Simultaneously, the FP and FN values were lower, showing that the model could successfully control both false alarms and missed detections. This consistent behavior is precisely the strength of ensemble learning, and Random Forest makes the most robust decisions when dealing with multi-source, nonlinear, and interaction-heavy data.

Table 5. Confusion Matrix Results of the evaluated different ML approaches

Model	**TN**	**FP**	**FN**	**TP**
RandomForest	2754	70	74	2751
ExtraTrees	2746	78	69	2756
XGBClassifier	2771	53	103	2722
DecisionTree	2736	88	81	2744
KNN	2728	96	105	2720
GradientBoosting	2621	203	255	2570
SVC_RBF	2237	587	795	2030
LogisticRegression	1758	1066	1114	1711
LinearSVC	1761	1063	1122	1703
GaussianNB	2079	45	1463	1362

The Extra Trees classifier exhibited similar tendencies. A low FN indicates a very low risk of missing positive cases in this study. Although the number of false positives was slightly higher, the model showed high sensitivity and reliability.

The Decision Tree model is naturally less powerful than ensemble models. However, TP and FN show that it performs better than the baseline, although it has not reached the best because of variance and overfitting issues.

In Figure 11, among the top five models, Random Forest and Extra Trees exhibited the best performance, achieving equally good scores across all metrics: Accuracy, Precision, Recall, and F1-Score. These models are accurate and balanced, providing low false-positive and false-negative rates. Second, the XGB Classifier shows a slightly lower recall; however, its accuracy and F1-Score are still high. Decision Tree is a simple model which performed correctly, but there is some variation between precision and recall, specifying some misclassifications. The KNN model gained good accuracy and F1-Score; however, the recall was slightly lower, suggesting the possibility of missing some positive cases. Therefore, Random Forest and Extra Trees are the strongest, but the other models are also quite effective, even though they show slightly lower performance.

Figure 11. Performance Comparison of ML Models Based on Accuracy, Precision, Recall, and F1-Score

4.4 Results With ML Models With Explainable AI (XAI)

Table 6 contrasts the performance metrics of several machine learning models with Explainable AI (XAI) to evaluate their effectiveness in predicting yield conditions. The performance metrics included accuracy, precision, recall, F1-score, false negative rate (FNR), and false positive rate (FPR). Gradient Boosting emerged as the best performer, achieving 97.86% accuracy, 97.94% precision, 97.77% recall, and a 97.86% F1-score. Furthermore, Random Forest displayed decent results with 97.57% accuracy and 97.76% precision. Moreover, the XGBClassifier gained good results, with 97.54% accuracy and 97.35% recall, maintaining a better balance between precision and recall. ExtraTrees and DecisionTree performed similarly, with high accuracy and recall but a slightly lower precision. KNN showed good accuracy (96.64%) and recall (96.85%), although some improvement in precision is needed. In contrast, SVC_RBF lagged significantly, with low accuracy and high false-positive and false-negative rates, indicating difficulty in correctly identifying low-yield events. Finally, LogisticRegression, LinearSVC, and GaussianNB performed the worst, particularly with low recall and high false-negative rates, leading these models to miss several crop issue events.

Table 6. Evaluating the performance of different ML approaches with Optuna across diverse metrics.

Model	Accuracy	Precision	Recall	F1-Score	FNR	FPR
GradientBoosting (Optuna)	0.978580	0.979433	0.977699	0.978565	0.022301	0.020538
RandomForest (Optuna)	0.975748	0.977612	0.973805	0.975705	0.026195	0.022309
XGBClassifier (Optuna)	0.975394	0.973554	0.977345	0.975446	0.022655	0.026558
ExtraTrees (Optuna)	0.975040	0.977580	0.972389	0.974978	0.027611	0.022309
DecisionTree (Optuna)	0.971853	0.974715	0.968850	0.971773	0.031150	0.025142
KNN (Optuna)	0.966366	0.964399	0.968496	0.966443	0.031504	0.035765
SVC_RBF (Optuna)	0.882811	0.884465	0.880708	0.882582	0.119292	0.115085
LogisticRegression (Optuna)	0.613560	0.616052	0.603186	0.609551	0.396814	0.376062
LinearSVC (Optuna)	0.613206	0.615691	0.602832	0.609193	0.397168	0.376416
GaussianNB (Optuna)	0.609134	0.646417	0.482124	0.552311	0.517876	0.263810

Table 7 interprets the performance of several ML models, showing the confusion metric results of models tuned with Explainable AI (XAI). It discusses each model's performance using metrics such as AUC_ROC, True Negative (TN), False Positive (FP), False Negative (FN), and True Positive (TP). The GradientBoosting model showed the best performance with the highest AUC_ROC (0.997242), TN (2766), and TP (2762), whereas the LogisticRegression model had a comparatively lower AUC_ROC (0.648169) and higher FP (1062) and FN (1121), indicating its lower effectiveness. This table helps analyze the models' accuracy, errors, and performance.

Table 7. Confusion Matrix Results of the evaluated different ML approaches with Optuna

Model	TN	FP	FN	TP
GradientBoosting (Optuna)	2766	58	63	2762
RandomForest (Optuna)	2761	63	74	2751
XGBClassifier (Optuna)	2749	75	64	2761
ExtraTrees (Optuna)	2761	63	78	2747
DecisionTree (Optuna)	2753	71	88	2737
KNN (Optuna)	2723	101	89	2736
SVC_RBF (Optuna)	2499	325	337	2488
LogisticRegression (Optuna)	1762	1062	1121	1704
LinearSVC (Optuna)	1761	1063	1122	1703
GaussianNB (Optuna)	2079	745	1463	1362

As shown in Table 8 and Figure 12, it can be observed how Optuna optimization affects the performance of many ML models, especially with respect to improvements in F1-Score and accuracy. Table 8 compares the base performance of the models with their performance after applying Optuna tuning. GradientBoosting, XGBClassifier, and ExtraTrees displayed strong performance across all metrics, with GradientBoosting showing the highest improvement in the Optuna F1-Score at 0.059936, which is clearly visible in Figure 3, where its Optuna F1-Score increased outstandingly compared to the base score. Similarly, XGBClassifier and ExtraTrees also showed some improvement in the F1-Score, although their improvements were relatively small.

Table 8. Comparison Between Base VS Optuna in ML approaches

Model	Base_F1	Optuna_F1	Improvement	Base_Accuracy	Optuna_Accuracy
GradientBoosting	0.918630	0.978565	0.059936	0.919278	0.978580
RandomForest	0.975912	0.975705	-0.000207	0.975925	0.975748
XGBClassifier	0.964305	0.975446	0.011141	0.964773	0.975394
ExtraTrees	0.973614	0.974978	0.001364	0.973624	0.975040
DecisionTree	0.970126	0.971773	0.001648	0.970083	0.971853
KNN	0.964368	0.966443	0.002075	0.964418	0.966366
SVC_RBF	0.746049	0.882582	0.136533	0.755355	0.882811
LogisticRegression	0.610853	0.609551	-0.001302	0.614091	0.613560
LinearSVC	0.609193	0.609193	0.000000	0.613206	0.613206
GaussianNB	0.552311	0.552311	0.000000	0.609134	0.609134

Figure 13 depicts the improvement in F1-Score after Optuna tuning, where SVC_RBF showed the greatest improvement, reaching approximately 0.136533, and the significant increase in its performance is clearly reflected in Figure 13. RandomForest and DecisionTree showed average improvements, resulting in some enhancements in their F1-Score and Accuracy, although they remained comparatively stable. LogisticRegression, LinearSVC, and KNN showed little to no improvement, and LogisticRegression even slightly decreased in F1-Score and is indicated by the negative bar in the second graph, indicating that Optuna optimization had less effect on these models.

Figure 12. Comparison Between Base Model and Optuna

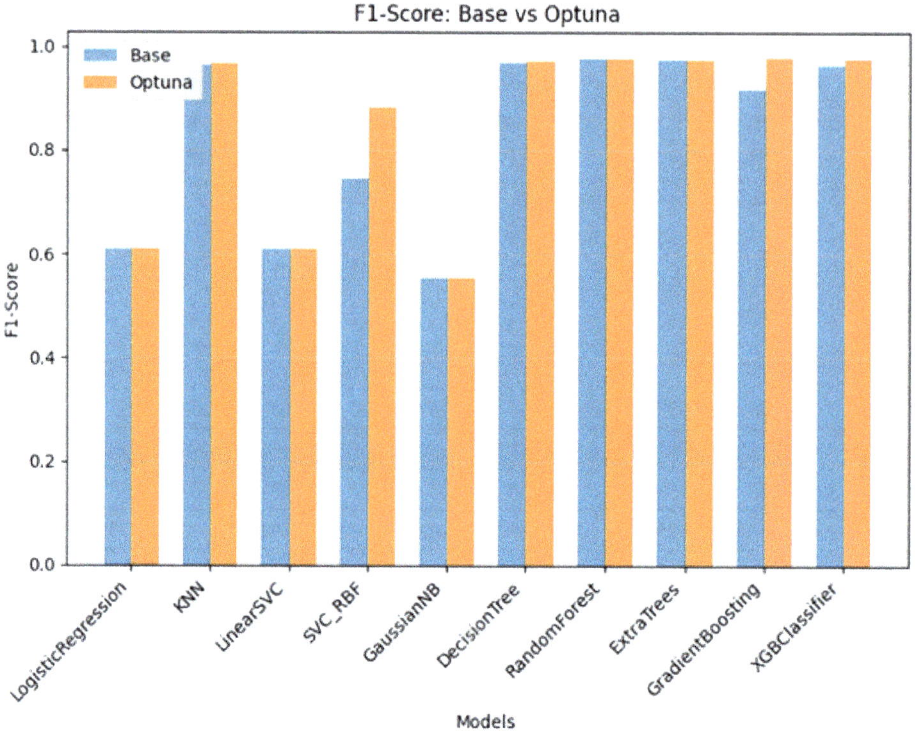

Figure 13. Improvement in F1-Score After Optuna Tuning

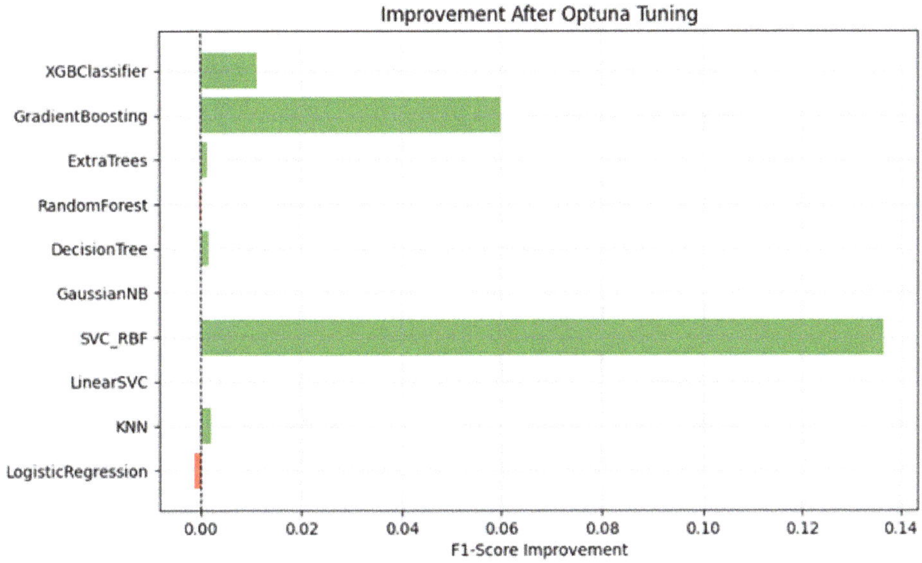

A comprehensive comparison of the classification performance of the assessed machine learning models using Receiver Operating Characteristic (ROC) curves and Precision–Recall (PR) curves, for both base models and Optuna-optimized models, is presented in Figure 14. From the ROC analysis revealed that tree-based ensemble models, such as Random Forest, Extra Trees, Gradient Boosting, and XGBoost indicated extreme performance even in their base state, with AUC values close to 0.99. In contrast, Logistic Regression, Linear SVC, and Gaussian Naive Bayes showed relatively weaker performance, with AUC values of approximately 0.64, indicating an inability to sufficiently capture the complex and nonlinear relationships present in the dataset. After hyperparameter optimization using Optuna, most models demonstrated better performance, particularly the Gradient Boosting and XGBoost models, which achieved almost flawless ROC performance. A similar trend was observed in the Precision–Recall curve analysis, where ensemble models were able to maintain high precision across a high recall range, resulting in high Average Precision scores. In particular, the Optuna-optimized Gradient Boosting model exhibited significant improvement, proving its effectiveness in detecting important positive cases while maintaining a low false negative rate. Overall, these results suggest that Optuna-optimized ensemble learning methods are highly effective for analyzing complex and multidimensional agriculture-based datasets, providing significantly better classification capabilities than conventional linear models.

Figure 14. Receiver Operating Characteristic (ROC) curves and Precision–Recall (PR) curves both for base models and Optuna-optimized models

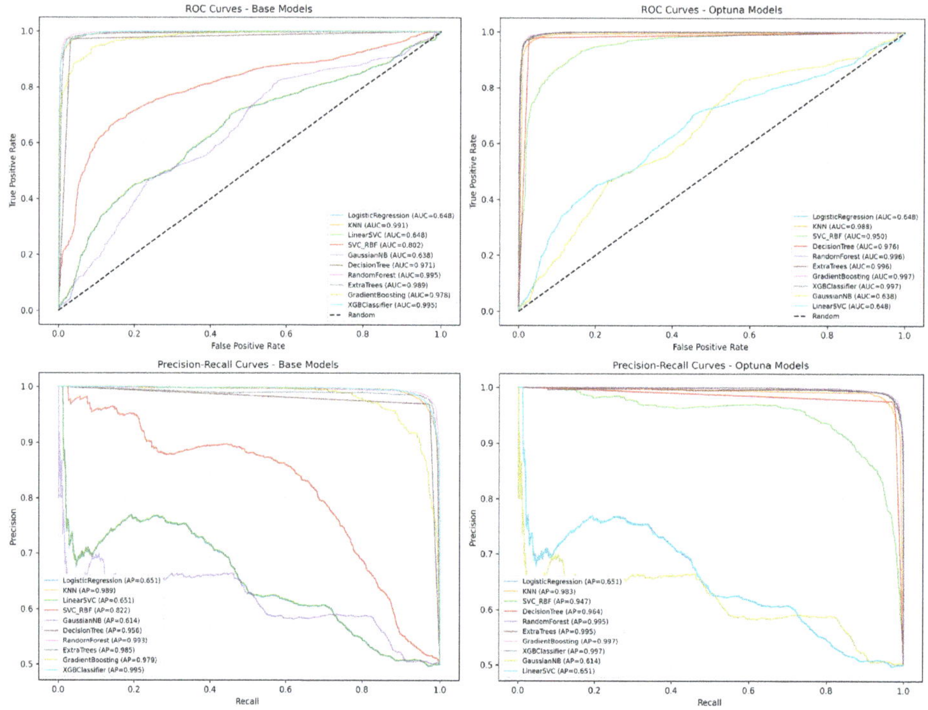

4.4.1 Decision Tree With Optuna

In Beeswarm Plot (Figure 15), the Item_Encoded feature is the most influential, with higher values pushing the Decision Tree model's predictions upward. Avg_Temp and Rainfall also play important roles, whereas Pesticides and Country_Encoded contribute less. Mean Absolute SHAP plot (Figure 16) shows that item_Encoded is also identified as the main feature in terms of feature importance in the Decision Tree, followed by Avg_Temp and Rainfall.

Figure 15. SHAP Beeswarm Summary of DecisionTree

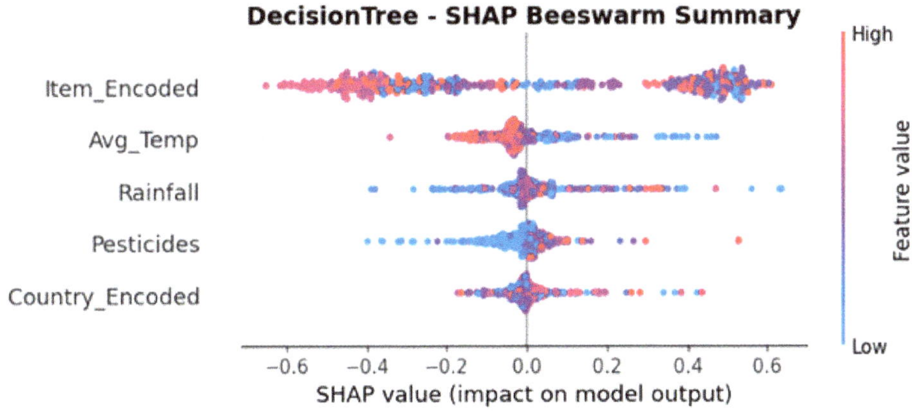

Figure 16. Mean Absolute SHAP in DecisionTree

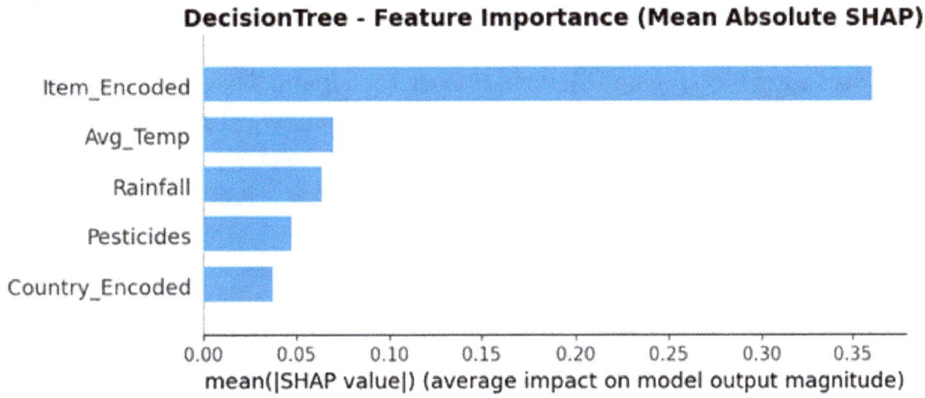

In the waterfall plot (Figure 17), Item_Encoded significantly decreased the prediction in the decision tree, while other features contributed slightly to the final prediction.

Figure 17. Waterfall Plot in Decision Tree

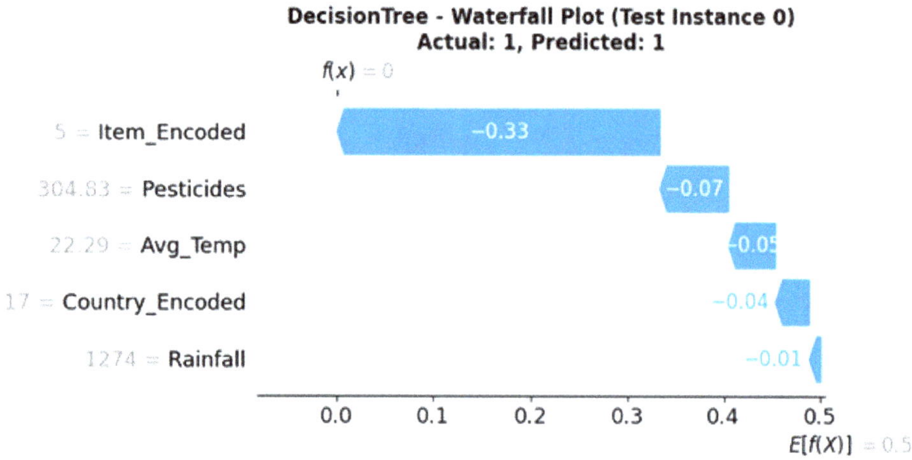

4.4.2 Random Forest With Optuna

In Beeswarm Plot (Figure 18), Item_Encoded and Rainfall have emerged such as the most influential features in the Random Forest model. Avg_Temp and Pesticides also have some effects, but the impact of Country_Encoded is low. Item_Encoded is again identified as the main feature, and Avg_Temp and Rainfall are in second place in Bar plot (Figure 19).

Figure 18. SHAP Beeswarm Summary of RandomForest

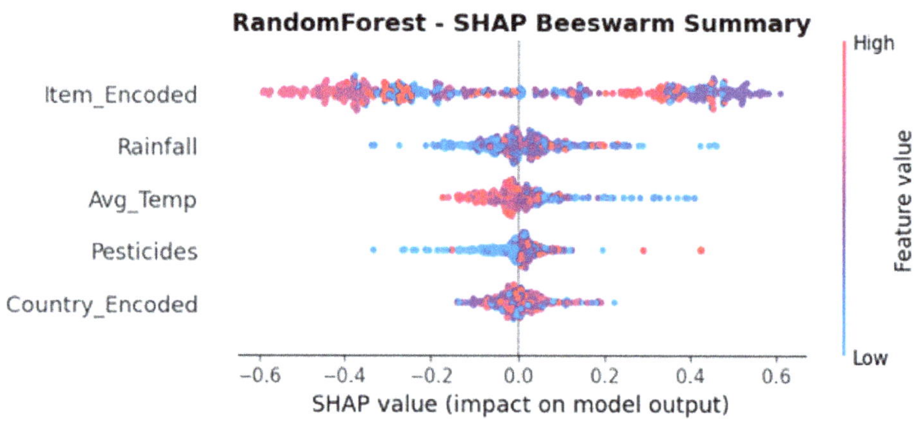

Figure 19. Mean Absolute SHAP in RandomForest

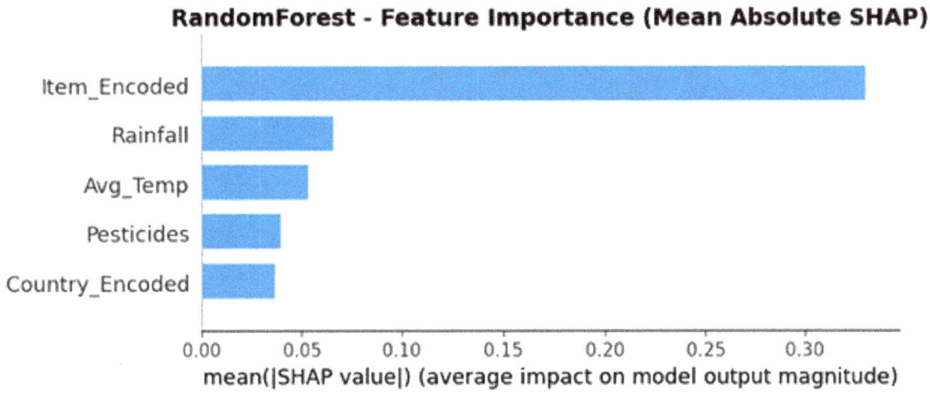

Through stacked force plot (Figure 20), we get a general idea of feature importance, where the impact of Item_Encoded is the highest and the impact of Country_Encoded or Pesticides is very low.

This waterfall plot (Figure 21) provides an in-depth analysis of predictions for a single instance in the Random Forest model. Despite this negative effect, the final predicted value (f(x) = 0.003) is still close to the original value, indicating that the combination of all features has not significantly altered the prediction outcome.

Figure 20. Stacked Force Plot in RandomForest

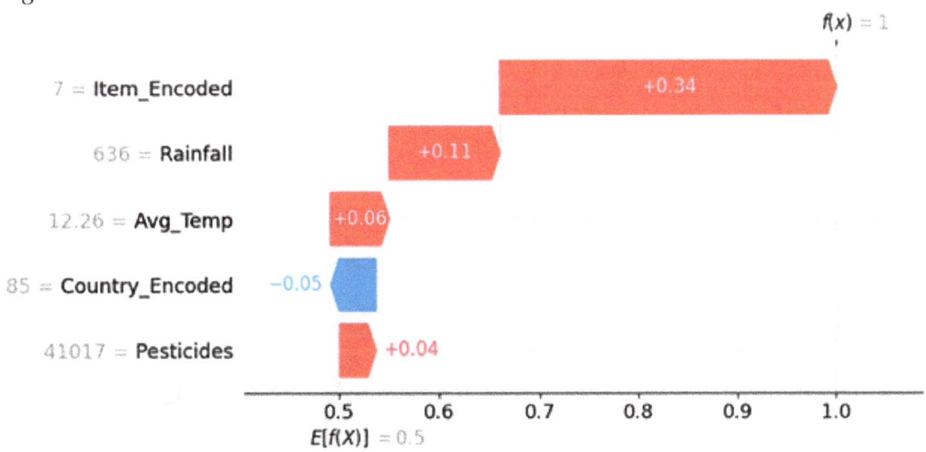

Figure 21. Waterfall Plot in RandomForest

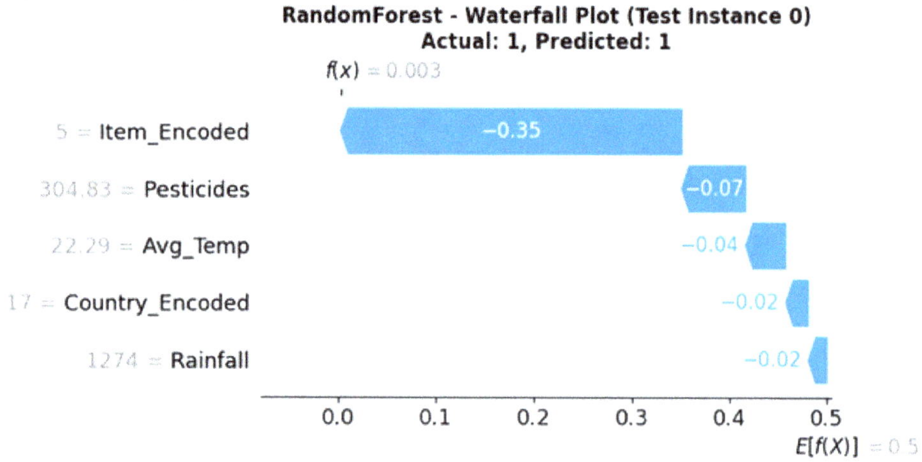

4.4.3 Extra Tree With Optuna

In Beeswarm Plot (Figure 22), Item_Encoded remains the main feature in Extra Trees, followed by the effects of Avg_Temp and Rainfall. Pesticides and Country_Encoded have less impact. Bar plot (Figure 23) shows that item_Encoded is again influential, and Avg_Temp and Rainfall contribute somewhat to the model's decision-making.

Figure 22. SHAP Beeswarm Summary of ExtraTrees

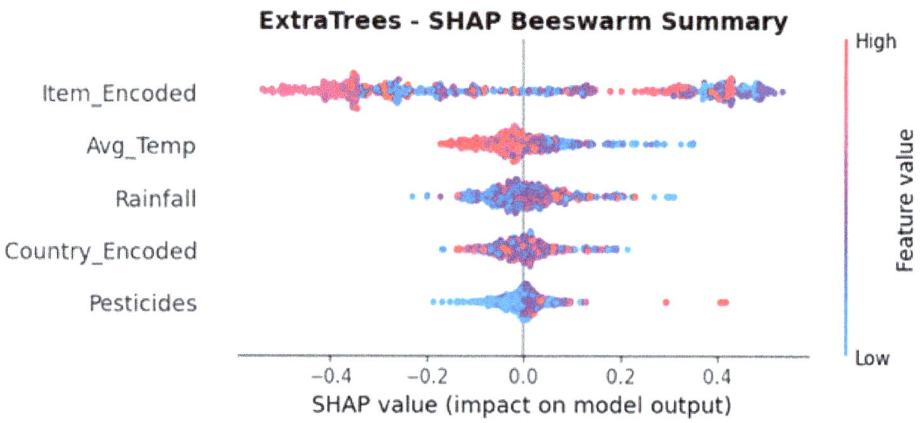

Figure 23. Mean Absolute SHAP in ExtraTrees

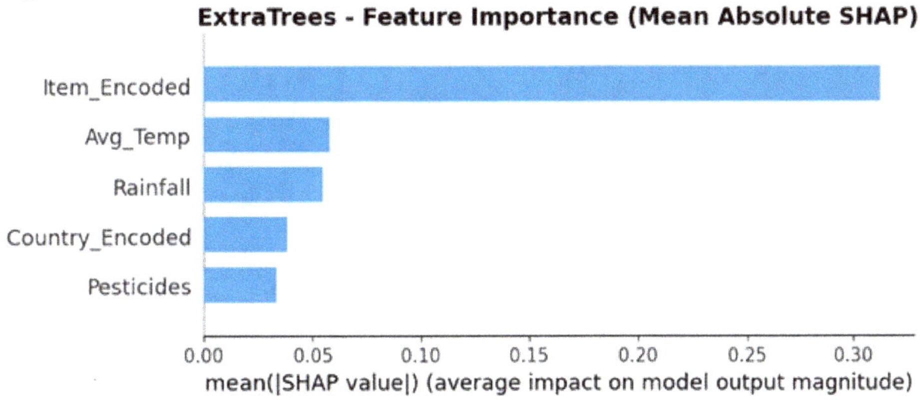

Stacked force plot (Figure 24) shows that Item_Encoded pushes the prediction towards higher values as the most influential feature. The waterfall plot (Figure 25) illustrates that while Item_Encoded has a significantly positive impact on the model, other features such as pesticides, average temperature, and rainfall generally pull the prediction down.

Figure 24. Stacked Force Plot in ExtraTrees

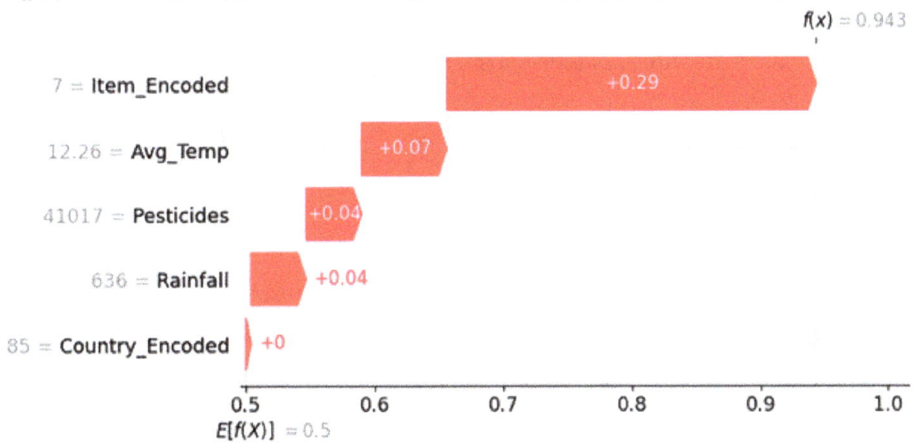

Figure 25. Waterfall Plot in ExtraTrees

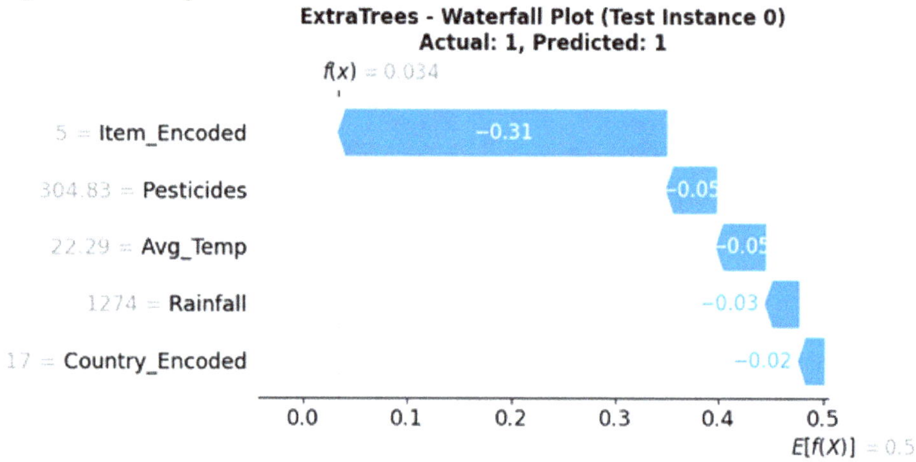

4.5 Lime Explanation

In Figure 26, item_Encoded (4.00 < Item_Encoded <= 7.00), this feature is the most impactful, having a large negative impact on the model's predictions. Avg_Temp (21.50 < Avg_Temp <= 26.00) and Pesticides (<= 1714.39), these features also have a negative impact. Rainfall (1083.00 < Rainfall <= 1668.00) and Country_Encoded (<= 23.00), these features also have a small but positive impact.

Figure 26. Local Feature impact on model predictions (4.00 < Item_Encoded <= 7.00).

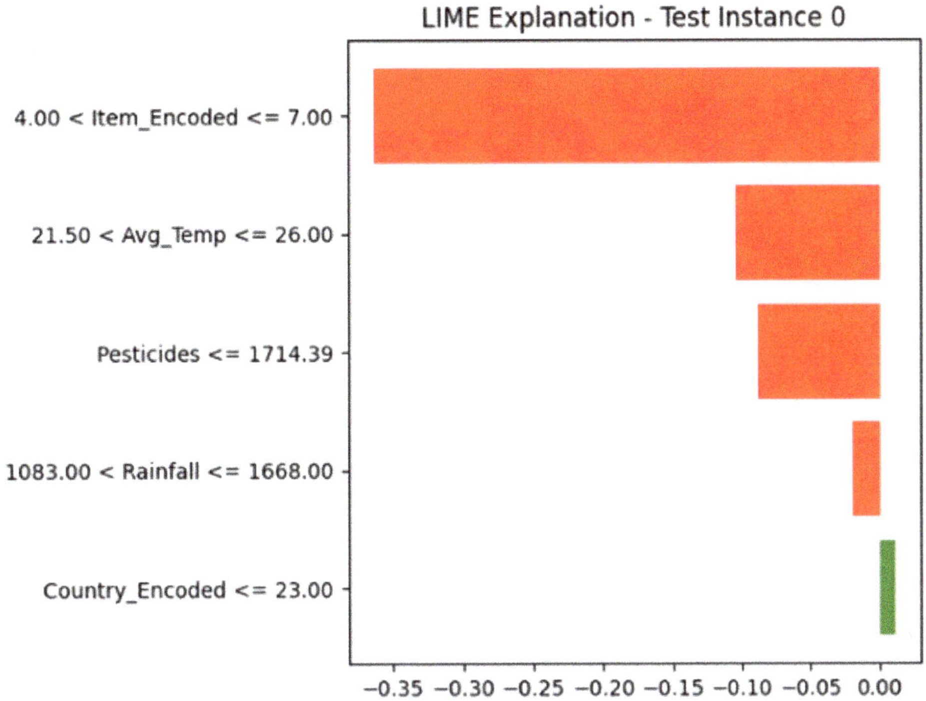

In Figure 27, item_Encoded (<= 3.00) has the largest positive impact. Rainfall (591.00 < Rainfall <=1083.00), this feature also has a positive impact. Country_ Encoded (23.00 <Country_Encoded <= 42.00), this feature has a small negative impact. Avg_Temp (16.72 < Avg_Temp <= 21.50) and Pesticides (<= 1714.39), these features also have a negative impact.

Figure 27. Local Feature impact on model predictions item_Encoded (<= 3.00)

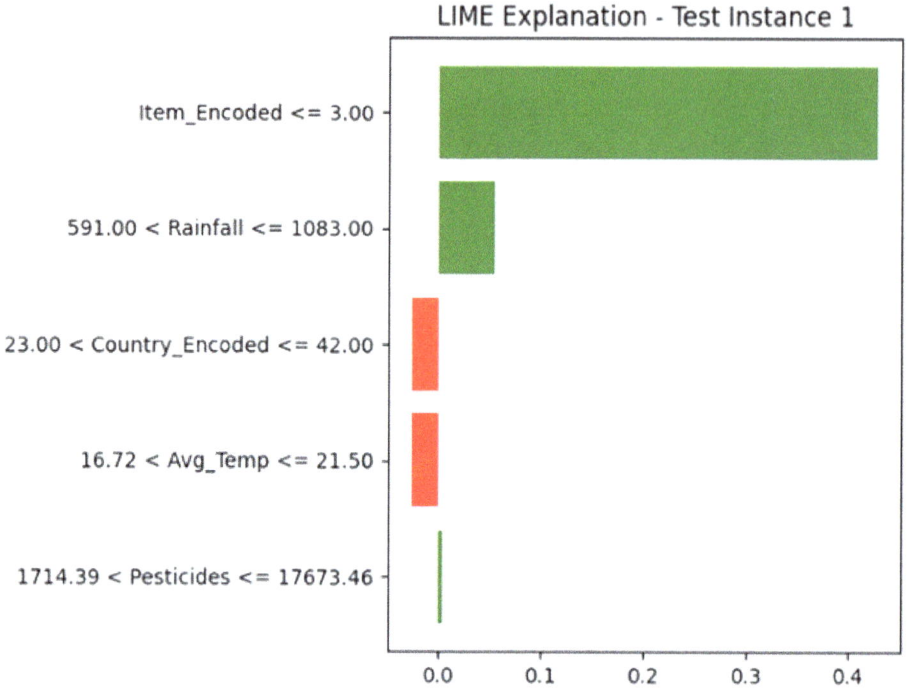

In this study, the predictive performance of the XGBoost classifier was evaluated using a force plot visualization, which highlights the model's decision-making process for each sample. The force plot is divided into two main components, where the blue lines represent the predicted probability of the positive class (1) and the red lines represent the predicted probability of the negative class (0) in Figure 28.

Figure 28. Predictive Performance of XGBClassifier using Force Plot Visualization

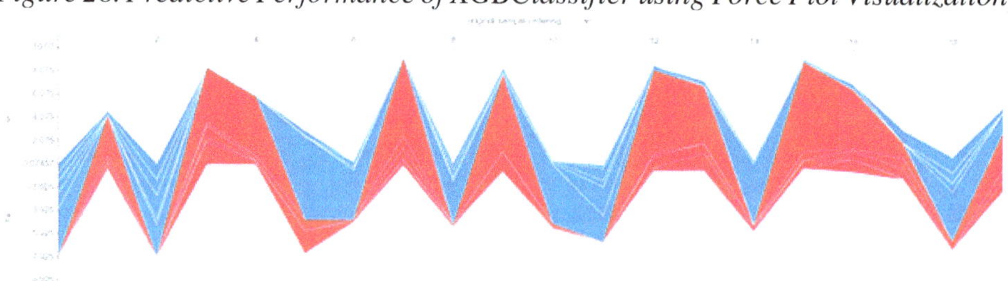

From Figure 28, each line represents a sample, and the varying heights indicate the level of confidence in the model's prediction. The fluctuations between the blue and red lines highlight the uncertainty and variation in the predicted probabilities, reflecting that the model has learned the complex relationships between the input features and the target class. Original sample ordering indicates that the samples are displayed according to their original sequence, which may highlight trends or patterns based on the order of the samples. This feature can be important in analyzing the model's dataset generalization capability or how it reacts to different feature sets for each sample.

The force plot also shows how different features interact with each other and influence the final prediction. For example, some features can cause significant changes toward the positive or negative class in the prediction. In our analysis, the model gives more importance to certain features (such as Item_Encoded, Avg_Temp), which is indicated by steep rises and falls in the graph.

4.6 Generalization Performance Under Realistic Scenarios

Models that perform well under random training–test splits may fail to generalize in variable environments, particularly in agricultural applications. This issue arises because random splitting often places observations from the same country and crop in both the training and testing sets. Consequently, models may learn patterns specific to particular countries or crops rather than general climate–yield relationships. This problem is exacerbated when country and crop identifiers are included as features, as models can effectively memorize location- or crop-specific behaviors that do not transfer to unseen contexts.

To evaluate how models perform under realistic conditions, we tested them using three stricter protocols: (i) country-holdout validation, which tests performance on entirely new geographic regions; (ii) crop-holdout validation, which tests performance on crop types never seen during training; and (iii) temporal holdout evaluation, which tests performance on future years using only historical training data.

4.6.1 Country-Holdout Validation (Unseen Regions)

We evaluated geographic generalization using five-fold GroupKFold cross-validation, in which all observations from the same country were placed in the same fold. This ensures that the model does not train on data from a country it will be tested on. The performance dropped considerably compared with random splitting. Gradient Boosting performed best under this protocol, achieving F1 = 0.8429, accuracy = 0.8437, and AUC = 0.9292. Extra Trees performed nearly as well (F1 = 0.8423). Although these scores remained acceptable, the error rates increased noticeably. Gradient Boosting had FNR = 0.1422 and FPR = 0.1573, meaning it missed roughly 14% of actual low-yield cases and incorrectly flagged 16% of high-yield cases when applied to new regions

The performance gap between random splitting and country-holdout reveals how much the random split overestimates real-world capability. Random Forest, for instance, dropped from F1 = 0.9757 under random splitting to F1 = 0.8314 under country-holdout. This large decline shows that much of the apparent accuracy in random splits comes from within-country data overlap rather than learned climate–yield relationships that transfer across regions.

4.6.2 Crop-Holdout Validation (Unseen Crops)

We tested crop transferability using five-fold GroupKFold cross-validation with crop type as the grouping variable. This means that the models were trained on some crops and tested on completely different crops that they had never encountered. All models performed poorly under this protocol. Decision Tree achieved the highest F1-score at 0.3925, while ensemble methods like Random Forest (F1 = 0.3896) and Extra Trees (F1 = 0.3879) scored similarly low. AUC values hovered near 0.44–0.54, barely better than random guessing. These results indicate that the climate-yield patterns learned from one set of crops do not reliably predict outcomes for different crop species.

This poor cross-crop performance reveals a fundamental limitation: yield responses to climate vary substantially across crops, and aggregated climate variables alone cannot capture these differences. Models essentially fail to generalize because wheat, maize, rice, and other crops respond differently to the same environmental conditions. Achieving robust cross-crop prediction would likely require crop-specific features or physiological parameters that account for these biological differences.

4.6.3 Temporal Holdout Evaluation (Future Years)

We assessed temporal stability by training models on data from 1990–2009 and testing on 2010–2013. This setup mirrors real-world use cases in which models built from historical records are applied to predict future growing seasons. Unlike the country and crop holdout protocols, temporal evaluation produced strong results. Random Forest achieved F1 = 0.9375, accuracy = 0.9333, and AUC = 0.9784. Extra Trees performed nearly identically with F1 = 0.9373 and accuracy = 0.9331. Although performance declined relative to random splitting, the drop was much smaller than in the country-holdout scenario. This suggests that climate–yield relationships remain fairly stable across consecutive time periods when the same countries and crops are involved.

However, the temporal evaluation revealed some concerning error patterns. Random Forest model had an FNR = 0.0979 and FPR = 0.0278. Although the false positive rate remained low, the false negative rate nearly tripled that of random splitting. This indicates that the model missed approximately 10% of actual low-yield years when making forward predictions, which is a significant issue for early warning systems, in which failing to detect poor harvests can have serious consequences.

4.6.4 Summary of Generalization Behavior Across Protocols

Table 9 compares F1-scores for all validation protocols. Tree-based ensemble methods consistently outperformed linear models across all scenarios; however, their robustness varied considerably depending on the type of generalization required. Models handled temporal shifts reasonably well but struggled more with new geographic regions and failed almost completely when applied to new crop types.

Table 9. Model F1-score across validation protocols

Model	**Random Split (F1)**	**Country-holdout GroupKFold (F1)**	**Crop-holdout GroupKFold (F1)**	**Temporal Holdout (F1)**
Random Forest	0.9757	0.8314	0.3896	0.9375
Extra Trees	0.9744	0.8423	0.3879	0.9373
Model	Random Split (F1)	Country-holdout GroupKFold (F1)	Crop-holdout GroupKFold (F1)	Temporal Holdout (F1)
XGBClassifier	0.9753	0.8270	0.3703	0.9327
Gradient Boosting	0.9432	0.8429	0.3781	0.9092
Decision Tree	0.9701	0.8014	0.3925	0.9302
KNN	0.9644	0.7152	0.3595	0.9249
Logistic Regression	0.6107	0.5885	0.3152	0.6347
Linear SVC	0.6092	0.5883	0.3145	0.6347
Gaussian NB	0.5523	0.4826	0.2491	0.5884
SVC (RBF)	0.7460	0.6785	0.3210	0.7661

These results underscore an important point: how you evaluate agricultural prediction models matters significantly. Random splitting produces overly optimistic performance estimates. Group-aware and temporal validation strategies provide more honest assessments of how models will actually perform when applied to new regions, new crops, or future growing seasons.

5. DISCUSSION

This study examined the performance of different machine learning models in predicting crop yields under both standard and realistic testing conditions. Random train-test splits showed excellent performance (Random Forest F1 $\approx$ 0.97); however, when the models were tested under conditions that better reflected actual use, performance dropped considerably. The central finding was the large gap between the random split results and the structured validation results. When the models were tested on countries they had not seen during training, F1-scores fell to approximately 0.83–0.86 for the best ensemble methods. This indicates that random splitting provides an overly optimistic picture of how these models will perform in practice. The performance drop occurs because random splits allow models to learn country-specific patterns that do not transfer to new regions. Crop-holdout validation revealed an even more serious limitation. When models trained on certain crops were tested on completely different crops, performance collapsed (best F1 $\approx$ 0.39), with several models producing AUC values that were barely better than random guessing. This shows that yield responses to climate differ fundamentally across crop species. A model that learns how maize responds to temperature and rainfall cannot reliably predict how rice or wheat will respond using the same aggregated climate variables. In contrast, temporal validation produced much better results (F1 $\approx$ 0.93–0.94 for Random Forest and Extra Trees). Models trained on historical data predicted future years reasonably well, suggesting that climate–yield relationships remain fairly stable over short time spans when applied to the same countries and crops.

Tree-based ensemble models outperformed linear classifiers in every validation scenario, with the largest gap under random splitting (approximately 35% in F1). This reflects their ability to capture complex interactions between temperature, rainfall, pesticide use, and yield outcome relationships that linear models cannot represent. Hyperparameter tuning with Optuna improved performance in some cases, particularly under country-aware cross-validation, although improvements were inconsistent across different validation protocols. SHAP and LIME explainability analyses consistently identified Item_Encoded (crop type) as the most influential feature, followed by rainfall and average temperature. This confirms that crop identity drives much of the prediction, which helps explain why models fail under crop-holdout validation without access to crop type, they lose their primary signal. Three main conclusions emerge from these results: how you evaluate agricultural prediction models has a major impact on the performance you observe, with random splits producing misleading results; ensemble methods offer strong nonlinear modeling capacity but their ability to generalize varies dramatically across different scenarios; and combining transparent model interpretation with rigorous validation protocols is necessary to understand whether models will actually work when deployed.

6. FUTURE WORK

The current study evaluated X-AMLF using historical aggregated data and did not include real-time sensing, crop stress labels, disease detection, or automated control systems. The results obtained in the current study are promising; however, further testing and development are needed for applications in real-world environments. Although the dataset spans many countries, coverage is uneven across regions and crops, and the applicability or limitations of the model are noticeable. To address this important limitation, future work will use data from different geographical regions, climatic regions, multiple seasons of crops, and multiple crop varieties. This will further enhance the dataset so that the models can learn more diverse patterns and increase their reliability.

To verify the effectiveness and practicality of this framework, testing it in a real agricultural environment is necessary. Future research will focus on field trials, where the X-AMLF framework will be used directly on farms or in fields. This will allow us to directly compare the model predictions with the actual conditions of the crop in the field. These tests will help us fine-tune the model and ensure that it meets the demands of everyday farming.

In the future, the interpretability of this model will be further improved, and more different languages will be added to it. This will make it easier for users to see a clear and complete picture of the entire work in their own language.

Future research may also consider including automated climate control systems, particularly for indoor or controlled-environment farming applications. In such settings, Internet of Things (IoT) devices can be used to monitor temperature, humidity, light intensity, and carbon dioxide (CO2) levels, whereas AI-based control mechanisms can dynamically adjust these parameters to maintain optimal growing conditions. In addition, improving support for low-power and mobile devices could enable real-time decision support, even in regions with limited Internet connectivity. Therefore, future research could need to be focused on effective hardware design and handy visualization tools, which could increase the adoption of sustainable systems by making them relatively inexpensive.

7. CONCLUSION

This study introduced the Explainable Agro-Intelligence Machine Learning Framework (X-AMLF) to evaluate the effectiveness of machine learning models in predicting crop yields using aggregated country-year-crop agricultural data. The framework combined climate indicators and pesticide use data with rigorous validation protocols and explainability techniques to assess both prediction accuracy and robustness under realistic conditions. Under standard random train-test splitting, tree-based ensemble models achieved strong classification performance (F1 $\approx$ 0.97). However, structured validation revealed substantial performance gaps. When tested on countries not seen during training, the performance dropped to F1 $\approx$ 0.83–0.86. Performance declined even further under crop-holdout validation (best F1 $\approx$ 0.39), where models trained on certain crops were tested on completely different crop types. These results show that random splitting can seriously overestimate real-world performance when the data contains geographic or crop-level groupings. Temporal validation, where models trained on 1990–2009 data were tested on 2010–2013, produced much better results (F1 $\approx$ 0.93–0.94), indicating that near-term forecasting within known regions and crops remains feasible. Across all evaluation protocols, ensemble methods like Random Forest, Extra Trees, Gradient Boosting, and XGBoost consistently outperformed linear models by capturing nonlinear interactions between temperature, rainfall, pesticide use, and yield outcomes. Hyperparameter tuning improved performance for some models, particularly Gradient Boosting, although the degree of improvement varied across validation scenarios.

An explainability analysis using SHAP and LIME revealed that crop identity (Item_Encoded) was the most influential feature, followed by rainfall and average temperature. This explains both why the models performed well under random splits and why they failed under crop-holdout validation. The heavy reliance on crop identity shows that the models learn crop-specific patterns rather than general climate–yield relationships that transfer across different crops. This finding underscores the importance of carefully designing features and choosing appropriate validation methods when evaluating model robustness. Three main insights emerged from this research: the evaluation strategy fundamentally shapes the reported performance in agricultural machine learning, with random splits producing misleading optimism; ensemble models offer strong nonlinear modeling but generalize differently across geographic, crop, and temporal dimensions; and combining structured validation with explainability provides a more honest assessment of whether models are ready for deployment than accuracy metrics alone. The X-AMLF framework contributes a rigorous evaluation methodology for agricultural machine learning systems by demonstrating that robustness analysis and interpretability matter as much as predictive accuracy when building tools intended for real agricultural decision-making.

REFERENCES

Akintuyi, O. B.Olabimpe Banke Akintuyi. (2024). Adaptive AI in precision agriculture: A review: Investigating the use of self-learning algorithms in optimizing farm operations based on real-time data. *Open Access Research Journal of Multidisciplinary Studies*, *7*(2), 16. DOI: 10.53022/oarjms.2024.7.2.0023

Alsanoosy, T., & Malik, J. A. (2025). Predicting plant stress using SAM-L: Novel self-adaptive-meta learner with XAI based on soil moisture and chlorophyll analysis. *Scientific Reports*, *15*(1), 42304. Advance online publication. DOI: 10.1038/s41598-025-26184-w PMID: 41309723

Amsalu, A., Garedew, W., Melka, G. A., Çelik, M. A., & Gemeda, D. O. (2026). Trends in climate extreme indices and implications for coffee farming in The Kafa Biosphere Reserve, southwest Ethiopia. *Climate Services*, *42*, 100637. DOI: 10.1016/j.cliser.2026.100637

Awad, M., & Khanna, R. (2015). Support Vector Regression. In *Apress eBooks* (p. 67). https://doi.org/DOI: 10.1007/978-1-4302-5990-9_4

Beckmann, L., Debener, J., & Kriebel, J. (2023). Understanding the determinants of bond excess returns using explainable AI. *Journal of Business Economics*, *93*(9), 1553–1590. DOI: 10.1007/s11573-023-01149-5

Begum, T., & Rahman, S.. (2024). *Machine learning-based crop recommendation system for precision agriculture*. IEEE Xplore., DOI: 10.1109/ICAIT61638.2024.10690764

Botero-Valencia, L., López, D., & García, L.. (2025). Machine learning in sustainable agriculture. *Agriculture*, *15*(4), 377. Advance online publication. DOI: 10.3390/agriculture15040377

Cambrin, D. R., Poeta, E., Pastor, E., Cerquitelli, T., Baralis, E., & Garza, P. (2024). KAN You See It? KANs and Sentinel for Effective and Explainable Crop Field Segmentation. *arXiv*. https://doi.org//arxiv.2408.07040DOI: 10.48550

Cartolano, A., Cuzzocrea, A., & Pilato, G. (2024). Analyzing and assessing explainable AI models for smart agriculture environments. *Multimedia Tools and Applications*, *83*(12), 37225–37246. DOI: 10.1007/s11042-023-17978-z

Cartolano, D., Fernandes, L., & Sato, K.. (2024). *Explainable AI model for enhancing transparency in smart agriculture*. Springer., DOI: 10.1007/s11042-023-17978-z

Çelik, M. A. (2026). Agro-climatic classification of Türkiye's Agricultural Basins using long-term ERA5-Land temperature data. *Physics and Chemistry of the Earth Parts A/B/C, 142*, 104307. DOI: 10.1016/j.pce.2026.104307

Çelik, M. A., Kıraç, A., & Gouveia, C. M. (2026). North Atlantic Oscillation modulation of drought and vegetation activity in Türkiye. *Theoretical and Applied Climatology, 157*(2), 74. Advance online publication. DOI: 10.1007/s00704-025-05971-5

Chen, Y.-H., Sharma, K., Sharma, C., & Sharma, S. (2023). Integrating explainable artificial intelligence and blockchain to smart agriculture: Research prospects for decision making and improved security. *Smart Agricultural Technology, 6*, 100350. DOI: 10.1016/j.atech.2023.100350

Fatimah, M., & Malik, J. A., et al. (2024). Integration of Agriculture 5.0 and Explainable AI (XAI) for Sustainable Development. *Environmental Science Journal.* https://doi.org/DOI: 10.28991/ESJ-2024-08-02-024

Geurts, P., Ernst, D., & Wehenkel, L. (2006). Extremely randomized trees. *Machine Learning, 63*(1), 3–42. DOI: 10.1007/s10994-006-6226-1

Hu, W., Zhao, F., ... (2023). Interpretable ML framework for crop-climate relationship analysis. *Elsevier.* https://www.elsevier.com/open-access/userlicense/1.0/

Huang, Y., Xu, T., Yang, Q., Pan, C., Lü, Z., Chen, H., Zhang, X., & Chen, C. (2023). Demand prediction of medical services in home and community-based services for older adults in China using machine learning. *Frontiers in Public Health, 11*, 1142794. Advance online publication. DOI: 10.3389/fpubh.2023.1142794 PMID: 37006569

Kumar, R., & Prasad, V.. (2025). *Machine learning models for sustainable agricultural development.* IEEE Xplore., DOI: 10.1109/AUTOCOM64127.2025.10956454

Laktionov, I., Diachenko, G., Rutkowska, D., & Kisiel-Dorohinicki, M. (2023). An Explainable AI Approach to Agrotechnical Monitoring and Crop Diseases Prediction in Dnipro Region of Ukraine. *Journal of Artificial Intelligence and Soft Computing Research, 13*(4), 247–272. DOI: 10.2478/jaiscr-2023-0018

Lapajne, J., & Yegorov, V.. (2025). UAV-based phenomics for potato crop analysis. *Computers and Electronics in Agriculture.* Advance online publication. DOI: 10.1016/j.compag.2023.107663

Liu, Y., Choi, T., & Liu, X. (2025). Towards Fully Automated Decision-Making Systems for Greenhouse Control: Challenges and Opportunities. *arXiv.* https://doi.org//arxiv.2503.21640DOI: 10.48550

Logeshwaran, J., Pradeep, R., Ramya, K., Siva, P., & Raghavan, R.. (2024). Improving crop production using an agro-deep learning framework in precision agriculture. *BMC Bioinformatics*, *25*(1), 341. Advance online publication. DOI: 10.1186/s12859-024-05970-9 PMID: 39487390

Mallik, S., Chakraborty, A., Podder, K., Talukdar, S., Rahman, A., & Mishra, U. (2025). Enhancing soil moisture prediction with explainable AI: Integrating IoT and multi-sensor remote sensing data through soft computing. *Applied Soft Computing*, *180*, 113406. DOI: 10.1016/j.asoc.2025.113406

Mansoor, S., Iqbal, S., Popescu, S. M., Kim, S. L., Chung, Y. S., & Baek, J.-H. (2025). Integration of smart sensors and IoT in precision agriculture: Trends, challenges and future perspectives. *Frontiers in Plant Science*, *16*, 1587869. DOI: 10.3389/fpls.2025.1587869 PMID: 40438737

Martin, A., Liu, Z., & Zhang, Y.. (2024). *XAI-powered smart agriculture framework for improving food production and sustainability*. IEEE., DOI: 10.1109/ACCESS.2024.3492973

Mohammed, S., & Yusuf, H.. (2025). *The role of AI in transforming precision agriculture*. Precision Agriculture and Sustainability., DOI: 10.14719/pst.6175

Mohan, R. N. V. J., Rayanoothala, P. S., & Sree, R. P. (2025). Next-gen agriculture: Integrating AI and XAI for precision crop yield predictions. *Frontiers in Plant Science*, *15*, 1451607. DOI: 10.3389/fpls.2024.1451607 PMID: 39845494

Nasir, N., Ramzan, S., Ali, B., Jabbar, S., Raza, A., Kim, C., Syafrudin, M., & Fitriyani, N. L. (2025). Interpretable deep multimodal-based tomato disease diagnosis and severity estimation. *Scientific Reports*, *15*(1), 37919. Advance online publication. DOI: 10.1038/s41598-025-21611-4 PMID: 41162432

Patil, Y., Ramachandran, H., Sundararajan, S., & Srideviponmalar, P. (2025). Comparative analysis of machine learning models for crop yield prediction across multiple crop types. *SN Computer Science*, *6*(1), 64. DOI: 10.1007/s42979-024-03602-w

Petropoulos, G., & Mantas, P.. (2025). *Interpretable machine learning framework for pulse crop yield prediction*. MDPI Applications., DOI: 10.3390/app15137074

Ponnurajan, M., & Iqbal, A.. (2025). *Federated Learning for crop yield prediction*. IEEE., DOI: 10.1109/ASSIC64892.2025.11158569

Qin, F., & Li, Y.. (2025). Predicting the effect of mechanical soil compaction on soybean crop yields. *Agricultural Systems*. Advance online publication. DOI: 10.1016/j.inpa.2025.09.002

Ramesh, V., & Kumaresan, P. (2025). Stacked ensemble model for accurate crop yield prediction using machine learning techniques. *Environmental Research Communications*, *7*(3), 035006. DOI: 10.1088/2515-7620/adb9c0

Rane, J., Kaya, Ö., Mallick, S. K., & Rane, N. L. (2024). Smart farming using artificial intelligence, machine learning, deep learning, and ChatGPT: Applications, opportunities, challenges, and future directions. https://doi.org/DOI: 10.70593/978-81-981271-7-4_6

Razak, A., Yogarayan, S., Sayeed, M. S., & Derafi, M. I. F. M. (2024). Agriculture 5.0 and Explainable AI for Smart Agriculture: A Scoping Review. *Emerging Science Journal*, *8*(2), 744–760. DOI: 10.28991/ESJ-2024-08-02-024

Sengupta, A., Guler, S. I., Gayah, V. V., & Warchol, S. (2023). Evaluating the reliability of automatically generated pedestrian and bicycle crash surrogates. *arXiv (Cornell University)*. https://doi.org//arxiv.2307.13178DOI: 10.48550

Shams, M. Y., Gamel, S. A., & Talaat, F. M. (2024). Enhancing crop recommendation systems with explainable artificial intelligence: A study on agricultural decision-making. *Neural Computing & Applications*, *36*(11), 5695–5714. DOI: 10.1007/s00521-023-09391-2

Shams, S., Chowdhury, T., ... (2024). XAI-CROP: Explainable AI-based Crop Recommendation System for Precision Agriculture. *Springer*. https://link.springer.com/article/10.1007/s00521-023-09391-2

Shastri, S., Kumar, S., Mansotra, V., & Salgotra, R. (2025). Advancing crop recommendation system with supervised machine learning and explainable artificial intelligence. *Scientific Reports*, *15*(1), 25498. DOI: 10.1038/s41598-025-07003-8 PMID: 40664775

Shawon, S. M., Ema, F. B., Mahi, A. K., Niha, F. L., & Zubair, H. T. (2025). Crop yield prediction using machine learning: An extensive and systematic literature review. *Smart Agricultural Technology*, *10*, 100718. DOI: 10.1016/j.atech.2024.100718

Syari, M., & Zhang, X.. (2025). *IoT-based smart farming for spinach cultivation*. IEEE Xplore., DOI: 10.1109/ICCIT65724.2025.11167551

Tahir, H. A., Alayed, W., & Hassan, W. U. (2025). A Federated Explainable AI Framework for Smart Agriculture: Enhancing Transparency, Efficiency, and Sustainability. *IEEE Access : Practical Innovations, Open Solutions*, *13*, 97567–97584. DOI: 10.1109/ACCESS.2025.3571340

Talukder, M., Islam, M. M., Uddin, M. A., Hasan, K. F., Sharmin, S., Alyami, S. A., & Moni, M. A. (2024). Machine learning-based network intrusion detection for big and imbalanced data using oversampling, stacking feature embedding and feature extraction. *Journal of Big Data*, *11*(1), 33. Advance online publication. DOI: 10.1186/s40537-024-00886-w

Tan, M., & Liu, Y.. (2025). Optimized Random Forest framework for rice yield prediction. *Agronomy Journal*. Advance online publication. DOI: 10.3390/agronomy15102273

Tariq, M., Ali, U., Abbas, S., Hassan, S., Naqvi, R. A., Khan, M. A., & Jeong, D. (2024). Corn leaf disease: Insightful diagnosis using VGG16 empowered by explainable AI. *Frontiers in Plant Science*, *15*, 1402835. Advance online publication. DOI: 10.3389/fpls.2024.1402835 PMID: 38988642

Turgut, O., Kok, I., & Ozdemir, S. (2024). AgroXAI: Explainable AI-Driven Crop Recommendation System for Agriculture 4.0. *arXiv*. https://doi.org/DOI: 10.1109/BigData62323.2024.10825771

Zou, L., & Li, C.. (2025). *Ensemble-quantile regression for crop yield prediction*. IEEE., DOI: 10.1109/ISSSR65654.2025.00057

Chapter 9
Advanced Artificial Intelligence and Multi-Source Remote Sensing for Agricultural Monitoring and Forecasting

S. Ramakrishnan
http://orcid.org/0009-0007-1488-7609
Erode Sengunthar Engineering College, India

S. Maheswari
http://orcid.org/0000-0003-0097-4887
Velalar College of Engineering and Technology, India

G. Pradeepa
Vivekanandha College of Engineering for Women, India

P. Brindha
http://orcid.org/0000-0002-0232-0572
Erode Sengunthar Engineering College, India

Brindha Samarasam
http://orcid.org/0000-0003-4179-4161
Nandha Engineering College, India

Parimala Devi M
http://orcid.org/0000-0003-3955-1339
Velalar College of Engineering and Technology, India

ABSTRACT

Agricultural systems are multifaceted adaptive systems that are constructed through dynamic processes of interaction between climate, soil, vegetation, and management activities, and which are progressively evidently, climate variability and extreme events. Conventional methods of monitoring have constraints of low spatial coverage,

DOI: 10.4018/979-8-2600-0778-5.ch009

expensive nature and low sensitivity to the spatiotemporal heterogeneity. Current developments in satellite remote sensing have made available large amounts of multisensory data including optical, SAR, and thermal data, that when combined with the meteorological and edaphic data, can be used to fully monitor the agriculture. This chapter reviews the methods of artificial intelligence (AI) and machine learning, deep learning, and spatiotemporal models to monitor crops, predict yields and assess food security. The focus is put on the time-series analysis, multi-source data fusion, and early-warning applications, including the drought detection and yield-gap analysis.

1. INTRODUCTION

Agricultural production systems represent a primary contact point between the Earth system and the human society. They are the source of world food security and are also sensitive to climatic, environmental, and anthropogenic processes. The interplay of atmospheric conditions, soil characteristics, water availability, crop physiology, and management practices is complex and controls crop growth and yield formation across multiple spatial and temporal scales. Over the past few decades, the interactions have been further destabilized by climate change, land-use acceleration, and the growing resource stress culminating in increased variability in crop productivity and increased vulnerability of food systems in the global context (Lobell et al., 2013; Rockström et al., 2017). In such circumstances, there is a strong need to monitor agriculture and estimate yields in time to assist adaptive management, stabilize food supply chains, and make informed policy interventions on regional to global levels.

The traditional agricultural monitoring methods such as field-based surveys, administrative statistics and empirical yield models still have a significant role in production assessment. These techniques are, however, usually limited in terms of spatial representativeness, operational cost, reporting delays, and data reliability. These limitations become more pronounced when applied to large and heterogeneous agricultural regions (Becker-Reshef et al., 2020). These restrictions are particularly strong in low-income and middle-income nations, whereby the sparse network of observation, and lack of the institutional capacity, limit the availability of quality and timely agricultural information. This means that traditional surveillance systems do not tend to record intra-season, spatial heterogeneity, and quick transformations caused by extreme weather conditions.

Remote sensing has evolved into satellite-based observation systems. These systems serve as a cornerstone for large-scale agricultural monitoring by providing systematic, repeatable, and spatially explicit measurements of land surface conditions. Modern Earth observation systems are designed in such a way that they can allow access to streams of multi-sensor data which encompass optical, Synthetic Aperture Radar (SAR), thermal and hyperspectral data. The data obtained can describe crop type distribution, vegetation vigor, canopy structure, and phenological growth. They also provide information on soil moisture conditions and crop responses to environmental stress across different agro-ecological zones (Atzberger, 2013). Satellite observations have become increasingly dense in space and time, enabling agricultural monitoring of near-real-time and interannual yield evaluation with a great deal of potential.

In spite of these developments, remote sensing data can be analytically exploited to serve agriculture, but this task has not been easy. Satellite data are high-dimensional and contain sensor-specific noise and data gaps. In addition, crop–environment interactions are highly nonlinear. These factors restrict the performance of traditional index-based and linear modeling approaches (Sebastianelli et al., 2025). Although computationally efficient, vegetation indices and rule-based classification methods are usually ineffective to generalize to other areas, other crop types, and climatic conditions. These issues have led to an increased popularity of data-driven analytical models that are able to learn complicated patterns with large heterogeneous data sets.

The recent advancements in artificial intelligence and especially in machine learning and deep learning have significantly enhanced the analysis of agricultural remote sensing. Data-driven models enable automatic feature extraction, nonlinear regression, and spatiotemporal pattern recognition. This improves the accuracy and scalability of crop monitoring and yield prediction tasks (Kamilaris & Prenafeta-Boldú, 2018; Liakos et al., 2018). Convolutional neural networks have shown good results in the prediction of the spatial structure and field-scale variability of satellite imagery, with recurrent and attention-based architectures now being used more frequently to predict seasonal crop dynamics using multi-temporal data. Such techniques are highly applicable to agricultural systems. In these systems, temporal dynamics and spatial heterogeneity are key determinants of crop yield.

AI-based agricultural models have been further enhanced due to the integration of satellite-observations and ancillary datasets. In conjunction with complementary data like meteorological observations, agronomic management data and edaphic properties, data-driven models have stronger physically coherent and agronomically interpretive depiction of crop growth processes (You et al., 2017). These integrated methods have demonstrated greater resilience to diverse climatic conditions and higher transferability to other regions which makes them especially useful to large-scale yield prediction and climate change impact.

In addition to monitoring the fields and estimating yields, remote sensing with the help of AI is increasingly used to evaluate food security and predictive systems. The indicators that can be identified by using satellite-derived data, such as the state of the vegetation, abnormal weather changes in moisture, and climatic stress, are analysed using the latest modeling technologies, which allows identifying the effects of the drought or crop failure at an early stage (Klompenburg et al., 2020). Such abilities are particularly essential in vulnerable and data derivate areas of the environment where timely information can direct proactive actions of the governments, development agencies, and humanitarian organisations.

Still, the practical use of artificial intelligence based agricultural monitoring systems is limited by the major challenges. There are still problems with the quality of training data, uncertainty of labelling, lack of transferability of models across agro-ecological regions, and low interpretability, which hinders scalability (Becker-Reshef et al., 2020). To overcome these problems, careful design of the models, strict validation, and the growing focus on the uncertainty-sensitive and interpretable models that can be used to facilitate a clear and credible decision-making process must be considered. Figure 1 depicts the integration process between satellite data and AI modeling support components within the proposed agricultural intelligence framework.

Figure 1. Conceptual Framework for AI-Enabled Agricultural Monitoring, Yield Estimation, and Food Security

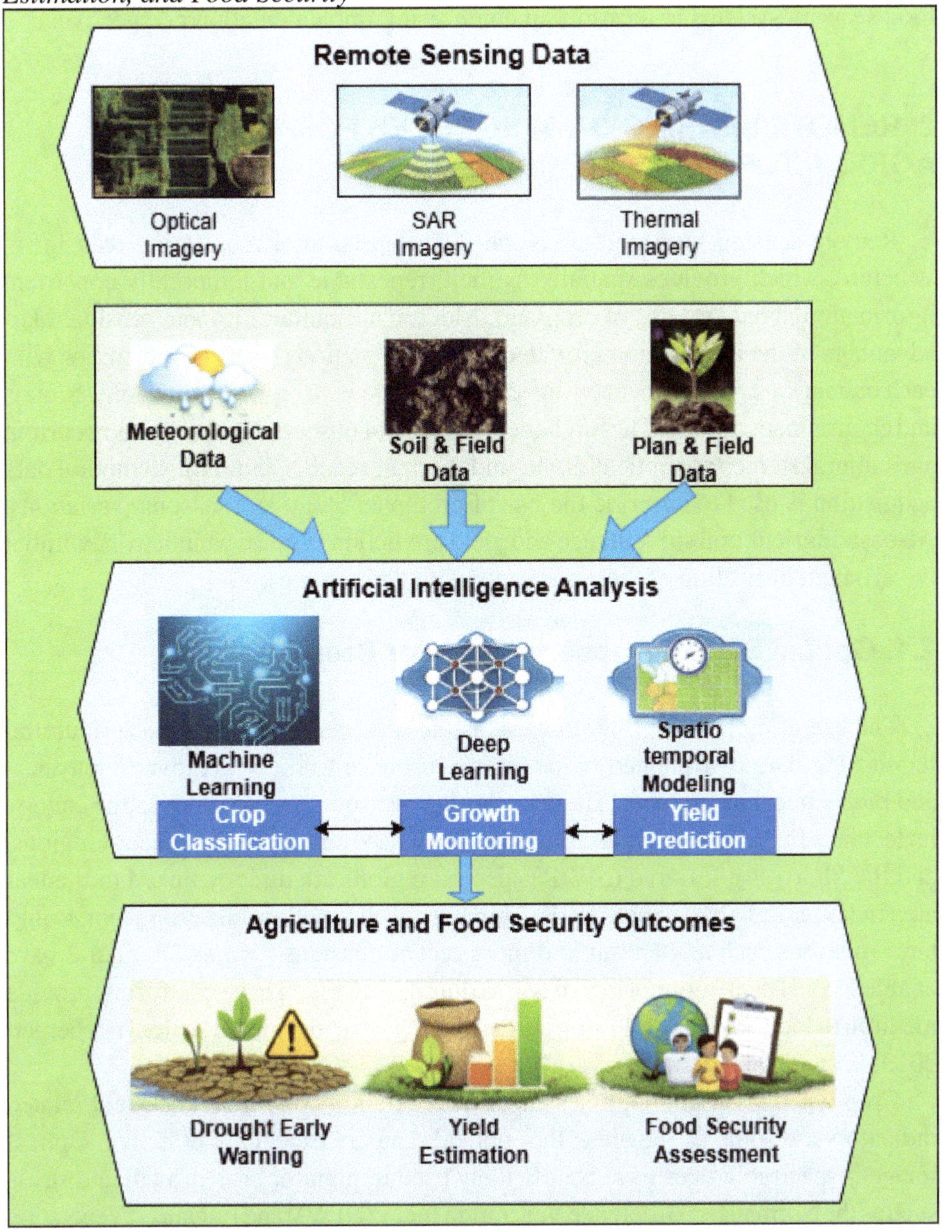

This chapter serves as a methodological synthesis with an operational overview. Its contribution brings a systematic approach to remote sensing analytics, climate-sensitive yield modeling, and uncertainty-sensitive decision structures into a consistent architecture that is applicable to real-life agricultural monitoring. In contrast to previous reviews, it focuses on scalability, rigor of the validation, and deployable food security systems in terms of an engineering implementation perspective.

2. REMOTE SENSING DATA SOURCES FOR AGRICULTURAL MONITORING

Remote sensing via satellites is the main agricultural monitoring data infrastructure, which provides spatially explicit, repeatable and temporally consistent information about systems of cropland. Modern agricultural remote sensing takes advantage of the information provided by a combination of sensor modalities with each one sensing complementary information relevant to crop structure, physiology, and environmental stress. The fundamental pillars of observational systems in current agricultural studies are optical, SAR, and thermal sensors, but multi-temporal data acquisition is used to describe the use of crop phenology and seasonal variations. AI-based agricultural surveillance and yield prediction systems cannot work without the efficient utilization of their corresponding data sources.

2.1. Optical Satellite Observations for Crop Analysis

The use of optical satellite imagery as the most popular source of agricultural monitoring data is explained by its strong response to the vegetative biophysical and biochemical parameters. The sensors that measure the variations in the canopy reflectance in the visible spectral regions, the Near-Infrared (NIR) spectral regions, and the Shortwave Infrared (SWIR) spectral regions are directly linked to the leaf area index, chlorophyll concentration, biomass accumulation, and crop vigor. Long-term missions such as Landsat and more recent missions such as Sentinel-2 have enabled consistent monitoring of agricultural systems. These platforms provide medium to high spatial resolution data across regional and global scales (Atzberger, 2013; Weiss et al., 2020).

Crop type classification, assessment of vegetation condition, and yield related indicators are some of the areas that optical data are especially effective. Optical imagery spectral indices have been widely used to monitor crop growth and stress such as the Normalized Difference Vegetation Index (NDVI) and Enhanced Vegetation Index (EVI). However, although these indices are useful as first-order indicators, they are generally less interpretable in complex situations where cropping patterns are mixed, the background effect of the soil and the inconsistent management practice have been observed. Due to this, AI-based methods have started to use the full multispectral reflectance information, as opposed to using a fixed set of indices, and learn task-specific representations with the data (Sebastianelli et al., 2025).

Though useful, the optical observations are constrained by the weather, cloud cover, and variation of the illumination. Continuous monitoring is hindered by constant cloud cover especially in areas that affect agriculture (especially in monsoon or rainy seasons) where cloud cover causes discontinuities in the data. Such constraints highlight the need to have complementary sensor modalities that would enable the provision of trustworthy observations in all weather conditions. As illustrated in Figure 2, how multispectral optical data capture vegetation and canopy characteristics used for crop monitoring and yield estimation.

Figure 2. Optical Satellite Observations for Crop Analysis

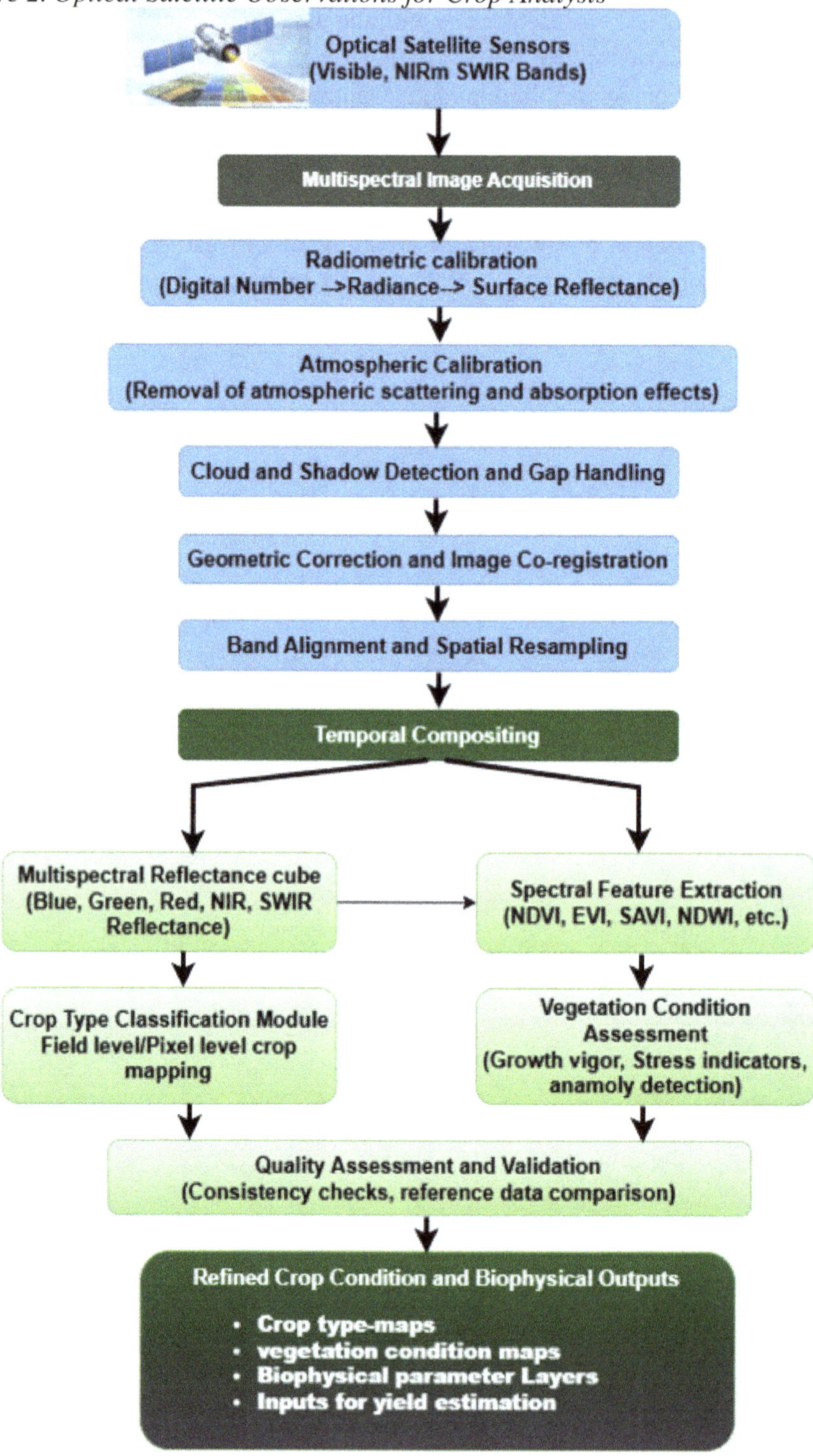

2.2. SAR and Thermal Data for All-Weather Monitoring

SAR represent an essential complement to optical imagery that observes the Earth surface during day and night time and they are cloud independent. SAR backscatter signals are sensitive to crop structure, surface roughness, canopy moisture, and soil moisture making them especially useful in agricultural monitoring in areas prone to cloud (Xu et al., 2018). SAR missions in space e.g. Sentinel-1 have enabled regular collection of multi-polarization data which can be used to map crops, determine their stage of growth and estimate the moisture content of soils.

The SAR observations will prove particularly beneficial in early development stages and in areas with high frequency of clouds, where optical data cannot be gathered. SAR data can be used as supplementary structural and moisture-related data when incorporated into AI-based models to increase the accuracy of classifying crops and to increase resilience to changing environmental factors (Orynbaikyzy et al., 2020). Nevertheless, the process of SAR data interpretation remains highly complicated as the speckle noise and the nonlinear scattering processes complicate the interpretation, which requires the use of advanced preprocessing and learning-based analysis methods. Figure 3, highlights the role of SAR and thermal data in enabling all-weather monitoring and crop stress assessment.

Figure 3. SAR and Thermal Data for All-Weather Monitoring

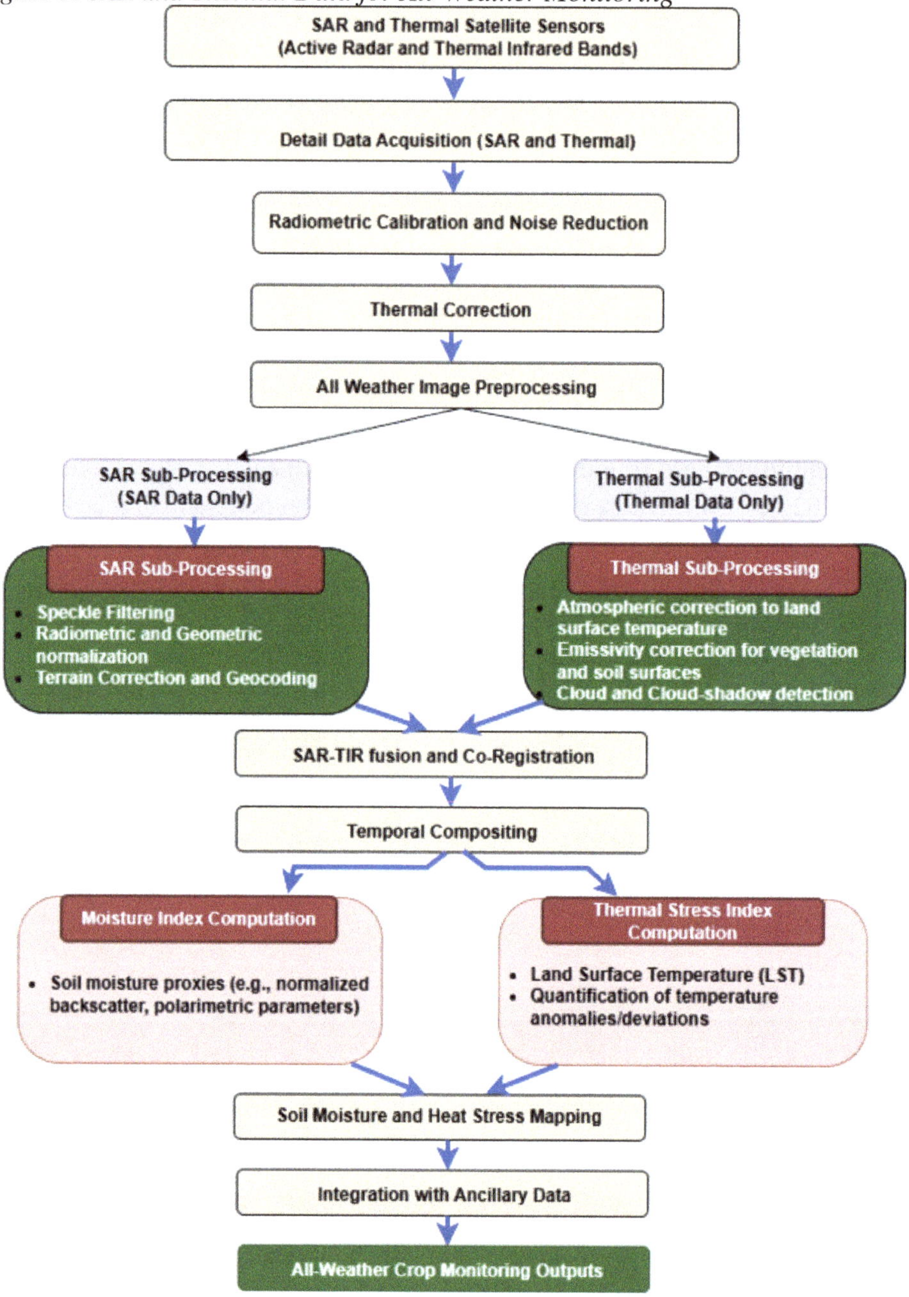

Thermal remote sensing can also give other information on the water status and stress of crops due to the variation of land surface temperatures. The degree of canopy temperature is directly connected to the rate of evapotranspiration and plant transpiration resulting in the thermal data having a special level of applicability in irrigation control, drought recognition and the evaluation of stress (Anderson et al., 2008). Even though thermal sensors generally have relatively coarse spatial resolutions relative to optical platforms, their combination with multispectral and SAR measurements can be used to make more complete measurements of crop conditions in AI-powered monitoring applications.

2.3. Multi-Temporal Observations and Crop Phenology

Farming systems are dynamic in nature and crop maturation is a successive phase of phenological changes through the growing season. Multi-temporal satellite data is thus imperative in depicting seasonal patterns of the vegetation development, stress reaction, and senescence. Satellite data can be analysed by a time-series to characterize phenological parameters of planting dates, maximum biomass timing and harvest periods that are highly predictive of end yields (Sakamoto et al., 2005). As shown in Figure 4, multi-temporal satellite observations capture seasonal crop dynamics for time-series analysis and yield modeling.

Figure 4. Multi-Temporal Observations and Crop Phenology

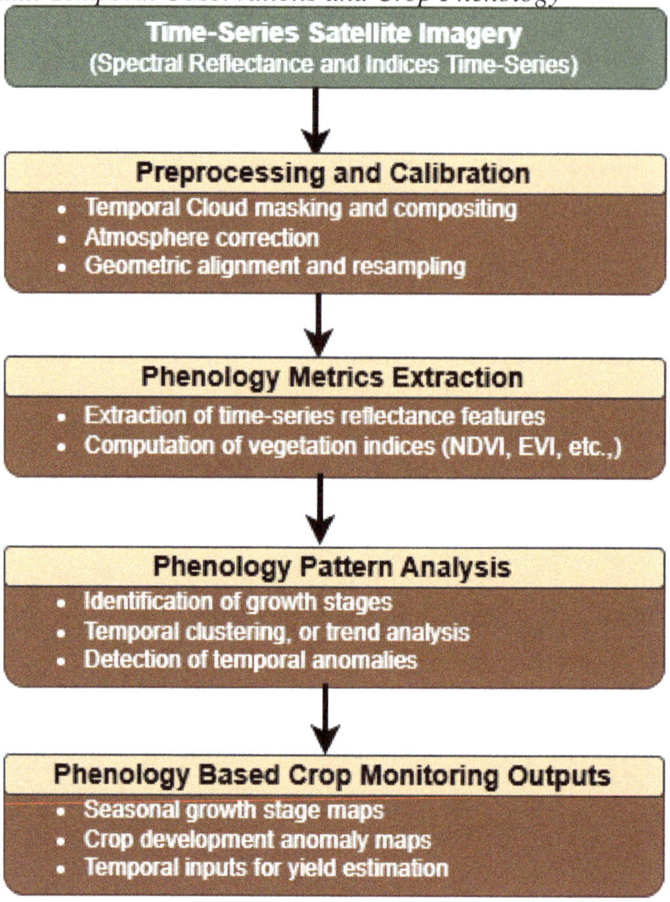

The recent developments in satellite revisit frequency and data access have enabled dense time series analysis, which has made it possible to use spatiotemporal AI models with explicit description of crop growth dynamics. Recurrent neural networks, time-dependent convolutional networks, and transformer-based systems have been shown to be very effective in learning phenological patterns based on a multi-year history of satellite time series (Xiong et al., 2022). These methods are beyond the impermanent snapshot analysis and the models consider the intra-seasonal variation, climate aberration and management induced disparities in crop growth.

Phenological monitoring is further improved through the combination of multi-temporal observations of optical, SAR, and thermal sensors that reduce the gaps between measurements and improve the continuity of observations over time. These combined time-series data would bring a solid base to AI-based yield prediction and food security evaluation especially during unpredictable climatic conditions.

3. ARTIFICIAL INTELLIGENCE TECHNIQUES FOR AGRICULTURAL MONITORING

The agricultural systems are more complex with high nonlinear interactions among climate, soil, vegetation and management practices and thus require the use of analytical methods that can manage high-dimensional, heterogeneous, and spatiotemporally structured data. Agricultural monitoring based on satellites continues to move towards Artificial Intelligence (AI) applications to go beyond index-based and parametric models and towards data-driven frameworks capable of learning discriminative representations directly on top of remote sensing observations. AI methods can be used to support scalable classification of crops, effective spatial feature identification and dynamic modelling of crop development processes in various agro-ecological conditions (Miralles et al., 2021; Reichstein et al., 2019).

In the agricultural remote sensing context, AI tools may be divided into three functional groups, namely

(i) Machine learning models of crop classification based on engineered features
(ii) Deep learning models of spatial representation learning, and
(iii) Spatiotemporal models that explicitly model crop dynamics throughout the growing season. Figure 5 depicts the various techniques to monitor agriculture.

Figure 5. Different techniques for Agriculture monitoring

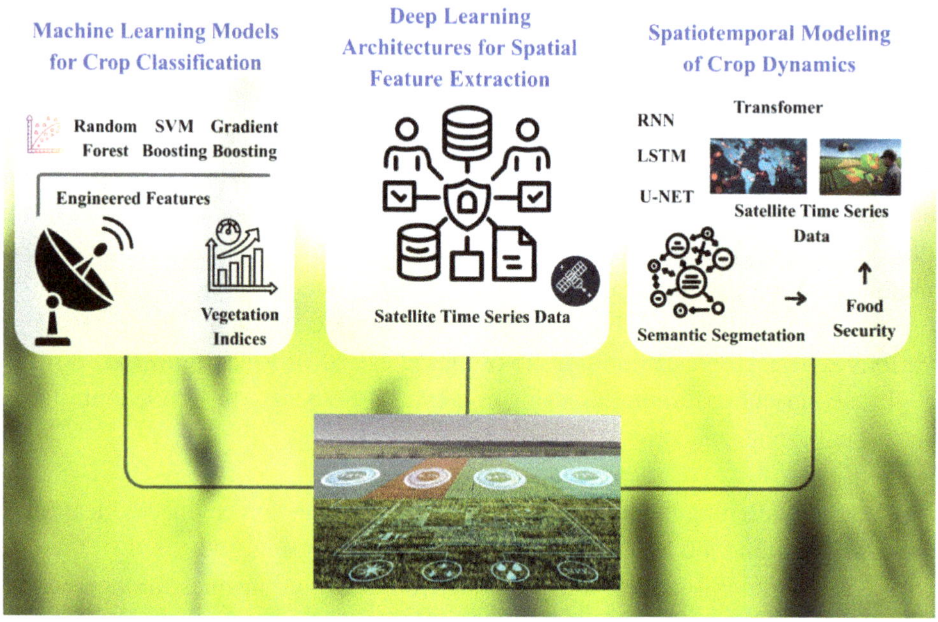

3.1. Machine Learning Models for Crop Classification

The wide application of Machine Learning (ML) models to crop-type classification can be explained by the ability to work successfully in high-dimensional feature space and nonlinear decision boundaries. The most common ML algorithms in use are Random Forests (RF), Support Vector Machines (SVM), gradient-boosting machines and ensemble-based classifiers, which are usually trained on spectral reflectance, vegetation indices, textural measures, and phenological measures based on satellite images (Liakos et al., 2018).

The use of random-forest classifiers in the agricultural field is especially favourable due to their resistance to noisy data, their capacity to deal with correlated data, and their natural anti-over-fitting capacity when applied to small amounts of labeled data (Inglada et al., 2017). Additionally, RF models offer measures of feature importance, which allow an interpretation of classification decisions partially, which is a significant aspect of operational agricultural monitoring. As computationally more challenging, but useful in separating spectrally similar classes of crops, SVM-based methods especially work well when the training data are sparse or unevenly distributed. Even though ML models can be computationally efficient and interpretable, the use of handcrafted features restricts their capability to utilize the spatial context and fine-scale structure of the images whereas high-resolution images have.

3.2. Deep Learning Architectures for Spatial Feature Extraction

Deep Learning (DL) models overcome these shortcomings of traditional models of ML by allowing hierarchical learning of features based on raw or only slightly processed satellite images. The basis of spatial deep learning in agricultural remote sensing is Convolutional Neural Networks (CNNs), which can use local spatial patterns, textures and contexts to identify crop types and evaluate canopy condition (Kamilaris & Prenafeta-Boldú, 2018) .

CNN in models use learnable convolutional filters on multispectral image patches, gradually changing spectral information of low level to high level representation of crop structure, field geometry and intra field variability. To be used in heterogeneous agricultural landscapes, fully convolutional models like U-Net and encoder-decoder networks have proven to be popular in pixel-level crop mapping and semantic segmentation problems (Ronneberger et al., 2015). The CNN-based methods are usually computationally intensive involving high training data volumes and demand high performance because of their high-performance levels. The recent works investigated transfer learning, self-supervised pretraining and weakly supervised learning schemes to improve the model generalization and to minimize the need of annotation (Zhu et al., 2022).

3.3. Spatiotemporal Modeling of Crop Dynamics

Agricultural practices are dynamic in nature and the growth of the crop goes through specific phenological stages that are highly affected by weather variability and management. The spatial models, as such, are not sufficient to describe the complete characteristics of the crop over the growing period. Explicit spatiotemporal AI models allow taking into account time-related data and studying the seasonal patterns, growth abnormalities, and interannual variability (You et al., 2017).

RNNs and Long Short-Term Memory (LSTM) networks have become popular models to predict multi-temporal satellite time series because they are capable of modeling temporal relationships and irregular observation periods due to cloud contamination. More recently, architectures based on transformers have become a point of interest, thanks to their capability to learn long-range temporal connections via attention mechanisms that provide models with opportunities to detect phenologically informative time steps not based on explicit temporal ordering (Garnot et al., 2020). Notably, spatiotemporal AI models allow predictive analysis at an earlier stage in the growing season, which facilitates the planning of agricultural management and food-security.

4. AI-BASED CROP YIELD ESTIMATION AND FORECASTING

The estimation of crop yields is considered one of the most difficult and significant issues in remote sensing and analytics of agriculture. Yield is not a direct observation but a cumulative effect of interactions between crop phenology, climatic variation, soil properties, and guidelines of management procedures throughout the entire growing season. Such processes are non-linear, spatially heterogeneous and time-varying making prediction of yield fundamentally more complex than crop classification or conditions assessment (Jabed & Murad, 2024; Kamilaris & Prenafeta-Boldú, 2018). Conventional yield-estimation techniques, such as empirical regression models, agro-meteorological indices or end-of-season surveys, are usually limited by simplified assumptions, lateness-availability and insufficient generalisation with unusual climatic conditions.

In recent years, the crop yield estimation and forecasting have become a prevailing paradigm of AI based methodologies. The ability to do joint modelling of the multi-source, multi-temporal data sets have shown superior predictive value and extended forecast lead times of AI approaches over traditional statistical and empirical models (Jabed & Murad, 2024; Kamilaris & Prenafeta-Boldú, 2018). These methods are especially useful in modelling non-linear crop-climate-soil interactions and also in making full use of dense satellite time series that capture the seasonal vegetation dynamics. In this regard, therefore AI-based yield models are gradually being employed at spatial scales including field-scale decision support, regional-scale production forecasting to support food security planning, and Nation-wide production forecasting (Reichstein et al., 2019).

4.1. Time-Series Analysis for Yield Prediction

Time-series based yield prediction takes advantage of the high empirical and physiological correlation between seasonal vegetation patterns and crop yield. The data available as multi-temporal satellite (e.g. surface reflectance, vegetation indices (e.g. NDVI, EVI), derived biophysical parameters (e.g. leaf area index, fraction of absorbed photosynthetically active radiation) and then arranged in a sequence over time, representing crop development between emergence and senescence. These sequences encode cumulative biomass production as well as the stress induced deviations which have direct effect on yield outcomes.

The Long Short-Term Memory networks and attention-based models have shown themselves to be very effective sequence-learning models used to model these temporal dependencies. The LSTM networks address the vanishing-gradient problem of the conventional recurrent neural networks, and therefore, they can store long-term temporal information over increasing seasons. Attention-based models also contribute to better performance by laying more emphasis on phenologically significant times, which are flowering and grain-filling stages, which has been known to have disproportionate effects on final yield (Fernandez-Beltran et al., 2021; Qiao et al., 2021).

Regular empirical analyses of large crops including maize, wheat, rice, and soybean are regularly reporting that models being trained on full-season or partial-season time series performed better than univariate single-date or peak-season models. It is reported that the Root Mean Square Error (RMSE) decreases by 10 to 30% when multi-temporal inputs are used in place of the static vegetation indices, with the largest improvement in this parameter in the case of early-season predictions (Singh et al., 2022). This is especially useful in early warning systems where it is possible to produce yield projections a few weeks or months before harvest.

Time-series models are further improved with phenology-aware features to predict them. Cumulative greenness, growth rate, timing of peak vegetation and slope of senescence metrics also make model learning consistent with agronomically significant growth stages. These representations make them less sensitive to noise and inter-annual deviation in the timing of observation, which make them yield stronger yield forecasts across regions and seasons (Ji et al., 2022).

Figure 6. Time-series–based AI workflow for crop yield estimation.

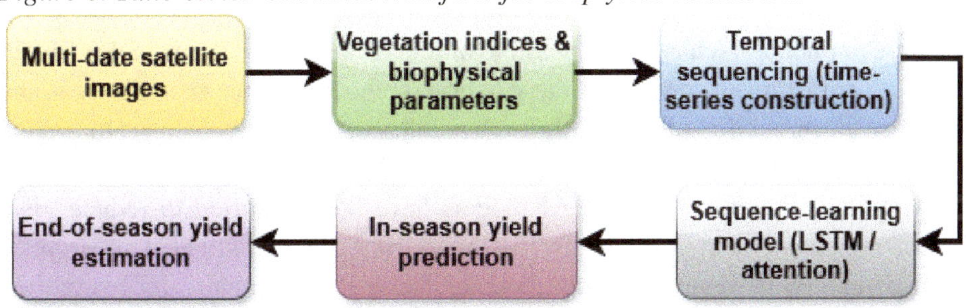

This time-series modelling paradigm is exemplified in Figure 6 and in this example, multi-date satellite data is converted to multi-temporally structured inputs which are fed through sequence-learning models to generate in-season yield predictions and end-of-season yield projections.

4.2. Climate-Aware Yield Modeling Using AI

Though vegetation time series provide integrated crop responses, they might not well explain the underlying cause of stress and inter-annual variation of yield. Climate extremes like heat waves, droughts, too much or too little rain, and radiation deficit, these extremities can significantly decrease the yield even where the indices of the vegetation seem normal at the early stages of growth. Therefore, yield modelling that takes climate into account has emerged as an urgent enhancement of AI-based models of forecasting.

The models of AI that are sensitive to climate directly combine meteorological drivers such as temperature extremes, precipitation anomalies, vapor pressure deficit, and solar radiation with the satellite-based vegetation indicators. This integration enables models to both directly describe the effects of climate on the physiology of crops, and indirectly describe the effects that are mediated by vegetation dynamics (Lobell et al., 2013). The heat stress at flowering will not necessarily result in less green but can by itself cause a great loss in the setting of grain and ultimate yield.

There are two overriding modelling strategies in the literature. The former is feature-level integration, where climatic and soil variables are both concatenated with remote sensing time series and processed together by machine learning or deep learning networks. This design allows models to acquire cross-domain interaction directly through data and this architecture has been demonstrated to increase yield prediction effectiveness especially in climatically variable areas (Sah et al., 2024).

The second approach is an approach of hybrid or process-informed modelling, which means that the results of crop-growth or water-balance models are introduced into the AI structures as either input or output. Axial features such as simulated soil moisture or phenological stage or possible biomass can be used as examples. Under non-stationary climate conditions, these hybrid methods enhance generalisation and minimise the possibility of physically implausible forecasts - a recognised limitation of entirely data-driven models (Reichstein et al., 2019).

Quantitatively, climate-aware AI models have been reported to reduce yield forecasting errors by approximately 15–25% during drought and heat-stress years when compared to vegetation-only models. These improvements underscore the importance of explicitly incorporating climate drivers for climate-resilient agricultural forecasting (Cheng et al., 2022).

4.3. Large-Scale and Regional Yield Forecasting

The extrapolation of scaling of yield estimation of local fields to regional and national scales brings up further complications associated with regional and national spatial heterogeneity, mixed farming practices and variability of data availability and reflectance. Massive yield forecasting systems have to balance between fine-scale predictions and the administrative reporting units and be consistent with the official agricultural statistics (Ji et al., 2022).

AI-based regional prediction systems usually combine field level or pixel level yield forecasts based on crop area maps obtained by satellite-based classification. The methodologies used to deal with spatial heterogeneity are usually hierarchical and ensemble modelling. The hierarchical models are used where local models are trained in the regions with available data and the higher-level models are used to combine information over the agro-ecological areas, thus leading to better transferability and robustness of predictions (Ji et al., 2022).

5. MULTI-SOURCE DATA FUSION FOR AGRICULTURAL INTELLIGENCE

Biophysical and climatic factors interact with management factors to cause changes to agricultural systems at both spatial and temporal scales. These interactions cannot be sufficiently portrayed using any single source of data especially in cases of climate variability and non-uniform land management. Consequently, there has been the emergence of multi-source data fusion, as a new paradigm in agricultural intelligence, which allows the enhancement of AI-based monitoring and decision-support systems, their reliability, robustness, and scalability. Fusion-based methods decrease uncertainty, minimize the data gaps and increase the explanatory capacity of predictive models by integrating satellite data with meteorological, soil, and ancillary data (Camps-Valls et al., 2011) (Belgiu & Drăguţ, 2016).

Analytically, data fusion can help AI models to go beyond correlation-based inference and furthermore provide more situated descriptions of the relationship between crops and the environment. This is of great significance to yield estimation, stress detection and early warning systems in which isolated data streams have the potential to miss cumulative or lagging effects of environmental drivers.

5.1. Integration of Satellite, Meteorological, and Soil Data

Agricultural data fusion is based on the combination of satellite remote sensing data with the meteorological and soil data. The soil datasets also add in the form of a dataset, the static and semi-dynamic parameters, such as soil texture, soil depth, organic matter content, and water-holding capacity, which control the crop response to climatic factors and management activities (Lobell et al., 2009).

This integrated perspective is advantageous in AI-based models as they are able to learn both in the short term (crop response to stress, e.g., stress-induced reflectance changes) and in the long term (cumulative effects, e.g. seasonal water availability). An example is that vegetation indices can be used to predict drought effects and yield losses by combining them with rainfall variations and soil moisture proxies, which is more successful than satellite-only methods (Lu et al., 2024). Further, the addition of soil characteristics enhances generalisation of models across those regions that have different edaphic conditions, which is an essential requirement to regional and national scale agricultural monitoring.

5.2. AI-Based Feature-Level and Decision-Level Fusion

The fusion strategies that are usually used in agriculture via AI are applicable at both the feature level and the decision level, each being beneficial in its own way. In feature-level fusion, several data sources are pre-fused together at feature level before model training. This method enables learning algorithms to directly learn the interactions across more than two or three modalities. More importantly, multimodal deep learning architectures, attention-based fusion systems, and joint latent embedding systems are actively used to promote feature-level combinations of satellite imagery, weather time series and soil characteristics (Gao et al., 2020).

In comparison, decision-level fusion retains separate models of each source of data and combines the outputs of the models using ensemble techniques or meta-models. The method is especially beneficial in cases where the sources of data vary widely in terms of each of the three mentioned factors. Decision-level fusion makes the redundancy of any particular data stream, thereby improving robustness and allowing models to be useful even when part of the data is unavailable, which is a common problem in practice in agricultural monitoring (Raza et al., 2025). Ensemble based decision fusion has been observed to enhance stability, as well as reduce prediction variance in seasonal and geographically varying contexts.

5.3. Cloud-Based Platforms for Large-Scale Agricultural Analytics

The process of multi-source data fusion on large scale requires massive computation and effective data-management systems (Gorelick et al., 2017). Cloud-native architectures allow directly incorporating the AI models into data-processing streams, which facilitates the minimization of data transfer and allows monitoring data in near-real-time. This is especially useful in terms of early warning systems, where it is essential that the stress or yield anomalies in crops are detected in time. In addition, cloud environments encourage reproducibility and transparency through standardisation of data access and analysis processes, which encourage collaborative research and operational implementation (Ozdogan et al., 2010).

5.4. Data Harmonization and Resolution Enhancement Techniques

Multi-source fusion involves careful harmonisation of multi-source datasets with different spatial resolutions, temporal frequency, spectral properties and with different coordinate systems. Standardized preprocessing and harmonization procedures are applied to ensure radiometric consistency, temporal continuity, and cross-sensor comparability prior to model integration (Song et al., 2001). The AI models can find spurious patterns in the presence of sensor artefacts instead of real biophysical signals without this kind of preprocessing (Belgiu & Drăguţ, 2016). These augmented datasets are better basis of performing downstream tasks like yield prediction, stressing, and food-security evaluation.

6. AI-ENABLED FOOD SECURITY ASSESSMENT AND EARLY WARNING SYSTEMS

Food-security assessment and early warning all represent essentially spatiotemporal inference problems: they require

(i) real-time monitoring of crop status and area,
(ii) to identify abnormal deviations of seasonal bestseller curves,
(iii) to measure exposure to climatic extremes, and
(iv) to use local production estimates to construct probabilistic regional and national predictions.

Traditional methods of monitoring such as household surveys, administrative statistics, and ad hoc professional judgments have latency, sampling bias, a poor spatial resolution, and, thus, their usefulness in predictive action during the evolving disasters is limited (Gorelick et al., 2017; Justice et al., 2019). Near-real-time, spatially explicit, uncertainty-sensitive food-security analytics based on the combination of satellite remote sensing, environmental data, and AI can provide a way forward to operate an operational early-warning system.

As a matter of fact, AI-based food-security systems combine multi-sensor remote sensing (optical, SAR, thermal) with gridded meteorological fields, soil and agronomic databases, past yield reports, and socio-economic indicators in end-to-end pipelines that undertake anomaly detection, impact quantification, and probabilistic prediction. In this description, technical methods, data flows, and model design options of four priority tasks are given, namely, crop-failure detection and yield-gap analysis; drought, flood, and extreme-event impact assessment; regional and nation-wide production forecasting; and policy and planning decision support. In both the subsections, there is a focus on algorithms, input preconditioning, uncertainty quantification as well as validation protocols that befits a technical chapter.

6.1. Crop Failure Detection and Yield Gap Analysis

a. **Problem framing:** It presents the change-point/anomaly-detection problem as based on multi-temporal vegetation signals. It aims at signaling the exit of the climatological envelope of agronomical importance (i.e. would lead to lower final yield). Yield-gap analysis decomposes observed yield shortfalls into climate-limited and management-limited components to guide targeted interventions. Figure 7 shows the crop failure detection and yield gap analysis.

Figure 7. Crop Failure and Yield Gap Analysis

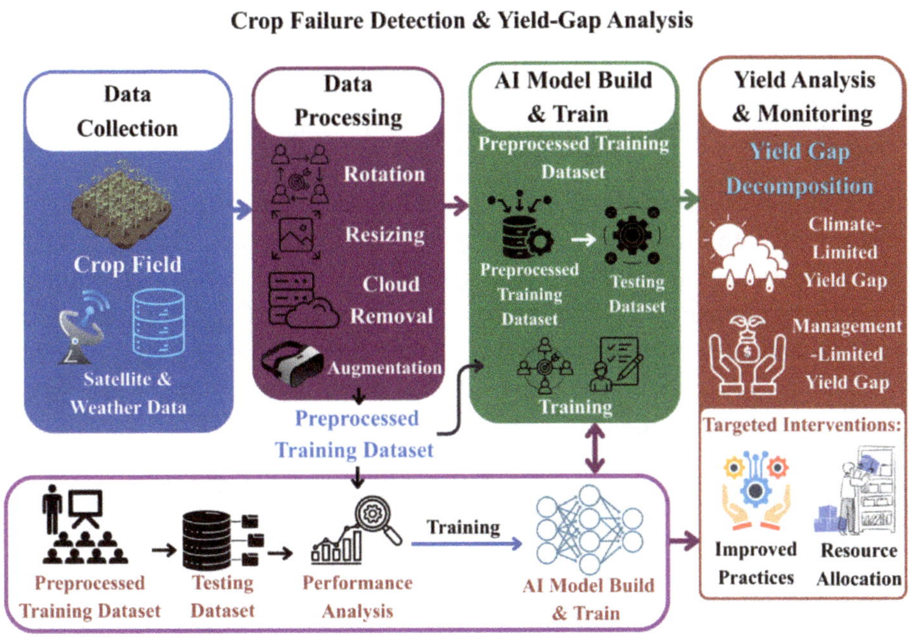

b. **Data architecture:** These inputs include surface reflectance and vegetation index data (e.g., NDVI, EVI) in dense time series (e.g., Sentinel -2 and Landsat). SAR backscatter (sentinel -1) during cloud-prone times. Addition inputs include: surface temperature of the land, gridded meteorological re-analyses (precipitation, temperature, radiation) and map of soil properties (Atzberger, 2013).

c. **Computer Algorithms and modeling strategies:** There are three common patterns of algorithms that are used:

   i. **Supervised anomaly classification:** Gradient-boosting arrays or random forests are trained with seasonal trajectory characteristics (where possible, including phenological dates, cumulative greenness, rate of decline) at the training locations on past failure events that have a historical label (Van Klompenburg et al., 2020). These models are very precise with respect to accurate labels but prone to label bias.

   ii. **Unsupervised or semi-supervised detection:** It applies autoencoders or Gaussian process–based outlier scoring applied to multi-year baseline trajectories. This method is suitable in operational monitoring when there is limited number of labelled failures (Reichstein et al., 2019).

iii. **Hybrid yield-gap decomposition:** The Process Informed Hybrid Models are based upon the outputs of crop models (e.g., water-limited potential yield or biological mass states) and machine-learning residual models of the relationship between obtainable and actual yields; the residual is the gap associated with management. Hybridization improves extrapolation in new regimes of climate compared to purely empirical models (You et al., 2017).

d. **Validation and metrics:** Generalization should be measured with the help of temporally stratified hold-outs and spatial cross-validation. Measures are precision of detection of failure events and recall of failure events and continuous measures of error (RMSE, MAE) of yield predictions. Operational users need the probabilities to be well-calibrated (Brier score, reliability diagrams) to such an extent that early warnings can set a threshold at which the probability of false-alarm is known.

6.2. Drought, Flood, and Extreme Event Impact Assessment

a. Scope and data fusion
   - Immediate mapping of the extent of the events and strong translation of the observed biophysical anomalies into anticipated production effects is required in impact assessment.
   - In the case of floods, SAR observations enable cloud-independent hazard extent mapping, while optical time series capture post-event vegetation impacts.
   - In the case of droughts and heat events, the linkage of precipitation losses and evapotranspiration stresses (provided by remote-sensing LST and meteorological forcings) and vegetation stress indices provides early predictors of loss in the yield (Van Wittenberghe et al., 2021; Zhu et al., 2022).

b. Modeling approaches
   - The mapping of both damage footprints and per cent yield loss have been focused on by Convolutional Neural Networks (CNNs) and spatio-temporal structures (CNN-LSTM, temporal attention), training to learn the mapping of pre-/post-event imagery and meteorological anomalies to historical loss indications (where available) such as crop insurance claims and official loss reports (Tupas et al., 2024). Multi-input networks which absorb both SAR and optical inputs and meteorological covariates are used to model nonlinear interactions and lagged responses of crops in the case of compound events. Figure 8 depicts the integra-

tion of hazard detection and impact modeling for assessing drought and flood effects on crop systems.

c. Uncertainty and timelines of operation
   - Uncertainty (ensembles, Bayesian neural nets, or quantile regression) should be explicitly propagated through models, to enable decision makers to have probabilistic forecasts of the patterns of predicted losses and recoveries

Figure 8. Drought, Flood and Extreme Event Impact Analysis

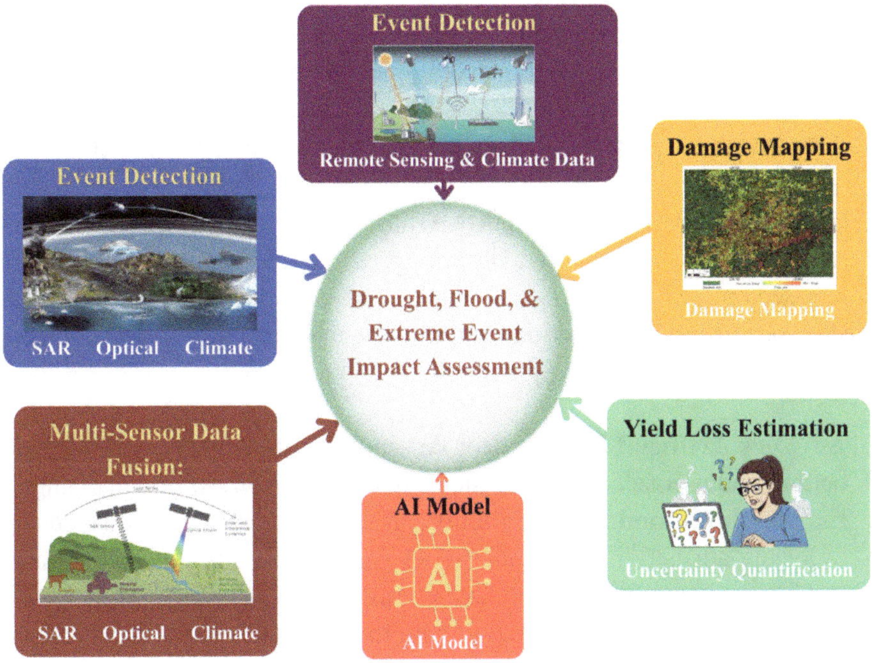

6.3 Regional and National Food Production Forecasting

Aggregation architecture: The yield predictions (or yield probability distributions) in field or pixel scale are summative to administrative units by use of crop area maps based on classification models.

There are two paradigms of structural modelling which make large-scale precision:

- **Hierarchical/ multi-level models:** The local models that have been trained in high quality labelled regions are embedded into a global model that shares information among regions via shared priors. Bayesian models enable the transfer of uncertainty among levels (Ji et al., 2022).

- **Phenology-grouped modeling:** The transferability is more enhanced by grouping pixels by phenological cluster (planting date, crop calendar) and training group-specific models since phenology is one of the mediators of climatic sensitivity (Inglada et al., 2017).

Predict performance and performance goal: The forecasts should be able to report proficiency at various spatial scales as well as lead times (e.g. in-season 1-3 months ahead of harvesting). The common measures of skills are the RMSE, the Mean Absolute Percentage Error (MAPE), and Continuous Ranking Probabilistic Scores (CRPS). Operation needs are different: national policy choices can be very tolerant, as opposed to localized intervention targeting.

6.4 Decision-Support Systems for Policy and Planning

Translation to action: The output of AI has to be converted into decision-relevant indicators: severity categories (early warning, watch, emergency), input distribution layers based on spatial targeting, and market and trade planning output projections based on probabilistic production shortfall. The main features of decision-support system are:

i. **Advisory Artificial Intelligence:** Presentation of complete predictive distributions and user selectable thresholds that are policy response mappings.
ii. **Explainability modules:** Credibility and actionable interpretation is supported by feature-attribution techniques (SHAP, integrated gradients), counterfactuals, and impact decomposition (climate versus management drivers).
iii. **Interoperability:** The integration with the national information systems and humanitarian platforms is possible because of structured APIs and compliance with the data standards (e.g., OGC). Figure 9 presents the structure of the AI-based early warning and decision-support system for food security applications.

Figure 9. AI-Driven Early Warning and Decision Support Framework

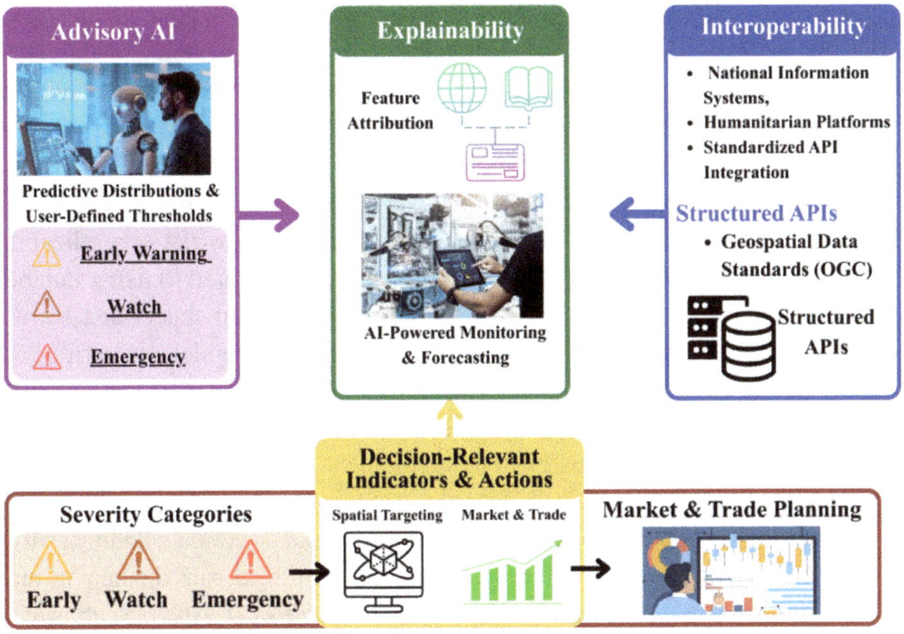

Institutionalisation challenges: The operational adoption requires the deployment of scalable data pipelines, development of human capacity to interpret, and governance arrangements that enable actions taken to respond to probabilistic signals without such requirements, technical accuracy will not be converted to yield improved food-security results (Weber & Cisneros, 2018).

7. CHALLENGES, EXPLAINABILITY, AND EMERGING RESEARCH DIRECTIONS

7.1. Data Scarcity and Label Uncertainty in Agricultural Datasets

Both lack of data and uncertainty of labels are inherent to the advancement of AI models in agricultural surveillance. Labeled datasets are essential to train strong AI systems and those that are of high quality; moreover, it is often unfeasibly expensive and logistically difficult to obtain such data in the agricultural setting. There is a high degree of heterogeneity in the agricultural setting due to the differences in the type of crops, the differences in soil behavior, management, and the differences in climatic condition and thus necessitates the use of different and representative data to train the models.

There are various sources of label uncertainty. Field values can also be invalid due to human error, uneven sampling guidelines or time differences with the satellite overpasses. As an example, ground-truth data obtained on a particular day might not be accurate relative to the satellite imagery obtained days or weeks before generating noisy or imprecise labels. Furthermore, there are differences in the resolution of space between plot level ground and satellite pixel that creates ambiguity in labeling. This uncertainty may corrupt the model performance and restrict generalizability.

In alleviating these problems, new approaches are oriented to using unlabeled or partially labeled data by using weak supervision, semi-supervised learning, and active learning models. Weak supervision allows the models to learn through the domain knowledge or heuristics rules by using the noisy or inaccurate labels. Semi-supervised learning takes advantage of the number of unlabeled data available through the satellites and uses them to supplement a few labelled data to improve representation learning.

Synthetic data generation presents another potential direction whereby artificial satellite images and scenarios of crop growth are generated with the help of mechanistic crop models or Generative Adversarial Networks (GANs). Such artificial data can be used to supplement empirical samples to enhance model robustness, and they can also be used to test their models under rare or extreme conditions. Domain adaptation methods also increase model transferability by adapting models trained on a specific region or type of crop to new domains that have different data distributions to reduce the impact of data scarcity and label noise.

7.2. Generalization of the Model across Regions and Crop Type.

Scalability of AI models between a diverse geographic area, agro-ecological zone, and the types of crops is a major limitation to large-scale operational use. The model used to train data in one particular area often becomes inaccurate when used in other areas because of changes in environmental factors, crop phenology, farming activities, and sensor properties. The spectral signature of the crops and their growth cycles vary significantly and make the creation of universal models difficult.

The inconsistency in satellite sensor characteristics; spatial resolution, spectral band and revisit frequency, also hinder model transferability. Moreover, the management activities of irrigation, fertilization, and density of planting bring about more heterogeneity which the model has to consider in order to generalize well.

To boost generalization, transfer learning methods re-train the pre-trained models with limited labeled data in new datasets to build upon already learned feature representations. The goal of meta-learning systems is to develop models that can quickly adapt to new tasks or domains with little data and be more flexible. Multi-task learning, whereby, the same model is trained on several related tasks or crop types at the same time encourages the representation of shared knowledge which enhances robustness.

Another very potent strategy is the incorporation of the knowledge of agronomic domains and physical constraints into AI models. The inclusion of crop development stages, crop phenological calendar, or soil-plant-atmosphere interactions as priors or regularisation terms result in the learning process being driven to agronomically viable predictions. This data-driven knowledge-driven hybridization of modeling is better at improving interpretability and generalization, especially when data are scarce or the environment is changing physics-informed and uncertainty-aware AI models.

7.3. Intelligible AI towards Reliable Agricultural Decision-Making

The fact that AI models (particularly, deep-learning structures) often make decisions in an obscure way is likely to hinder trust and acceptance among the stakeholders in the agricultural sector, including farmers, policymakers, and extension employees. To overcome this gap, explainable AI (XAI) methods are necessary to offer interpretable information about model predictions, as well as how model input features contribute to them.

The XAI techniques, such as SHapley Additive exPlanations (SHAP), provide an opportunity to measure the value of each input feature to a particular prediction and allow stakeholders to identify the role of specific climatic variables, soil parameters, and satellite-based indices in the model outcomes. Integrated gradients and counterfactual explanations also explain how changes in inputs can adjust predictions and are therefore useful towards supporting scenario analysis and sensitivity testing. Explainability leads to transparency, which allows the users to identify possible biases, data quality concerns, or model failures. It can also be used to perform analysis of errors and refine the model, by showing unanticipated dependencies between features.

In addition to this, explainability will help in regulatory conformity and ethical issues because it offers accountability to AI-based agricultural systems. It allows sharing AI insights with non-technical stakeholders, eliminating the gap between the complex models, and real-life applications in policy and resource management.

7.4. Physics-inspired and Uncertainty-aware AI Models

Combining physical process knowledge and AI models is a potential future area of improving model reliability, interpretability, and robustness. Physics-informed AI incorporates mechanistic knowledge of crop development, soil moisture interactions, and hydrology into the learning system, as explicit constraints, biases, or a hybrid system based on the combination of process-based and data-driven systems. This combination helps and avert the danger of physically implausible forecasts that are common in the purely empirical models, particularly in new climatic or management regimes. Physics-informed methods enhance predictions of extrapolation by providing valuable insights into causal processes based on prior principles of agronomics and the environment. In order to explicitly incorporate the agronomic constraints in neural networks, the physics-informed loss functions may be brought into use. The predictive model output is by definition $\hat{y}$ (yield or risk score) and the target of observation be y. The total loss may be stated in the form of

$$L_{total} = L_{total} + \lambda L_{physics}$$

where:

- $\mathcal{L}_{data} = \| y - \hat{y} \|^2$represents the empirical prediction error,
- $\mathcal{L}_{physics}$ encodes agronomic constraints such as water balance consistency, crop growth stage progression, or temperature stress thresholds,
- λ controls the strength of physical regularization.

It is also important to be uncertain about modeling because agricultural decision-making is also associated with risk assessment and incomplete or noisy data most of the time. The measurement of prediction uncertainty allows the decision-makers to estimate the level of confidence, balance risks, and rank interventions in the order of priorities. The predictive uncertainty in Bayesian neural networks, ensemble modeling, and quantile regression models is explicitly modeled and hence predictable on probabilistic forecasts as opposed to point estimates.

Uncertainty quantification is used in early warning to set thresholds in alerting and resource mobilization in situations involving drought, pest infestations, or crop yield reductions. It is also useful in supporting adaptive management because it identifies areas or times with great uncertainty in the forecast that might require further monitoring or data gathering. Such a combination of physics-informed and uncertainty-aware AI models, therefore, provides a way to improve to a more transparent, reliable, and actionable system of agricultural intelligence.

7.5. Future Research Objectives

- Multi-Scale and Multi-Modal Fusion: Scientists are coming up with architectures that combine the use of disparate data sources; optical imagery, Synthetic Aperture Radar (SAR), thermal measurements, meteorological observations, and soil characteristics at spatial and temporal scales. Such an integrative method is meant to forge complex interactions between biophysical processes thus contributing to the predictive power.
- Few-Shot and Self-Supervised Learning: With the help of self-supervised pre-training on large amounts of unlabeled satellite imagery, it is possible to obtain generalizable representations using very little labelled data to fine-tune.
- Human-in-the-loop Systems and explainability: The attempts to increase model interpretability are accompanied by the use of human knowledge during training, validation, and deployment stages. The approach will enhance the credibility and relevance of the models.
- Real-Time Analytics and Edge Computing: Near-real-time agricultural monitoring and early warning systems can be implemented on lightweight AI models deployed to edge devices or cloud platforms, especially in resource-constrained environments.
- Sustainable AI and Ethical AI: The discussion on the issues of data privacy, fair access, and environmental issues is necessary to assure responsible use of AI technologies by various agricultural communities

8. CASE STUDY: AI-DRIVEN REAL-TIME PEST OUTBREAK PREDICTION AND MANAGEMENT IN RARE CROP SYSTEMS IN INDIA

The case study is based on the usage of artificial intelligence combined with remote sensing and weather observation to allow earlier identification and control of pest outbreak of rare and underused crops grown in India. They focus on crops like foxtail millet, black rice (Oryza conuleum indica) and kodo millet that are produced in niche agro-ecological regions with little pest surveillance infrastructure and ground truth data (Setiyono et al., 2018).

8.1 Sources of Data and Pre-processing

The monitoring system is able to streamline various multi-modal satellite data, such as optical images in Sentinel 2, SAR in Sentinel 1, and thermal infrared data, as well as local weather conditions such as temperature, humidity and rain. These datasets are supplemented by soil property maps that are specific to the cultivation regions. Atmospheric correction, cloud masking, shadow masking, and temporal smoothing are the pre-processing steps to produce continuous and high-quality time series that can be used as phenological indices of the rare crops. This multi-source fusion eliminates data gaps due to frequent cloud covers during monsoons and non-homogeneous terrain to provide the AI modelling with strong inputs (Asam et al., 2022).

8.2 Domain Knowledge and Model Architecture Integration

A custom deep-learning structure that is used is a multi-task model that combines both spatial feature extraction using a CNN and a temporal attention model that captures dynamic changes in the pest population in response to changes in micro-climatic conditions and crop-specific developmental stages. Physical priors, such as the incorporation of agronomic domain knowledge (pest life cycles of these rare crops, phenology calendars) within the learning process leads to the regularisation and restriction of the learning process. This hybridisation produces better biological plausibility and reduces over-fitting. Semi-supervised learning is an approach that takes advantage of the large amounts of unlabeled satellite data recorded at any point during the growing season, surmounting the limitation of limited labelled data of pest occurrence by enhancing feature representation and model generalisation (Chen et al., 2021).

8.3 Agricultural AI of Uncertainty-Aware Modelling and Explainability

Bayesian neural networks also measure predictive uncertainty, which results in probabilistic predictions of the risk of pest outbreaks. This quantification of uncertainty allows decision-makers to make prioritisation of interventions using confidence-increased resource allocation, which is optimal in the case of risk and variability. The explainability is realized through SHapley Additive exPlanations (SHAP) values that define important environmental and phenological contributors to risk of outbreak, including localized temperature abnormalities, soil moisture losses, and certain stages of crop development, which affect the risk of outbreak. These interpretable understandings can be supported with agronomic knowledge, which will promote trust between the stakeholders and will open up clear communication between stakeholders and the farmers and extension workers (Khan et al., 2025).

8.4 Real-Time Implementation and Testing

The AI system was introduced in Maharashtra and Odisha in 2023 in isolated areas where crops were growing. The constant consumption of close-to-real-time satellite and meteorological streams of data allowed to detect the signs of the pest infestation 2-3 weeks before the symptoms are reported to the farmer. Quantitatively, the detection of pest outbreaks with a precision of 0.87 and a recall of 0.81 with an F1-score of 0.84 were observed, when historical pest occurrence records were used to validate it. Combination of SAR and thermal images minimized the false-positives rates by 12% relative to baselines by vegetation-index only. Bayesian uncertainty calibration produced a Brier of 0.11 which shows well-calibrated probabilistic results which can be used in operational thresholding. These indicators show that the suggested AI framework enhances the reliability of detection as well as early warning lead time (Tzachor et al., 2022).

8.5 Impact and Outcomes

Timely integrated pest management (IPM) strategies were adopted by farmers and local extension workers through the use of the AI-driven early warning system, thus reducing an estimated 20 per cent of the use of pesticide and preventing crop damage of less than 12 per cent in areas where the system was used. Clear results of AI served to make farmers more confident in their decision-making process, and it facilitated the sustainable use of pests in rare crop systems that conventional monitoring systems tend to overlook. This method will help to preserve biodiversity and achieve sustainable farming through the reduction of chemical use and the health of crops (Hajjar et al., 2023).

8.6 Difficulties and Adaptive Strategies

The main difficulties included a lack of data and ambiguity of labels that is characteristic of rare crop systems that have little ground truth data. Some mitigation tactics were semi-supervised learning and domain adaptation that was specific to the pest biology and phenology of the crops. The incorporation of crop specific biological knowledge and physical limitations were essential in avoiding over-fitting and also in making relevant predictions in changing climatic conditions. Data fusion over multiple sources and uncertainty-sensitive modelling was needed in the context of operational monitoring in resource-constrained and heterogeneous settings as is common in the context of rare crop farming. The case highlights the role of combining agronomic knowledge and high-level AI practices in overcoming the constraints of sparse data and environmental changeability (Gadiraju & Vatsavai, 2023).

8.7 Global Relevance and Climate Variability Integration

Even though the case study revolves around rare crop systems in India, the concept of AI-based framework is naturally flexible to other climatic and agro-ecological settings. As an illustration, research regarding the control of the drought and vegetation functionality by the global climate drivers like the North Atlantic Oscillation (NAO) in Turkey has underscored the significance of incorporating the indexes of teleconnections into forecasting agricultural activities (Çelik et al., 2026). On the same note, studies of climate extreme indices in Kafa Biosphere Reserve in southwestern Ethiopia show that the rise in temperature variability and precipitation anomalies have a direct impact on crop system like coffee agriculture (Amsalu et al., 2026). The role of the spatial climatic heterogeneity in the definition of crops suitability and seasonal productivity is also supported by long-term agro-climatic classification with the help of ERA5-Land temperature datasets in the context of agricultural basins of Türkiye (Çelik, 2026). The proposed framework can be re-calibrated to different regions such as Sub-Saharan Africa, Mediterranean climates as well as South American agricultural systems by adding regional climatic oscillations, extreme event predictors, and agro-cleavage zoning layers into the modeling framework, making it more global in its applicability and calibration to data-variable systems.

9. CONCLUSION AND FUTURE WORK

Artificial intelligence combined with multi-source remote sensing and meteorological data provides a radical potential of agricultural surveillance, especially in complex settings like the infrequent crop regime where monitoring facilities are scarce. The chapter clarifies how state of the art AI practices such as semi-supervised learning, domain adaptation, and physics-informed modeling are useful in addressing the critical problems due to data scarcity, label uncertainty, and environmental heterogeneity. The implementation of uncertainty-conscious models and explainable AI boosts transparency and builds trust among the stakeholders, thus, allowing risk-based decision-making and resource allocation optimization. Despite the significant developments, there are still a number of obstacles that stand in the way of embracing operations on a broad basis. Research problems that must be tackled moving forward would focus on the creation of standardized, large scale, and varieties of data that includes both rare and under-utilized crops to enable meaningful model training and benchmarking. It is hoped that additional knowledge in the agronomic domain to be built into physical processes models would further increase model generalization and biological plausibility. Cultivation of the explainability frameworks to meet the interests of the agricultural stakeholders will be essential to support trust and pragmatic application. In addition, increased real-time analytics and edge-computing will support the ability to intervene in resource-constrained environments in time. The deployment of AI must be implemented in a sustainable and ethical manner, whereby careful consideration is placed on data governance, fair access, and environmental control, so that in the course of technological progress, there can be beneficial outcomes of technology, which can be applied to biodiversity conservation and food security implementation. Research, policymakers, extension, and farming communities should also work collaboratively to transfer AI innovations into implementable resources that can help to have a resilient, sustainable, and inclusive agricultural system at the global level.

REFERENCES

Amsalu, A., Garedew, W., Melka, G. A., Çelik, M. A., & Gemeda, D. O. (2026). Trends in climate extreme indices and implications for coffee farming in The Kafa Biosphere Reserve, southwest Ethiopia. *Climate Services*, *42*, 100637. DOI: 10.1016/j.cliser.2026.100637

Anderson, M., Norman, J., Kustas, W., Houborg, R., Starks, P., & Agam, N. (2008). A thermal-based remote sensing technique for routine mapping of land-surface carbon, water and energy fluxes from field to regional scales. *Remote Sensing of Environment*, *112*(12), 4227–4241. DOI: 10.1016/j.rse.2008.07.009

Atzberger, C. (2013). Advances in remote sensing of agriculture: Context description, existing operational monitoring systems and major information needs. *Remote Sensing (Basel)*, *5*(2), 949–981. DOI: 10.3390/rs5020949

Becker-Reshef, I., Justice, C., Barker, B., Humber, M., Rembold, F., Bonifacio, R., & Shitote, C. (2020). Strengthening agricultural decisions in countries at risk of food insecurity: The GEOGLAM Crop Monitor for Early Warning. *Remote Sensing of Environment*, *237*, 111553. DOI: 10.1016/j.rse.2019.111553

Belgiu, M., & Drăguţ, L. (2016). Random forest in remote sensing: A review of applications and future directions. *ISPRS Journal of Photogrammetry and Remote Sensing*, *114*, 24–31. DOI: 10.1016/j.isprsjprs.2016.01.011

Camps-Valls, G., Tuia, D., Gómez-Chova, L., Jiménez, S., & Malo, J. (2011). Remote sensing image processing.

Çelik, M. A. (2026). Agro-climatic Classification of Türkiye's Agricultural Basins Using Long-Term ERA5-Land Temperature Data. *Physics and Chemistry of the Earth Parts A/B/C*, *142*, 104307. DOI: 10.1016/j.pce.2026.104307

Çelik, M. A., Kıraç, A., & Gouveia, C. M. (2026). North Atlantic Oscillation modulation of drought and vegetation activity in Türkiye. *Theoretical and Applied Climatology*, *157*(2), 74. DOI: 10.1007/s00704-025-05971-5

Cheng, E., Zhang, B., Peng, D., Zhong, L., Yu, L., Liu, Y., & Chen, Y. (2022). Wheat yield estimation using remote sensing data based on machine learning approaches. *Frontiers in Plant Science*, *13*, 1090970. DOI: 10.3389/fpls.2022.1090970 PMID: 36618627

Fernandez-Beltran, R., Baidar, T., Kang, J., & Pla, F. (2021). Rice-yield prediction with multi-temporal sentinel-2 data and 3D CNN: A case study in Nepal. *Remote Sensing (Basel)*, *13*(7), 1391. DOI: 10.3390/rs13071391

Gao, J., Li, P., Chen, Z., & Zhang, J. (2020). A survey on deep learning for multimodal data fusion. *Neural Computation, 32*(5), 829–864. DOI: 10.1162/neco_a_01273 PMID: 32186998

Garnot, V. S. F., Landrieu, L., Giordano, S., & Chehata, N. (2020). Satellite image time series classification with pixel-set encoders and temporal self-attention. Proceedings of the IEEE/CVF conference on computer vision and pattern recognition, Gorelick, N., Hancher, M., Dixon, M., Ilyushchenko, S., Thau, D., & Moore, R. (2017). Google Earth Engine: Planetary-scale geospatial analysis for everyone. *Remote Sensing of Environment, 202*, 18–27.

Inglada, J., Vincent, A., Arias, M., Tardy, B., Morin, D., & Rodes, I. (2017). Operational high resolution land cover map production at the country scale using satellite image time series. *Remote sensing, 9*(1), 95.

Jabed, M. A., & Murad, M. A. A. (2024). Crop yield prediction in agriculture: A comprehensive review of machine learning and deep learning approaches, with insights for future research and sustainability. *Heliyon, 10*(24), e40836. DOI: 10.1016/j.heliyon.2024.e40836 PMID: 39720079

Ji, Z., Pan, Y., Zhu, X., Zhang, D., & Wang, J. (2022). A generalized model to predict large-scale crop yields integrating satellite-based vegetation index time series and phenology metrics. *Ecological Indicators, 137*, 108759. DOI: 10.1016/j.ecolind.2022.108759

Justice, C. J., Becker-Reshef, I., Barker, B., Humber, M. L., Sanchez, A., Nakalembe, C., & Sobue, S. (2019). Strengthening agricultural decisions in countries at risk of food insecurity: The GEOGLAM Crop Monitor for Early Warning. AGU Fall Meeting Abstracts, Kamilaris, A., & Prenafeta-Boldú, F. X. (2018). Deep learning in agriculture: A survey. *Computers and Electronics in Agriculture, 147*, 70–90.

Klompenburg, T., Kassahun, A., & Catal, C. (2020). Crop yield prediction using machine learning. *Computers and Electronics in Agriculture.*

Liakos, K. G., Busato, P., Moshou, D., Pearson, S., & Bochtis, D. (2018). Machine learning in agriculture: A review. *Sensors (Basel), 18*(8), 2674. DOI: 10.3390/s18082674 PMID: 30110960

Lobell, D. B., Cassman, K. G., & Field, C. B. (2009). Crop yield gaps: Their importance, magnitudes, and causes. *Annual Review of Environment and Resources, 34*(1), 179–204. DOI: 10.1146/annurev.environ.041008.093740

Lobell, D. B., Hammer, G. L., McLean, G., Messina, C., Roberts, M. J., & Schlenker, W. (2013). The critical role of extreme heat for maize production in the United States. *Nature Climate Change*, *3*(5), 497–501. DOI: 10.1038/nclimate1832

Lu, J., Li, J., Fu, H., Tang, X., Liu, Z., Chen, H., Sun, Y., & Ning, X. (2024). Deep learning for multi-source data-driven crop yield prediction in northeast China. *Agriculture*, *14*(6), 794. DOI: 10.3390/agriculture14060794

Miralles, P., Scannapieco, A. F., Jagadam, N., Baranwal, P., Faldu, B., Abhang, R., & Batul, B. (2021). Machine learning in earth observation operations: A review. 72nd International Astronautical Congress (IAC), Orynbaikyzy, A., Gessner, U., Mack, B., & Conrad, C. (2020). Crop type classification using fusion of sentinel-1 and sentinel-2 data: Assessing the impact of feature selection, optical data availability, and parcel sizes on the accuracies. *Remote Sensing*, *12*(17), 2779.

Ozdogan, M., Yang, Y., Allez, G., & Cervantes, C. (2010). Remote sensing of irrigated agriculture: Opportunities and challenges. *Remote Sensing (Basel)*, *2*(9), 2274–2304. DOI: 10.3390/rs2092274

Qiao, M., He, X., Cheng, X., Li, P., Luo, H., Zhang, L., & Tian, Z. (2021). Crop yield prediction from multi-spectral, multi-temporal remotely sensed imagery using recurrent 3D convolutional neural networks. *International Journal of Applied Earth Observation and Geoinformation*, *102*, 102436. DOI: 10.1016/j.jag.2021.102436

Raza, A., Shahid, M. A., Zaman, M., Miao, Y., Huang, Y., Safdar, M., Maqbool, S., & Muhammad, N. E. (2025). Improving Wheat Yield Prediction with Multi-Source Remote Sensing Data and Machine Learning in Arid Regions. *Remote Sensing (Basel)*, *17*(5), 774. DOI: 10.3390/rs17050774

Reichstein, M., Camps-Valls, G., Stevens, B., Jung, M., Denzler, J., Carvalhais, N., & Prabhat, F. (2019). Deep learning and process understanding for data-driven Earth system science. *Nature*, *566*(7743), 195–204. DOI: 10.1038/s41586-019-0912-1 PMID: 30760912

Rockström, J., Williams, J., Daily, G., Noble, A., Matthews, N., Gordon, L., & Steduto, P. (2017). Sustainable intensification of agriculture for human prosperity and global sustainability. *Ambio*, *46*(1), 4–17. DOI: 10.1007/s13280-016-0793-6 PMID: 27405653

Ronneberger, O., Fischer, P., & Brox, T. (2015). U-net: Convolutional networks for biomedical image segmentation. *International Conference on Medical image computing and computer-assisted intervention*

Sah, S., Haldar, D., Singh, R., Das, B., & Nain, A. S. (2024). Rice yield prediction through integration of biophysical parameters with SAR and optical remote sensing data using machine learning models. *Scientific Reports, 14*(1), 21674.

Sakamoto, T., Yokozawa, M., Toritani, H., Shibayama, M., Ishitsuka, N., & Ohno, H. (2005). A crop phenology detection method using time-series MODIS data. *Remote Sensing of Environment, 96*(3-4), 366–374. DOI: 10.1016/j.rse.2005.03.008

Sebastianelli, A., Mauro, F., Delilbasic, A., Di Stasio, P., Fan, F., Meoni, G.,... Ullo, S. (2025). Quantum Machine Learning for Earth Observation: A Review and Future Prospects.

Singh, G., Singh, S., Sethi, G., & Sood, V. (2022). Deep learning in the mapping of agricultural land use using Sentinel-2 satellite data. *Geographies, 2*(4), 691–700. DOI: 10.3390/geographies2040042

Song, C., Woodcock, C. E., Seto, K. C., Lenney, M. P., & Macomber, S. A. (2001). Classification and change detection using Landsat TM data: When and how to correct atmospheric effects? *Remote Sensing of Environment, 75*(2), 230–244. DOI: 10.1016/S0034-4257(00)00169-3

Tupas, M. E., Roth, F., Bauer-Marschallinger, B., & Wagner, W. (2024). Assessment of time-series-derived no-flood references for sar-based Bayesian flood mapping. *GIScience & Remote Sensing, 61*(1), 2427304. DOI: 10.1080/15481603.2024.2427304

Van Klompenburg, T., Kassahun, A., & Catal, C. (2020). Crop yield prediction using machine learning: A systematic literature review. *Computers and Electronics in Agriculture, 177*, 105709. DOI: 10.1016/j.compag.2020.105709

Van Wittenberghe, S., Sabater Medina, N., Cendrero Mateo, M. P., Tenjo Gil, N. C., Moncholi Estornell, A., Alonso Chordá, L., & Moreno Méndez, J. F. (2021). Towards the quantitative and physically-based interpretation of solar-induced vegetation fluorescence retrieved from global imaging.

Weber, R., & Cisneros, E. (2018). Assessing Monitoring and Early Warning Systems for Food Security Risk.

Weiss, M., Jacob, F., & Duveiller, G. (2020). Remote sensing for agricultural applications: A meta-review. *Remote Sensing of Environment, 236*, 111402. DOI: 10.1016/j.rse.2019.111402

Xiong, X., Tian, Q., Rodriguez, L. F., & Lin, T. (2022). Data-Driven Modeling for Crop Mapping and Yield Estimation. *Sensing, Data Managing, and Control Technologies for Agricultural Systems*, 131-158.

Xu, L., Zhang, H., Wang, C., Zhang, B., & Liu, M. (2018). Crop classification based on temporal information using sentinel-1 SAR time-series data. *Remote Sensing (Basel), 11*(1), 53. DOI: 10.3390/rs11010053

You, J., Li, X., Low, M., Lobell, D., & Ermon, S. (2017). Deep gaussian process for crop yield prediction based on remote sensing data. Proceedings of the AAAI conference on artificial intelligence, Zhu, Y., Wu, S., Qin, M., Fu, Z., Gao, Y., Wang, Y., & Du, Z. (2022). A deep learning crop model for adaptive yield estimation in large areas. *International Journal of Applied Earth Observation and Geoinformation, 110*, 102828.

Compilation of References

Abegunde, V. O., Sibanda, M., & Obi, A. (2019). The Dynamics of Climate Change Adaptation in Sub-Saharan Africa: A Review of Climate-Smart Agriculture among Small-Scale Farmers. *Climate (Basel)*, *7*(11), 132. DOI: 10.3390/cli7110132

Abraham, K., Abdelwahab, M., & Abo-Zahhad, M. (2023). Classification and detection of natural disasters using machine learning and deep learning techniques: A review. *Earth Science Informatics*, *17*(2), 869–891. DOI: 10.1007/s12145-023-01205-2

Adaawen, S., Rademacher-Schulz, C., Schraven, B., & Segadlo, N. (2019). Drought, migration, and conflict in sub-Saharan Africa: what are the links and policy options? Drought Challenges, 15–31. https://doi.org/DOI: 10.1016/B978-0-12-814820-4.00002-X

Addaney, M. (2025). Addressing the impacts of climate change on human rights through adaptation-focused litigation: Cases before the African Union's human rights bodies. *Journal of Human Rights*, *24*(3), 353–369. DOI: 10.1080/14754835.2025.2507588

Adebisi, Y. A., Akande-Sholabi, W., Jimoh, N. D., Lali, H., Adebowale, K. A., Ouachhou, A., Egwu, K. C., Nachibwede, L. V., Shomuyiwa, D. O., Gilbert, A., Olpengs, D., Sesay, N., & Oladunjoye, I. O. (2025). Leveraging a people-centred approach to combat antimicrobial resistance in Africa. *Tropical Medicine and Health*, *53*(1), 177. Advance online publication. DOI: 10.1186/s41182-025-00863-w PMID: 41316445

Adger, W. N. (2006). Vulnerability. *Global Environmental Change*, *16*(3), 268–281. DOI: 10.1016/j.gloenvcha.2006.02.006

Adisa, O. M., Masinde, M., Botai, J. O., & Botai, C. M. (2020). Bibliometric analysis of methods and tools for drought monitoring and prediction in Africa. *Sustainability (Basel)*, *12*(16), 6516. DOI: 10.3390/su12166516

Agrawal, J., & Arafat, M. (2024). Transforming Farming: A Review of AI-Powered UAV Technologies in Precision Agriculture. *Drones (Basel)*, *8*(11), 664. Advance online publication. DOI: 10.3390/drones8110664

Akintuyi, O. B.Olabimpe Banke Akintuyi. (2024). Adaptive AI in precision agriculture: A review: Investigating the use of self-learning algorithms in optimizing farm operations based on real-time data. *Open Access Research Journal of Multidisciplinary Studies*, *7*(2), 16. DOI: 10.53022/oarjms.2024.7.2.0023

Akkem, Y., Biswas, S., & Varanasi, A. (2023). Smart farming using artificial intelligence: A review. *Engineering Applications of Artificial Intelligence*, *120*, 105899. DOI: 10.1016/j.engappai.2023.105899

Alahmad, T., Neményi, M., & Nyéki, A. (2023). Applying IoT Sensors and Big Data to Improve Precision Crop Production: A Review. *Agronomy (Basel)*, *13*(10), 2603. Advance online publication. DOI: 10.3390/agronomy13102603

Alder, N., Ebert, K., Herbrich, R., & Hacker, P. (2024). AI, Climate, and Transparency: Operationalizing and Improving the AI Act. DOI: 10.2139/ssrn.4941352

Alestra, C., Cette, G., Chouard, V., & Lecat, R. (2020). Long-term Growth Impact of Climate Change and Policies: The Advanced Climate Change Long-term (ACCL) Scenario Building Model. SSRN Electronic Journal. DOI: 10.2139/ssrn.3569499

Alsanoosy, T., & Malik, J. A. (2025). Predicting plant stress using SAM-L: Novel self-adaptive-meta learner with XAI based on soil moisture and chlorophyll analysis. *Scientific Reports*, *15*(1), 42304. Advance online publication. DOI: 10.1038/s41598-025-26184-w PMID: 41309723

Althoff, D., Rodrigues, L. N., & Bazame, H. C. (2021). Uncertainty quantification for hydrological models based on neural networks: The dropout ensemble. *Stochastic Environmental Research and Risk Assessment*, *35*(5), 1051–1067. DOI: 10.1007/s00477-021-01980-8

Alyemiş, H. (2025). *The Impacts of Climate Change on Social Risks: An Evaluation within the Framework of Green Social Policy*. TAM Akademi Journal., DOI: 10.58239/tamde.2025.02.001.x

Amini Toosi, H., Famiglietti, J., & Motta, M. (2026). *Climate Change and Clean Energy Solution Nexus. Global Nexus Handbook*. Portico., DOI: 10.1002/9781394248087.ch21

Amsalu, A., Garedew, W., Melka, G. A., Çelik, M. A., & Gemeda, D. O. (2026). Trends in climate extreme indices and implications for coffee farming in The Kafa Biosphere Reserve, southwest Ethiopia. *Climate Services, 42*, 100637. DOI: 10.1016/j.cliser.2026.100637

Anderson, K., Ryan, B., Sonntag, W., Kavvada, A., & Friedl, L. (2017). Earth observation in service of the 2030 Agenda for Sustainable Development. *Geo-Spatial Information Science, 20*(2), 77–96. DOI: 10.1080/10095020.2017.1333230

Anderson, M., Norman, J., Kustas, W., Houborg, R., Starks, P., & Agam, N. (2008). A thermal-based remote sensing technique for routine mapping of land-surface carbon, water and energy fluxes from field to regional scales. *Remote Sensing of Environment, 112*(12), 4227–4241. DOI: 10.1016/j.rse.2008.07.009

Anguelovski, I., Connolly, J. J. T., Masip, L., & Pearsall, H. (2017). Assessing green gentrification in historically disenfranchised neighborhoods: A longitudinal and spatial analysis of Barcelona. *Urban Geography, 39*(3), 458–491. DOI: 10.1080/02723638.2017.1349987

Apraku, A., Gyampoh, B. A., Morton, J. F., & Mariwah, S. (2025). Managing climate change through social justice in Africa: Key lessons from the Libode area of Eastern Cape, South Africa. *Scientific African, 27*, e02534. DOI: 10.1016/j.sciaf.2025.e02534

Araújo, S., Peres, R., Barata, J., Lidon, F., & Ramalho, J. (2021). Characterising the Agriculture 4.0 Landscape—Emerging Trends, Challenges and Opportunities. *Agronomy (Basel), 11*(4), 667. DOI: 10.3390/agronomy11040667

Arruda, H., Silva, E. R., Lessa, M., Proença, D.Jr, & Bartholo, R. (2022). VOSviewer and Bibliometrix. *Journal of the Medical Library Association: JMLA, 110*(3), 392–395. DOI: 10.5195/jmla.2022.1434 PMID: 36589296

Ashique, S., Raikar, A., Jamil, S., Lakshminarayana, L., Gajbhiye, S. A., De, S., & Kumar, S. (2025). Artificial intelligence integration with nanotechnology: A new frontier for sustainable and precision agriculture. *Current Nanoscience, 21*(2), 242–273. DOI: 10.2174/0115734137275111231206072049

Aslan, M., Sabancı, K., & Aslan, B. (2024). Artificial Intelligence Techniques in Crop Yield Estimation Based on Sentinel-2 Data: A Comprehensive Survey. *Sustainability (Basel), 16*(18), 8277. Advance online publication. DOI: 10.3390/su16188277

Atzberger, C. (2013). Advances in remote sensing of agriculture: Context description, existing operational monitoring systems and major information needs. *Remote Sensing (Basel), 5*(2), 949–981. DOI: 10.3390/rs5020949

Awad, M., & Khanna, R. (2015). Support Vector Regression. In *Apress eBooks* (p. 67). https://doi.org/DOI: 10.1007/978-1-4302-5990-9_4

Awais, M., Zaigham, S., Naqvi, A., Zhang, H., Li, L., Zhang, W., Awwad, F., Ismail, E., Khan, M., Raghavan, V., & Hu, J. (2023). AI and machine learning for soil analysis: An assessment of sustainable agricultural practices. *Bioresources and Bioprocessing*, *10*(1), 90. Advance online publication. DOI: 10.1186/s40643-023-00710-y PMID: 38647622

Ayadi, R., Forouheshfar, Y., & Moghadas, O. (2025). Enhancing system resilience to climate change through artificial intelligence: A systematic literature review. *Frontiers in Climate*, *7*, 1585331. Advance online publication. DOI: 10.3389/fclim.2025.1585331

Babacan, H. (2025). From Crisis to Policy Action: Tackling Climate Change Through SDG 13. *The International Journal of Community and Social Development*, *7*(2), 205–221. DOI: 10.1177/25166026251351370

Bailey, K., & Okoduwa, S. I. R. (2025). Leveraging social media for public health nutrition in low- and middle-income countries: Addressing malnutrition, enhancing food security, and promoting health equity. *World Nutrition*, *16*(2), 109–121. DOI: 10.26596/wn.2025162109-121

Balan, S., Vorotynskyy, V., Patalakha, V., Rybak, I., & Tarasiuk, V. (2025). The Role of State Information Policy in Shaping Democratic Transformation Amid Hybrid Threats. DOI: 10.1590/1678-98732433e015

Baltazar, R., Florencio, B., Vicente, A., & Belizario, P. (2024). The Role of Artificial Intelligence in Disaster Prediction, Mitigation, and Response in the Philippines: Challenges and Opportunities. *International Journal of Artificial Intelligence*, *11*(1), 37–51. DOI: 10.36079/lamintang.ijai-01101.675

Barbedo, J. (2025). A Review of Artificial Intelligence Techniques for Wheat Crop Monitoring and Management. *Agronomy (Basel)*, *15*(5), 1157. Advance online publication. DOI: 10.3390/agronomy15051157

Bazi, Y., Bashmal, L., Al Rahhal, M. M., Al Dayil, R., & Al Ajlan, N. (2021). Vision transformers for remote sensing image classification. *Remote Sensing (Basel)*, *13*(3), 516. Advance online publication. DOI: 10.3390/rs13030516

Becker-Reshef, I., Justice, C., Barker, B., Humber, M., Rembold, F., Bonifacio, R., & Shitote, C. (2020). Strengthening agricultural decisions in countries at risk of food insecurity: The GEOGLAM Crop Monitor for Early Warning. *Remote Sensing of Environment*, *237*, 111553. DOI: 10.1016/j.rse.2019.111553

Beckmann, L., Debener, J., & Kriebel, J. (2023). Understanding the determinants of bond excess returns using explainable AI. *Journal of Business Economics, 93*(9), 1553–1590. DOI: 10.1007/s11573-023-01149-5

Begum, T., & Rahman, S.. (2024). *Machine learning-based crop recommendation system for precision agriculture.* IEEE Xplore., DOI: 10.1109/ICAIT61638.2024.10690764

Belgiu, M., & Drăguţ, L. (2016). Random forest in remote sensing: A review of applications and future directions. *ISPRS Journal of Photogrammetry and Remote Sensing, 114*, 24–31. DOI: 10.1016/j.isprsjprs.2016.01.011

Benedetto, M., Sousa, A., Reis, J. L., Pratas, J., & Pimenta, P. (2025). Unleashing Open Government Data: Implications for Urban Sustainability and Public Management.

Benos, L., Tagarakis, A., Dolias, G., Berruto, R., Kateris, D., & Bochtis, D. (2021). Machine Learning in Agriculture: A Comprehensive Updated Review. *Sensors (Basel), 21*(11), 3758. Advance online publication. DOI: 10.3390/s21113758 PMID: 34071553

Betancourt, C., Stomberg, T., Roscher, R., Schultz, M. G., & Stadtler, S. (2021). AQ-Bench: A benchmark dataset for machine learning on global air quality metrics. *Earth System Science Data, 13*(6), 3013–3033. DOI: 10.5194/essd-13-3013-2021

Biazar, S., Golmohammadi, G., Nedhunuri, R., Shaghaghi, S., & Mohammadi, K. (2025). Artificial Intelligence in Hydrology: Advancements in Soil, Water Resource Management, and Sustainable Development. *Sustainability (Basel), 17*(5), 2250. Advance online publication. DOI: 10.3390/su17052250

Birgani, R., Takian, A., Djazayery, A., Kianirad, A., & Pouraram, H. (2022). Climate Change and Food Security Prioritizing Indices: Applying Analytical Hierarchy Process (AHP) and Social Network Analysis (SNA). *Sustainability (Basel), 14*(14), 8494. Advance online publication. DOI: 10.3390/su14148494

Biswas, U., Mahato, S., & Joshi, P. K. (2025). Spatial prediction of forest fires in India: A machine learning approach for improved risk assessment and early warning systems. *Environmental Science and Pollution Research International, 32*(8), 4856–4878. DOI: 10.1007/s11356-025-35982-8 PMID: 39891812

Boers, N., Goswami, B., Rheinwalt, A., Bookhagen, B., Hoskins, B., & Kurths, J. (2019). Complex networks reveal global pattern of extreme-rainfall teleconnections. *Nature, 566*(7744), 373–377. DOI: 10.1038/s41586-018-0872-x PMID: 30700912

Bojer, A. K., Abshare, M. W., Mesfin, F., & Al-Quraishi, A. M. F. (2025). Assessing climate and land use impacts on surface water yield using remote sensing and machine learning. *Scientific Reports*, *15*(1), 18477. Advance online publication. DOI: 10.1038/s41598-025-03493-8 PMID: 40425713

Bonan, G. B., & Doney, S. C. (2018). Climate, ecosystems, and planetary futures: The challenge to predict life in Earth system models. *Science*, *359*(6375), eaam8328. Advance online publication. DOI: 10.1126/science.aam8328 PMID: 29420265

Bond, P. (2013). Water Rights, Commons and Advocacy Narratives. *South African Journal on Human Rights*, *29*(1), 125–143. DOI: 10.1080/19962126.2013.11865068

Boonpook, W., Tan, Y., Nardkulpat, A., Torsri, K., Torteeka, P., Kamsing, P., Sawangwit, U., Pena, J., & Jainaen, M. (2023). Deep learning semantic segmentation for land use and land cover types using Landsat 8 imagery. *ISPRS International Journal of Geo-Information*, *12*(1), 14. Advance online publication. DOI: 10.3390/ijgi12010014

Botero-Valencia, L., López, D., & García, L.. (2025). Machine learning in sustainable agriculture. *Agriculture*, *15*(4), 377. Advance online publication. DOI: 10.3390/agriculture15040377

Bouanani, N., Laamrani, A., Hajji, H., Bourriz, M., Bourzeix, F., Abdelali, H., El-Battay, A., Amazirh, A., & Chehbouni, A. (2025). Estimating Soil Attributes for Yield Gap Reduction in Africa Using Hyperspectral Remote Sensing Data with Artificial Intelligence Methods: An Extensive Review and Synthesis. *Remote Sensing (Basel)*, *17*(9), 1597. Advance online publication. DOI: 10.3390/rs17091597

Bouquet, P., Jackson, I., Nick, M., & Kaboli, A. (2023). AI-based forecasting for optimised solar energy management and smart grid efficiency. *International Journal of Production Research*, *62*(13), 4623–4644. DOI: 10.1080/00207543.2023.2269565

Brenowitz, N. D., & Bretherton, C. S. (2018). Prognostic Validation of a Neural Network Unified Physics Parameterization. Geophysical Research Letters, 45(12), 6289–6298. Portico. DOI: 10.1029/2018GL078510

Bukar, U. A., Sayeed, M. S., Razak, S. F. A., Yogarayan, S., Amodu, O. A., & Mahmood, R. A. R. (2023). A method for analyzing text using VOSviewer. *MethodsX*, *11*, 102339. DOI: 10.1016/j.mex.2023.102339 PMID: 37693657

Cambrin, D. R., Poeta, E., Pastor, E., Cerquitelli, T., Baralis, E., & Garza, P. (2024). KAN You See It? KANs and Sentinel for Effective and Explainable Crop Field Segmentation. *arXiv*. https://doi.org//arxiv.2408.07040DOI: 10.48550

Campiglio, E. (2016). Beyond carbon pricing: The role of banking and monetary policy in financing the transition to a low-carbon economy. *Ecological Economics*, *121*, 220–230. DOI: 10.1016/j.ecolecon.2015.03.020

Camps-Valls, G., Tuia, D., Gómez-Chova, L., Jiménez, S., & Malo, J. (2011). Remote sensing image processing.

Camps-Valls, G., Reichstein, M., & Mahecha, M. D.. (2021). A perspective on machine learning for Earth system modeling. *Nature Reviews. Earth & Environment*, *2*(11), 719–737. DOI: 10.1038/s43017-021-00192-7

Cardellicchio, A., Renò, V., Guadagno, C. R., Cellini, F., & Amitrano, C. (2025). Editorial: Agricultural innovation in the age of climate change: a 4.0 approach. *Frontiers in Plant Science*, *16*, 1715677. Advance online publication. DOI: 10.3389/fpls.2025.1715677 PMID: 41158684

Carneiro, B., Resce, G., Caravaggio, N., Santangelo, A., Ruscica, G., Tucci, G., Pacillo, G., Eilerts, G., Läderach, P., & Coffey, K. (2025). Text mining and machine learning reveal global determinants of food insecurity. *Scientific Reports*, *15*(1), 36709. Advance online publication. DOI: 10.1038/s41598-025-20670-x PMID: 41120495

Carr, E. R., Goble, R., Rosko, H. M., Vaughan, C., & Hansen, J. (2019). Identifying climate information services users and their needs in Sub-Saharan Africa: A review and learning agenda. *Climate and Development*, *12*(1), 23–41. DOI: 10.1080/17565529.2019.1596061

Cartolano, A., Cuzzocrea, A., & Pilato, G. (2024). Analyzing and assessing explainable AI models for smart agriculture environments. *Multimedia Tools and Applications*, *83*(12), 37225–37246. DOI: 10.1007/s11042-023-17978-z

Cash, D. W., Clark, W. C., Alcock, F., Dickson, N. M., Eckley, N., Guston, D. H., Jäger, J., & Mitchell, R. B. (2003). Knowledge systems for sustainable development. *Proceedings of the National Academy of Sciences of the United States of America*, *100*(14), 8086–8091. DOI: 10.1073/pnas.1231332100 PMID: 12777623

Castañeda Rodriguez, L., Galvez-Nieto, A., Aguilar Chunga, Y. A., Ccalla Chusho, J. A., & Salinas Romero, M. E. (2026). Transformative Urban Resilience and Collaborative Participation in Public Spaces: A Systematic Review of Theoretical and Methodological Insights. *Urban Science (Basel, Switzerland)*, *10*(1), 51. DOI: 10.3390/urbansci10010051

Celik, M. A. (2020). Bibliometric network analysis on new tendencies, techniques and terms used in drought research. [IGGE]. *International Journal of Geography and Geography Education*, *42*, 602–630. DOI: 10.32003/igge.721487

Çelik, M. A. (2026). Agro-climatic classification of Türkiye's agricultural basins using long-term ERA5-Land temperature data. *Physics and Chemistry of the Earth Parts A/B/C*, *142*, 104307. DOI: 10.1016/j.pce.2026.104307

Çelik, M. A., Ağırkaya, M. B., & Gemeda, D. O. (2026a). Bridging data scarcity with AI methodology: Global research trends in machine and deep learning for air quality prediction. *Journal of Atmospheric and Solar-Terrestrial Physics*, *281*, 106777. DOI: 10.1016/j.jastp.2026.106777

Çelik, M. A., & Çelik, E. (2024). Are Urbanisation and Biodiversity Antithetical? A Bibliometric Analysis. Coğrafya Dergisi. *The Journal of Geography*, *0*(48), 121–135. DOI: 10.26650/JGEOG2024-1350047

Çelik, M. A., Kıraç, A., & Gouveia, C. M. (2026b). North Atlantic Oscillation modulation of drought and vegetation activity in Türkiye. *Theoretical and Applied Climatology*, *157*(2), 74. Advance online publication. DOI: 10.1007/s00704-025-05971-5

Chapman, W. E., Delle Monache, L., Alessandrini, S., Subramanian, A. C., Ralph, F. M., Xie, S.-P., Lerch, S., & Hayatbini, N. (2022). Probabilistic predictions from deterministic atmospheric river forecasts with deep learning. *Monthly Weather Review*, *150*(1), 215–234. DOI: 10.1175/MWR-D-21-0106.1

Chari, T., & Munoriyarwa, A. (2025). *Climate Change Communication in Africa*. Routledge., DOI: 10.4324/9781003660217

Chattopadhyay, A., Nabizadeh, E., & Hassanzadeh, P. (2020). Analog Forecasting of Extreme-Causing Weather Patterns Using Deep Learning. Journal of Advances in Modeling Earth Systems, 12(2). Portico. DOI: 10.1029/2019MS001958

Chavula, P., & Kayusi, F. (2025). Systematic Review on the Application of Nanotechnology and Artificial Intelligence in Agricultural Economics. *LatIA*, *3*, 322. DOI: 10.62486/latia2025322

Chavula, P., Kayusi, F., & Juma, L. (2024). Leveraging Artificial Intelligence for Enhancing Wheat Yield Resilience Amidst Climate Change in Sub-Saharan Africa. *LatIA*, *2*, 88. DOI: 10.62486/latia202588

Cheng, S., Quilodrán-Casas, C., Ouala, S., Farchi, A., Liu, C., & Tandeo, P. (2023). Machine learning with data assimilation and uncertainty quantification for dynamical systems: A review. IEEE/CAA Journal of Automatica Sinica, 10(6), 1361–1387. https://doi.org/DOI: 10.1109/JAS.2023.123537

Cheng, E., Zhang, B., Peng, D., Zhong, L., Yu, L., Liu, Y., & Chen, Y. (2022). Wheat yield estimation using remote sensing data based on machine learning approaches. *Frontiers in Plant Science, 13*, 1090970. DOI: 10.3389/fpls.2022.1090970 PMID: 36618627

Cheng, Y., Shen, L., Zhang, J., He, H., Gu, X., Wang, S., & Ma, T. (2025). A Multi-Stage Deep Learning Framework for Multi-Source Cloud Top Height Retrieval from FY-4A/AGRI Data. *Atmosphere (Basel), 16*(11), 1288. DOI: 10.3390/atmos16111288

Chen, H., Qi, Z., & Shi, Z. (2022). Remote sensing image change detection with transformers. *IEEE Transactions on Geoscience and Remote Sensing, 60*, 1–14. DOI: 10.1109/TGRS.2021.3095166

Chen, H., Song, J., Han, C., Xia, J., & Yokoya, N. (2024). ChangeMamba: Remote sensing change detection with spatiotemporal state space model. *IEEE Transactions on Geoscience and Remote Sensing, 62*, 1–20. DOI: 10.1109/TGRS.2024.3417253

Chen, L., Msigwa, G., Yang, M., Osman, A. I., Fawzy, S., Rooney, D. W., & Yap, P.-S. (2022). Strategies to achieve a carbon neutral society: A review. *Environmental Chemistry Letters, 20*(4), 2277–2310. DOI: 10.1007/s10311-022-01435-8 PMID: 35431715

Chen, P., Gao, J., Ji, Z., Liang, H., & Peng, Y. (2022). Do Artificial Intelligence Applications Affect Carbon Emission Performance?—Evidence from Panel Data Analysis of Chinese Cities. *Energies, 15*(15), 5730. DOI: 10.3390/en15155730

Chen, Y.-H., Sharma, K., Sharma, C., & Sharma, S. (2023). Integrating explainable artificial intelligence and blockchain to smart agriculture: Research prospects for decision making and improved security. *Smart Agricultural Technology, 6*, 100350. DOI: 10.1016/j.atech.2023.100350

Choi, J., Hidayat, M., Cho, S., Hwang, W., Lee, H., Cho, B., Kim, M., Baek, I., & Kim, G. (2025). Recent Trends in Machine Learning, Deep Learning, Ensemble Learning, and Explainable Artificial Intelligence Techniques for Evaluating Crop Yields Under Abnormal Climate Conditions. *Plants, 14*(18), 2841. Advance online publication. DOI: 10.3390/plants14182841 PMID: 41011993

Cho, S., Soleh, H., Choi, J., Hwang, W., Lee, H., Cho, Y., Cho, B., Kim, M., Baek, I., & Kim, G. (2024). Recent Methods for Evaluating Crop Water Stress Using AI Techniques: A Review. *Sensors (Basel), 24*(19), 6313. Advance online publication. DOI: 10.3390/s24196313 PMID: 39409355

Chukwuma, E., Afolabi, O., Okonkwo, C., Olamigoke, O., & Okonkwo, C. (2024). Application of geospatial technology and decision model in the development of an improved food security index. *Scientific Reports*, *14*(1), 30204. Advance online publication. DOI: 10.1038/s41598-024-77845-1 PMID: 39632976

Cicilio, P., Glennon, D., Mate, A., Barnes, A., Chalishazar, V., Cotilla-Sanchez, E., Vaagensmith, B., Gentle, J., Rieger, C., Wies, R., & Kapourchali, M. H. (2021). Resilience in an Evolving Electrical Grid. *Energies*, *14*(3), 694. DOI: 10.3390/en14030694

Clark, M., Fröhner, C., Jørgensen, A., Pienkowski, T., Yekela, S., Isacs, A., Crowe, O., Andrews, J., Smaldino, P., Gallizioli, I., Arena, G., & Mills, M. (2025). Eliminating climate change mitigation trade-offs in African rangelands. DOI: 10.21203/rs.3.rs-7750461/v1

Cohen, S. J., & Waddell, M. W. (2009). Climate Change in the 21st Century. https://doi.org/DOI: 10.1515/9780773581296

Cortez, F. (2023). Artificial Intelligence, Climate Change and Innovative Democratic Governance. *European Journal of Risk Regulation*, *14*(3), 484–503. DOI: 10.1017/err.2023.60

Coumou, D., & Rahmstorf, S. (2012). A decade of weather extremes. *Nature Climate Change*, *2*(7), 491–496. DOI: 10.1038/nclimate1452

Cramer, W., Guiot, J., Fader, M., Garrabou, J., Gattuso, J.-P., Iglesias, A., Lange, M. A., Lionello, P., Llasat, M. C., Paz, S., Peñuelas, J., Snoussi, M., Toreti, A., Tsimplis, M. N., & Xoplaki, E. (2018). Climate change and interconnected risks to sustainable development in the Mediterranean. *Nature Climate Change*, *8*(11), 972–980. DOI: 10.1038/s41558-018-0299-2

Creutzig, F., Roy, J., Lamb, W. F., Azevedo, I. M. L., Bruine de Bruin, W., Dalkmann, H., Edelenbosch, O. Y., Geels, F. W., Grubler, A., Hepburn, C., Hertwich, E. G., Khosla, R., Mattauch, L., Minx, J. C., Ramakrishnan, A., Rao, N. D., Steinberger, J. K., Tavoni, M., Ürge-Vorsatz, D., & Weber, E. U. (2018). Towards demand-side solutions for mitigating climate change. *Nature Climate Change*, *8*(4), 260–263. DOI: 10.1038/s41558-018-0121-1

d'Andrimont, R., Yordanov, M., Sedano, F., Verhegghen, A., Strobl, P., Zachariadis, S., Camilleri, F., Palmieri, A., Eiselt, B., Rubio Iglesias, J. M., & van der Velde, M. (2024). Advances in LUCAS Copernicus 2022: Enhancing Earth observations with comprehensive in situ data on EU land cover and use. *Earth System Science Data*, *16*(12), 5723–5735. DOI: 10.5194/essd-16-5723-2024

Danilevicz, M., Upadhyaya, S., Batley, J., Bennamoun, M., Bayer, P., & Edwards, D. (2025). Understanding plant phenotypes in crop breeding through explainable AI. *Plant Biotechnology Journal, 23*(10), 4200–4213. DOI: 10.1111/pbi.70208 PMID: 40574309

Das, P., Posch, A., Barber, N., van Werkhoven, K., & Ganguly, A. R. (2024). Hybrid physics–AI outperforms numerical weather prediction for extreme precipitation nowcasting. *npj Climate and Atmospheric Science, 7*, 282. DOI: 10.1038/s41612-024-00834-8

Davidsdottir, B., Ásgeirsson, E. I., Fazeli, R., Gunnarsdottir, I., Leaver, J., Shafiei, E., & Stefánsson, H. (2024). Integrated Energy Systems Modeling with Multi-Criteria Decision Analysis and Stakeholder Engagement for Identifying a Sustainable Energy Transition. *Energies, 17*(17), 4266. DOI: 10.3390/en17174266

D, C. S., & Devarajan, Y. (2025). D, C.S., Devarajan, Y. Investigation of Emerging Technologies in Agriculture: An In-depth Look at Smart Farming, Nano-agriculture, AI, and Big Data. *Journal of Biosystems Engineering, 50*(2), 170–192. DOI: 10.1007/s42853-025-00258-z

Dechezleprêtre, A., Glachant, M., & Ménière, Y. (2009). Technology transfer by CDM projects: A comparison of Brazil, China, India and Mexico. *Energy Policy, 37*(2), 703–711. DOI: 10.1016/j.enpol.2008.10.007

Degen, D., Cacace, M., & Wellmann, F. (2022). 3D multi-physics uncertainty quantification using physics-based machine learning. *Scientific Reports, 12*, 17491. DOI: 10.1038/s41598-022-21739-7

Demaeyer, J., Bhend, J., Lerch, S., Primo, C., Van Schaeybroeck, B., Atencia, A., Ben Bouallègue, Z., Chen, J., Dabernig, M., Evans, G., Faganeli Pucer, J., Hooper, B., Horat, N., Jobst, D., Merše, J., Mlakar, P., Möller, A., Mestre, O., Taillardat, M., & Vannitsem, S. (2023). The EUPPBench postprocessing benchmark dataset v1.0. *Earth System Science Data, 15*(6), 2635–2653. DOI: 10.5194/essd-15-2635-2023

Dessai, S., & Hulme, M. (2004). Does climate adaptation policy need probabilities? *Climate Policy, 4*(2), 107–128. DOI: 10.1080/14693062.2004.9685515

Dhal, S., & Kar, D. (2024). Transforming Agricultural Productivity with AI-Driven Forecasting: Innovations in Food Security and Supply Chain Optimization. *Forecasting, 6*(4), 925–951. Advance online publication. DOI: 10.3390/forecast6040046

Dhanya, V., Subeesh, A., Kushwaha, N., Vishwakarma, D., Kumar, N., Ritika, G., & Singh, A. (2022). *Deep learning based computer vision approaches for smart agricultural applications.* Artificial Intelligence in Agriculture., DOI: 10.1016/j.aiia.2022.09.007

Dhinakaran, D., Edwin Raja, S., Ramathilagam, A., Vennila, G., & Alagulakshmi, A. (2025). Ethical and legal challenges with IoT in home digital twins. *MethodsX, 14*, 103409. DOI: 10.1016/j.mex.2025.103409 PMID: 40852565

Dhinakaran, D., Jagadish Kumar, N., Raja Brundha, A., Subiksha, N., & Karpagam, T. (2024). Precision Maintenance with PARM and Augmented Reality for Asset Optimization. *Navigating the Augmented and Virtual Frontiers in Engineering, Pages, 28.* Advance online publication. DOI: 10.4018/979-8-3693-5613-5.ch002

Dhinakaran, D., Udhaya Sankar, S. M., Latha, B. C., Anns, A. E. J., & Sri, V. K. "Dam Management and Disaster Monitoring System using IoT," *2023 International Conference on Sustainable Computing and Data Communication Systems (ICSCDS)*, Erode, India, 2023, pp. 1197-1201, DOI: 10.1109/ICSCDS56580.2023.10105132

Diffenbaugh, N. S., & Burke, M. (2019). Global warming has increased global economic inequality. *Proceedings of the National Academy of Sciences of the United States of America, 116*(20), 9808–9813. DOI: 10.1073/pnas.1816020116 PMID: 31010922

Dinku, T., Block, P., Sharoff, J., Hailemariam, K., Osgood, D., del Corral, J., Cousin, R., & Thomson, M. C. (2014). Bridging critical gaps in climate services and applications in africa. *Earth Perspectives, 1*(1), 15. DOI: 10.1186/2194-6434-1-15

Drakopulos, L., Havice, E., & Campbell, L. (2022). Architecture, agency and ocean data science initiatives: Data-driven transformation of oceans governance. *Earth System Governance, 12*, 100140. DOI: 10.1016/j.esg.2022.100140

Dueben, P. D., Schultz, M. G., Chantry, M., Gagne, D. J., Hall, D. M., & McGovern, A. (2022). Challenges and benchmark datasets for machine learning in the atmospheric sciences: Definition, status, and outlook. *Artificial Intelligence for the Earth Systems, 1*(3), e210002. DOI: 10.1175/AIES-D-21-0002.1

Dumenu, W. K., & Obeng, E. A. (2016). Climate change and rural communities in Ghana: Social vulnerability, impacts, adaptations and policy implications. Environmental Science &amp. *Environmental Science & Policy, 55*, 208–217. DOI: 10.1016/j.envsci.2015.10.010

Dutta, S., Chakraborty, S., Goswami, R., Banerjee, H., Majumdar, K., Li, B., & Jat, M. (2020). Maize yield in smallholder agriculture system—An approach integrating socio-economic and crop management factors. *PLoS One*, *15*(2), e0229100. Advance online publication. DOI: 10.1371/journal.pone.0229100 PMID: 32092077

Dzebo, A., & Stripple, J. (2015). Transnational adaptation governance: An emerging fourth era of adaptation. *Global Environmental Change*, *35*, 423–435. DOI: 10.1016/j.gloenvcha.2015.10.006

Ejiyi, C. J., Cai, D., Thomas, D., Obiora, S., Osei-Mensah, E., Acen, C., Eze, F. O., Sam, F., Zhang, Q., & Bamisile, O. O. (2025). Comprehensive review of artificial intelligence applications in renewable energy systems: Current implementations and emerging trends. *Journal of Big Data*, *12*(1), 169. Advance online publication. DOI: 10.1186/s40537-025-01178-7

Elbasi, E., Mostafa, N., AlArnaout, Z., Zreikat, A., Cina, E., Varghese, G., Shdefat, A., Topcu, A., Abdelbaki, W., Mathew, S., & Zaki, C. (2023). Artificial Intelligence Technology in the Agricultural Sector: A Systematic Literature Review. *IEEE Access : Practical Innovations, Open Solutions*, *11*, 171–202. DOI: 10.1109/ACCESS.2022.3232485

Erfanmanesh, M. (2017). Status and quality of open access journals in Scopus. *Collection Building*, *36*(4), 155–162. DOI: 10.1108/CB-02-2017-0007

Espinel, R., Herrera-Franco, G., García, J., & Escandón-Panchana, P. (2024). Artificial Intelligence in Agricultural Mapping. *Agriculture*, *14*(7), 1071. Advance online publication. DOI: 10.3390/agriculture14071071

Fathi Qarachal, J., & Alizadeh, M. (2025). Harnessing precision agriculture and nanotechnology for sustainable farming: A review. *Nanotechnol. Environ. Eng.*, *10*(2), 37. DOI: 10.1007/s41204-025-00429-5

Fathi, M., Shah-Hosseini, R., & Moghimi, A. (2023). 3D-ResNet-BiLSTM model: A deep learning model for county-level soybean yield prediction with time-series Sentinel-1, Sentinel-2 imagery, and Daymet data. *Remote Sensing (Basel)*, *15*(23), 5551. Advance online publication. DOI: 10.3390/rs15235551

Fatimah, M., & Malik, J. A., et al. (2024). Integration of Agriculture 5.0 and Explainable AI (XAI) for Sustainable Development. *Environmental Science Journal*. https://doi.org/DOI: 10.28991/ESJ-2024-08-02-024

Fauchald, P., Christensen, T. R., & Christensen, T. (2026). Climate change impacts on Arctic ecosystems and associated feedbacks. *Frontiers in Environmental Science*, *13*, 1747632. Advance online publication. DOI: 10.3389/fenvs.2025.1747632

Favour, N.Favour N. EzeAdepeju Nafisat SanusiIsrael Jonathan IheomaChijioke Cyriacus EkechiMicheal Adeolu OlatunbosunFavour Chizurum UkasoanyaMuhdawwal Aremu Eleshin. (2025). Artificial Intelligence in Climate Change Mitigation and Adaptation: A Review of Emerging Technologies and Real-World Applications. *Global Journal of Engineering and Technology Advances*, *24*(2), 235–250. DOI: 10.30574/gjeta.2025.24.2.0247

Feng, N., & Chandra, Y. (2025). Accountability in Government Use of AI: Citizen Concerns and Preferences. *Public Administration*, padm.70030. Advance online publication. DOI: 10.1111/padm.70030

Fernandez-Beltran, R., Baidar, T., Kang, J., & Pla, F. (2021). Rice-yield prediction with multi-temporal sentinel-2 data and 3D CNN: A case study in Nepal. *Remote Sensing (Basel)*, *13*(7), 1391. DOI: 10.3390/rs13071391

Flato, G., Marotzke, J., & Abiodun, B.. (2019). Evaluation of climate models. In *IPCC Special Report on Climate Change*. Cambridge University Press., DOI: 10.1017/9781009157896.009

Galaz, V., Centeno, M. A., Callahan, P. W., Causevic, A., Patterson, T., Brass, I., Baum, S., Farber, D., Fischer, J., Garcia, D., McPhearson, T., Jimenez, D., King, B., Larcey, P., & Levy, K. (2021). Artificial intelligence, systemic risks, and sustainability. *Technology in Society*, *67*, 101741. DOI: 10.1016/j.techsoc.2021.101741

Gao, Z., Luo, Z., Zhang, W., Lv, Z., & Xu, Y. (2020). Deep Learning Application in Plant Stress Imaging: A Review. *AgriEngineering*. https://doi.org/DOI: 10.3390/agriengineering2030029

Gao, J., Li, P., Chen, Z., & Zhang, J. (2020). A survey on deep learning for multimodal data fusion. *Neural Computation*, *32*(5), 829–864. DOI: 10.1162/neco_a_01273 PMID: 32186998

Garnot, V. S. F., Landrieu, L., Giordano, S., & Chehata, N. (2020). Satellite image time series classification with pixel-set encoders and temporal self-attention. Proceedings of the IEEE/CVF conference on computer vision and pattern recognition, Gorelick, N., Hancher, M., Dixon, M., Ilyushchenko, S., Thau, D., & Moore, R. (2017). Google Earth Engine: Planetary-scale geospatial analysis for everyone. *Remote Sensing of Environment*, *202*, 18–27.

Gavasso-Rita, Y., Papalexiou, S., Li, Y., Elshorbagy, A., Li, Z., & Schuster-Wallace, C. (2023). Crop models and their use in assessing crop production and food security: A review. *Food and Energy Security*. Advance online publication. DOI: 10.1002/fes3.503

Gawlikowski, J., Tassi, C. R. N., Ali, M., Lee, J., Humt, M., Feng, J., Kruspe, A., Triebel, R., Jung, P., Roscher, R., Shahzad, M., Yang, W., Bamler, R., & Zhu, X. X. (2023). A survey of uncertainty in deep neural networks. *Artificial Intelligence Review*, *56*(Suppl 1), 1513–1589. DOI: 10.1007/s10462-023-10562-9

Gebresenbet, G., Bosona, T., Patterson, D., Persson, H., Fischer, B., Mandaluniz, N., Chirici, G., Zacepins, A., Komašilovs, V., Pitulac, T., & Nasirahmadi, A. (2023). A concept for the application of integrated digital technologies to enhance future smart agricultural systems. *Smart Agricultural Technology*, *5*, 100255. Advance online publication. DOI: 10.1016/j.atech.2023.100255

Geurts, P., Ernst, D., & Wehenkel, L. (2006). Extremely randomized trees. *Machine Learning*, *63*(1), 3–42. DOI: 10.1007/s10994-006-6226-1

Gevaert, C. M. (2022). Explainable AI for earth observation: A review including societal and regulatory perspectives. *International Journal of Applied Earth Observation and Geoinformation*, *112*, 102869. DOI: 10.1016/j.jag.2022.102869

Gong, H., Mu, T., Li, Q., Dai, H., Li, C., He, Z., Wang, W., Han, F., Tuniyazi, A., Li, H., Lang, X., Li, Z., & Wang, B. (2022). Swin-Transformer-enabled YOLOv5 with attention mechanism for small object detection on satellite images. *Remote Sensing (Basel)*, *14*(12), 2861. Advance online publication. DOI: 10.3390/rs14122861

Gorelick, N., Hancher, M., Dixon, M., Ilyushchenko, S., Thau, D., & Moore, R. (2017). Google Earth Engine: Planetary-scale geospatial analysis for everyone. *Remote Sensing of Environment*, *202*, 18–27. DOI: 10.1016/j.rse.2017.06.031

Goyal, V., Rani, D., Mehrotra, S., Deng, C., & Wang, Y. (2023). Unlocking the Potential of Nano-Enabled Precision Agriculture for Efficient and Sustainable Farming. *Plants*, *12*(21), 3744. Advance online publication. DOI: 10.3390/plants12213744 PMID: 37960100

Grassi, G., House, J., Kurz, W. A., Cescatti, A., Houghton, R. A., Peters, G. P., Sanz, M. J., Viñas, R. A., Alkama, R., Arneth, A., Bondeau, A., Dentener, F., Fader, M., Federici, S., Friedlingstein, P., Jain, A. K., Kato, E., Koven, C. D., Lee, D., & Zaehle, S. (2018). Reconciling global-model estimates and country reporting of anthropogenic forest CO2 sinks. *Nature Climate Change*, *8*(10), 914–920. DOI: 10.1038/s41558-018-0283-x

Grimm, N. B., Faeth, S. H., Golubiewski, N. E., Redman, C. L., Wu, J., Bai, X., & Briggs, J. M. (2008). Global Change and the Ecology of Cities. *Science*, *319*(5864), 756–760. DOI: 10.1126/science.1150195 PMID: 18258902

Gund, R., Badgujar, C., Samiappan, S., & Jagadamma, S. (2025). Application of Digital Twin Technology in Smart Agriculture: A Bibliometric Review. *Agriculture*, *15*(17), 1799. Advance online publication. DOI: 10.3390/agriculture15171799

Gupta, J., Termeer, C., Klostermann, J., Meijerink, S., van den Brink, M., Jong, P., Nooteboom, S., & Bergsma, E. (2010). The Adaptive Capacity Wheel: A method to assess the inherent characteristics of institutions to enable the adaptive capacity of society. Environmental Science &amp. *Environmental Science & Policy*, *13*(6), 459–471. DOI: 10.1016/j.envsci.2010.05.006

Gu, Z., & Zeng, M. (2023). The Use of Artificial Intelligence and Satellite Remote Sensing in Land Cover Change Detection: Review and Perspectives. *Sustainability (Basel)*, *16*(1), 274. DOI: 10.3390/su16010274

Hadley, K., Wheat, S., Rogers, H. H., Balakumar, A., Gonzales-Pacheco, D., Davis, S. S., Linstadt, H., Cushing, T., Ziska, L. H., Piper, C., & Sorensen, C. (2023). Mechanisms underlying food insecurity in the aftermath of climate-related shocks: A systematic review. *The Lancet. Planetary Health*, *7*(3), e242–e250. DOI: 10.1016/S2542-5196(23)00003-7 PMID: 36774944

Hall, J. W., Lempert, R. J., Keller, K., Hackbarth, A., Mijere, C., & McInerney, D. J. (2012). Robust Climate Policies Under Uncertainty: A Comparison of Robust Decision Making and Info-Gap Methods. Risk Analysis, 32(10), 1657–1672. Portico. DOI: 10.1111/j.1539-6924.2012.01802.x

Ham, Y.-G., Kim, J.-H., & Luo, J.-J. (2019). Deep learning for multi-year ENSO forecasts. *Nature*, *573*(7775), 568–572. DOI: 10.1038/s41586-019-1559-7 PMID: 31534218

Harini, M., Dhinakaran, D., Prabhu, D., Udhaya Sankar, S. M., Pooja, V., & Kokila Sruthi, P. (2023). "Levarging Blockchain for Transparency in Agriculture Supply Chain Management Using IoT and Machine Learning," 2023 World Conference on Communication & Computing (WCONF), RAIPUR, India, pp. 1-6, DOI: 10.1109/WCONF58270.2023.10235156

Haynes, K., Lagerquist, R., McGraw, M., Musgrave, K., & Ebert-Uphoff, I. (2023). Creating and evaluating uncertainty estimates with neural networks for environmental-science applications. *Artificial Intelligence for the Earth Systems*, *2*(2), e220061. DOI: 10.1175/AIES-D-22-0061.1

Held, I. M. (2005). The Gap between Simulation and Understanding in Climate Modeling. *Bulletin of the American Meteorological Society*, *86*(11), 1609–1614. DOI: 10.1175/BAMS-86-11-1609

He, M., & Rozos, E. (2025). Editorial: Harnessing artificial intelligence to address climate-induced challenges in water resources management. *Frontiers in Water*, *7*, 1637341. Advance online publication. DOI: 10.3389/frwa.2025.1637341

Hoffmann, J., Bauer, P., Sandu, I., Wedi, N., Geenen, T., & Thiemert, D. (2023). Destination Earth – A digital twin in support of climate services. *Climate Services*, *30*, 100394. DOI: 10.1016/j.cliser.2023.100394

Hong, D., Han, Z., Yao, J., Gao, L., Zhang, B., & Plaza, A. (2022). SpectralFormer: Rethinking hyperspectral image classification with transformers. *IEEE Transactions on Geoscience and Remote Sensing*, *60*, 1–15. DOI: 10.1109/TGRS.2022.3172371

Hope, K. R.Sr. (2009). Climate change and poverty in Africa. International Journal of Sustainable Development &amp. *International Journal of Sustainable Development and World Ecology*, *16*(6), 451–461. DOI: 10.1080/13504500903354424

Hoseinzadeh, S., & Garcia, D. (2024). Ai-driven innovations in greenhouse agriculture: Reanalysis of sustainability and energy efficiency impacts. *Energy Conversion and Management*, *X*, 100701. Advance online publication. DOI: 10.1016/j.ecmx.2024.100701

Hsu, A., Höhne, N., Kuramochi, T., Roelfsema, M., Weinfurter, A., Xie, Y., Lütkehermöller, K., Chan, S., Corfee-Morlot, J., Drost, P., Faria, P., Gardiner, A., Gordon, D. J., Hale, T., Hultman, N. E., Moorhead, J., Reuvers, S., Setzer, J., Singh, N., & Widerberg, O. (2018). A research roadmap for quantifying non-state and subnational climate mitigation action. *Nature Climate Change*, *9*(1), 11–17. DOI: 10.1038/s41558-018-0338-z

Hu, W., Zhao, F., ... (2023). Interpretable ML framework for crop-climate relationship analysis. *Elsevier*. https://www.elsevier.com/open-access/userlicense/1.0/

Huang, X., Zhou, Y., Yang, X., Zhu, X., & Wang, K. (2023). SS-TMNet: Spatial–spectral transformer network with multi-scale convolution for hyperspectral image classification. *Remote Sensing (Basel)*, *15*(5), 1206. Advance online publication. DOI: 10.3390/rs15051206

Huang, Y., Xu, T., Yang, Q., Pan, C., Lü, Z., Chen, H., Zhang, X., & Chen, C. (2023). Demand prediction of medical services in home and community-based services for older adults in China using machine learning. *Frontiers in Public Health*, *11*, 1142794. Advance online publication. DOI: 10.3389/fpubh.2023.1142794 PMID: 37006569

Huang, Z., Wang, J., Bing, L., Qiu, Y., Guo, R., Yu, Y., Ma, M., Niu, L., Tong, D., Andrew, R. M., Friedlingstein, P., Canadell, J. G., Xi, F., & Liu, Z. (2023). Global carbon uptake of cement carbonation accounts 1930–2021. *Earth System Science Data*, *15*(11), 4947–4958. DOI: 10.5194/essd-15-4947-2023

Huntingford, C., Jeffers, E. S., Bonsall, M. B., Christensen, H. M., Lees, T., & Yang, H. (2019). Machine learning and artificial intelligence to aid climate change research and preparedness. *Environmental Research Letters*, *14*(12), 124007. DOI: 10.1088/1748-9326/ab4e55

Huo, C., Chen, K., Zhang, S., Wang, Z., Yan, H., Shen, J., Hong, Y., Qi, G., Fang, H., & Wang, Z. (2025). When remote sensing meets foundation model: A survey and beyond. *Remote Sensing (Basel)*, *17*(2), 179. Advance online publication. DOI: 10.3390/rs17020179

Hurlimann, A., Moosavi, S., & Browne, G. R. (2021). Urban planning policy must do more to integrate climate change adaptation and mitigation actions. *Land Use Policy*, *101*, 105188. DOI: 10.1016/j.landusepol.2020.105188

Hu, W., Ghazvinian, M., Chapman, W. E., Sengupta, A., Ralph, F. M., & Delle Monache, L. (2023). Deep learning forecast uncertainty for precipitation over the western United States. *Monthly Weather Review*, *151*(6), 1367–1385. DOI: 10.1175/MWR-D-22-0268.1

Huynh, N. N. T., Garambois, P.-A., Colleoni, F., & Monnier, J. (2026). A hybrid physics–AI approach using universal differential equations with state-dependent neural networks for learnable, regionalisable, spatially distributed hydrological modelling. *Geoscientific Model Development*, *19*(3), 1055–1074. DOI: 10.5194/gmd-19-1055-2026

Huynh, N. N. T., Garambois, P.-A., Renard, B., Colleoni, F., Monnier, J., & Roux, H. (2025). A distributed hybrid physics–AI framework for learning corrections of internal hydrological fluxes and enhancing high-resolution regionalised flood modelling. *Hydrology and Earth System Sciences*, *29*(15), 3589–3613. DOI: 10.5194/hess-29-3589-2025

Ibrahim, R. E., Motshegwa, T., Benouar, D., Khodja, M., Ead, H., Zhang, L., Crawley, F., Mchunu, N. P., Hachigonta, S., Hodson, S., Mcbride, V., & Yohannes, T. (2025). Sustainable North-South Africa Collaboration for Disaster and Crisis Management: A Strategic Capacity Development Framework using Open Science, Artificial Intelligence and Geoinformatics. *Data Science Journal*, *24*, 11. Advance online publication. DOI: 10.5334/dsj-2025-011

Ilyas, M., Liaqat, U., Eqan, M., Boškailo, E., Safdar, M., Ajanovic, T., Yaqoob, H., Chovancová, J., & Onifade, S. T. (2026). Policy frameworks and regulations for net zero. Net ZERO: Foundations, Technologies and Strategic Innovation, 189–247. DOI: 10.1016/B978-0-443-36426-6.00010-1

Inglada, J., Vincent, A., Arias, M., Tardy, B., Morin, D., & Rodes, I. (2017). Operational high resolution land cover map production at the country scale using satellite image time series. *Remote sensing, 9*(1), 95.

Intergovernmental Panel on Climate Change (IPCC). (2023). Climate Change 2022 – Impacts, Adaptation and Vulnerability. https://doi.org/DOI: 10.1017/9781009325844

Ioannou, K., Karasmanaki, E., & Tsantopoulos, G. (2025). Environmental Policy as a Tool for Sustainable Development. *Sustainability (Basel)*, *17*(8), 3674. DOI: 10.3390/su17083674

Iqbal, B. A., & Ghauri, F. N. (2011). Climate change: The biggest challenge in 21st century. *Mediterranean Journal of Social Sciences*, *2*(6), 41–41.

Irrgang, C., Boers, N., Sonnewald, M., Barnes, E. A., Kadow, C., Staneva, J., & Saynisch-Wagner, J. (2021). Towards neural Earth system modelling by integrating artificial intelligence in Earth system science. *Nature Machine Intelligence*, *3*(8), 667–674. DOI: 10.1038/s42256-021-00374-3

Islam, F. A. S. (2025). Artificial Intelligence-driven Optimization of Nature-based Carbon Sequestration: A Scalable Architecture for Urban Climate Resilience. *International Journal of Environment and Climate Change*, *15*(7), 252–277. DOI: 10.9734/ijecc/2025/v15i74928

Islam, F. A. S., & Islam, M. A. N. (2025). AI-Driven Integration of Nanotechnology and Green Nanotechnology for Sustainable Energy and Environmental Remediation. *Journal of Engineering Research and Reports*, *27*(7), 260–311. DOI: 10.9734/jerr/2025/v27i71574

Jabed, M. A., & Murad, M. A. A. (2024). Crop yield prediction in agriculture: A comprehensive review of machine learning and deep learning approaches, with insights for future research and sustainability. *Heliyon*, *10*(24), e40836. DOI: 10.1016/j.heliyon.2024.e40836 PMID: 39720079

Jafar, A., Bibi, N., Naqvi, R., Sadeghi-Niaraki, A., & Jeong, D. (2024). Revolutionizing agriculture with artificial intelligence: Plant disease detection methods, applications, and their limitations. *Frontiers in Plant Science*, *15*, 1356260. Advance online publication. DOI: 10.3389/fpls.2024.1356260 PMID: 38545388

Janetschek, H., Brandi, C., Dzebo, A., & Hackmann, B. (2019). The 2030 Agenda and the Paris Agreement: Voluntary contributions towards thematic policy coherence. *Climate Policy*, *20*(4), 430–442. DOI: 10.1080/14693062.2019.1677549

Janga, B., Asamani, G. P., Sun, Z., & Cristea, N. (2023). A review of practical AI for remote sensing in Earth sciences. *Remote Sensing (Basel)*, *15*(16), 4112. DOI: 10.3390/rs15164112

Jasanoff, S. (2010). A New Climate for Society. Theory, Culture &amp. *Theory, Culture & Society*, *27*(2–3), 233–253. DOI: 10.1177/0263276409361497

Jiang, C. M., Duangthip, D., Chan, A. K. Y., Tamrakar, M., Lo, E. C. M., & Chu, C. H. (2021). Global research interest regarding silver diamine fluoride in dentistry: A bibliometric analysis. *Journal of Dentistry*, *113*, 103778. DOI: 10.1016/j.jdent.2021.103778 PMID: 34391874

Ji, Z., Pan, Y., Zhu, X., Zhang, D., & Wang, J. (2022). A generalized model to predict large-scale crop yields integrating satellite-based vegetation index time series and phenology metrics. *Ecological Indicators*, *137*, 108759. DOI: 10.1016/j.ecolind.2022.108759

José de Oliveira, O., Francisco da Silva, F., Juliani, F., César Ferreira Motta Barbosa, L., & Vieira Nunhes, T. (2019). *Bibliometric Method for Mapping the State-of-the-Art and Identifying Research Gaps and Trends in Literature: An Essential Instrument to Support the Development of Scientific Projects*. Scientometrics Recent Advances., DOI: 10.5772/intechopen.85856

Juel, R., Whitmee, S., Green, R., Davies, M., Milner, J., & Turcu, C. (2026). Integrated indicators for tracking climate mitigation and adaptation: A protocol for a rapid review with a focus on health co-benefits and trade-offs. *Wellcome Open Research*, *11*, 30. DOI: 10.12688/wellcomeopenres.25619.1

Justice, C. J., Becker-Reshef, I., Barker, B., Humber, M. L., Sanchez, A., Nakalembe, C., & Sobue, S. (2019). Strengthening agricultural decisions in countries at risk of food insecurity: The GEOGLAM Crop Monitor for Early Warning. AGU Fall Meeting Abstracts, Kamilaris, A., & Prenafeta-Boldú, F. X. (2018). Deep learning in agriculture: A survey. *Computers and Electronics in Agriculture*, *147*, 70–90.

Karakosta, C., & Papathanasiou, J. (2024). Climate-Driven Sustainable Energy Investments: Key Decision Factors for a Low-Carbon Transition Using a Multi-Criteria Approach. *Energies*, *17*(21), 5515. DOI: 10.3390/en17215515

Karniadakis, G. E., Kevrekidis, I. G., Lu, L., Perdikaris, P., Wang, S., & Yang, L. (2021). Physics-informed machine learning. *Nature Reviews. Physics*, *3*(6), 422–440. DOI: 10.1038/s42254-021-00314-5

Kaya, C. (2025). Intelligent Environmental Control in Plant Factories: Integrating Sensors, Automation, and AI for Optimal Crop Production. *Food and Energy Security*, *14*(1), e70026. Advance online publication. DOI: 10.1002/fes3.70026

Keerthana, M., Dhinakaran, D., Ananthi, M., Harish, R., Udhaya Sankar, S. M., & Sree, M. S. (2023). "IoT Based Automated Irrigation System for Agricultural Activities," 2023 12th International Conference on Advanced Computing (ICoAC), Chennai, India, pp. 1-6, DOI: 10.1109/ICoAC59537.2023.10249426

Kerle, N., Nex, F., Gerke, M., Duarte, D., & Vetrivel, A. (2019). UAV-Based Structural Damage Mapping: A Review. *ISPRS International Journal of Geo-Information*, *9*(1), 14. DOI: 10.3390/ijgi9010014

Khan, M., Mfitumukiza, D., & Huq, S. (2021). Capacity building for implementation of Nationally Determined Contributions under the Paris Agreement. Making Climate Action More Effective, 95–106. .DOI: 10.4324/9781003162407-8

Khan, A. (2025). AI-Optimized Renewable Energy Management: Enhancing Efficiency and Grid Stability Using Machine Learning. *International Journal of Research & Technology*, *13*(1), 24–32.

Khan, M., Hanan, A., Kenzhebay, M., Gazzea, M., & Arghandeh, R. (2024). Transformer-based land use and land cover classification with explainability using satellite imagery. *Scientific Reports*, *14*(1), 16744. Advance online publication. DOI: 10.1038/s41598-024-67186-4 PMID: 39033183

Kikon, A., & Deka, P. (2021). Artificial intelligence application in drought assessment, monitoring, and forecasting: A review. *Stochastic Environmental Research and Risk Assessment*, *36*(5), 1197–1214. DOI: 10.1007/s00477-021-02129-3

Kirkwood, C., Economou, T., Pugeault, N., & Odbert, H. (2022). Bayesian deep learning for spatial interpolation in the presence of auxiliary information. *Mathematical Geosciences*, *54*(3), 507–531. DOI: 10.1007/s11004-021-09988-0

Kishore, K. V., Dhinakaran, D., Sankar, S. U., Chandu, K., & Visuwasam, L. M. M. (2023, November). Fish farm monitoring system using IoT. In *2023 International Conference on System, Computation, Automation and Networking (ICSCAN)* (pp. 1-6). IEEE. doi:DOI: 10.1109/ICSCAN58655.2023.10395013

Klenk, N., Fiume, A., Meehan, K., & Gibbes, C. (2017). Local knowledge in climate adaptation research: moving knowledge frameworks from extraction to co-production. WIREs Climate Change, 8(5). Portico. https://doi.org/DOI: 10.1002/wcc.475

Klinsky, S. (2017). An initial scoping of transitional justice for global climate governance. *Climate Policy*, *18*(6), 752–765. DOI: 10.1080/14693062.2017.1377594

Klompenburg, T., Kassahun, A., & Catal, C. (2020). Crop yield prediction using machine learning. *Computers and Electronics in Agriculture*.

Klotz, D., Kratzert, F., Gauch, M., Sampson, A. K., Brandstetter, J., Klambauer, G., Hochreiter, S., & Nearing, G. (2022). Uncertainty estimation with deep learning for rainfall–runoff modeling. *Hydrology and Earth System Sciences*, *26*(6), 1673–1693. DOI: 10.5194/hess-26-1673-2022

Kouziokas, G. N., & Perakis, K. (2017). Decision Support System Based On Artificial Intelligence, Gis And Remote Sensing For Sustainable Public And Judicial Management. *European Journal of Sustainable Development*, *6*(3). Advance online publication. DOI: 10.14207/ejsd.2017.v6n3p397

Kratzert, F., Klotz, D., Brenner, C., Schulz, K., & Herrnegger, M. (2019). Rainfall–runoff modelling using long short-term memory networks. *Water Resources Research*, *55*(8), 6503–6522. DOI: 10.1029/2018WR024567

Kumar, D., Shekhar, S., & Tewary, T. (2025). Editorial: AI and data analytics for climate data management. *Frontiers in Environmental Science*, *13*, 1679608. Advance online publication. DOI: 10.3389/fenvs.2025.1679608

Kumari, P., Lavania, P., Dobriyal, M., Srivastav, M., & Kumar, P. (2026). Artificial intelligence and forest management. Advanced Geospatial and Ground Based Techniques in Forest Monitoring, 255–272. .DOI: 10.1016/B978-0-443-18949-4.00002-6

Kumar, M. S., Palanichamy, N. V., Shivakumar, K. M., Chandrakumar, M., Kalpana, M., & Murugananthi, D. (2025). Impact of climate change on global economy: A comprehensive review. *Regional Sustainability*, *6*(6), 100274. DOI: 10.1016/j.regsus.2025.100274

Kumar, R., & Prasad, V.. (2025). *Machine learning models for sustainable agricultural development*. IEEE Xplore., DOI: 10.1109/AUTOCOM64127.2025.10956454

Kuziboev, B., Çatık, A. N., Pícha, K., Makhmudov, S., Matyakubova, A., & Balli, E. (2025). Time-varying dynamics of energy and sustainability uncertainty. *Research in Globalization*, *11*, 100325. DOI: 10.1016/j.resglo.2025.100325

Lacoste, A., Lehmann, N., Rodriguez, P., Sherwin, E. D., Kerner, H., Lütjens, B., Irvin, J. A., Dao, D., Alemohammad, H., Drouin, A., Gunturkun, M., Huang, G., Vazquez, D., Newman, D., Bengio, Y., Ermon, S., & Zhu, X. X. (2023). GEO-Bench: Toward foundation models for Earth monitoring. In Advances in Neural Information Processing Systems 36 (NeurIPS 2023) Datasets and Benchmarks Track (pp. 51080–51093). https://doi.org/DOI: 10.48550/arXiv.2306.03831

Lahn, B., Sognnaes, I., & Hermansen, E. A. T. (2025). Analysing the use of emissions scenarios in practice. *Npj Climate Action*, *4*(1), 117. Advance online publication. DOI: 10.1038/s44168-025-00320-8 PMID: 41450564

Laktionov, I., Diachenko, G., Rutkowska, D., & Kisiel-Dorohinicki, M. (2023). An Explainable AI Approach to Agrotechnical Monitoring and Crop Diseases Prediction in Dnipro Region of Ukraine. *Journal of Artificial Intelligence and Soft Computing Research*, *13*(4), 247–272. DOI: 10.2478/jaiscr-2023-0018

Lapajne, J., & Yegorov, V.. (2025). UAV-based phenomics for potato crop analysis. *Computers and Electronics in Agriculture*. Advance online publication. DOI: 10.1016/j.compag.2023.107663

Leiserowitz, A. A. (2005). American Risk Perceptions: Is Climate Change Dangerous? Risk Analysis, 25(6), 1433–1442. Portico. https://doi.org/DOI: 10.1111/j.1540-6261.2005.00690.x

Lemos, M. C., Kirchhoff, C. J., & Ramprasad, V. (2012). Narrowing the climate information usability gap. *Nature Climate Change*, *2*(11), 789–794. DOI: 10.1038/nclimate1614

Leong, W. Y. (2025). Artificial Intelligence-Driven Analytics for Monitoring and Mitigating Climate Change Impacts. *IEEE ICEIB*, *7*, 7. Advance online publication. DOI: 10.3390/engproc2025108007

Lesk, C., Rowhani, P., & Ramankutty, N. (2016). Influence of extreme weather disasters on global crop production. *Nature*, *529*(7584), 84–87. DOI: 10.1038/nature16467 PMID: 26738594

Liakos, K. G., Busato, P., Moshou, D., Pearson, S., & Bochtis, D. (2018). Machine learning in agriculture: A review. *Sensors (Basel)*, *18*(8), 2674. DOI: 10.3390/s18082674 PMID: 30110960

Li, X., & Li, J. (2024). MFCA-Net: A deep learning method for semantic segmentation of remote sensing images. *Scientific Reports*, *14*(1), 5745. Advance online publication. DOI: 10.1038/s41598-024-56211-1 PMID: 38459115

Li, Z., Yuan, J., Li, G., Wang, H., Li, X., Li, D., & Wang, X. (2023). RSI-YOLO: Object detection method for remote sensing images based on improved YOLO. *Sensors (Basel), 23*(14), 6414. Advance online publication. DOI: 10.3390/s23146414 PMID: 37514708

Lobell, D. B., Cassman, K. G., & Field, C. B. (2009). Crop yield gaps: Their importance, magnitudes, and causes. *Annual Review of Environment and Resources, 34*(1), 179–204. DOI: 10.1146/annurev.environ.041008.093740

Lobell, D. B., Hammer, G. L., McLean, G., Messina, C., Roberts, M. J., & Schlenker, W. (2013). The critical role of extreme heat for maize production in the United States. *Nature Climate Change, 3*(5), 497–501. DOI: 10.1038/nclimate1832

Lobell, D. B., Schlenker, W., & Costa-Roberts, J. (2011). Climate Trends and Global Crop Production Since 1980. *Science, 333*(6042), 616–620. DOI: 10.1126/science.1204531 PMID: 21551030

Logeshwaran, J., Pradeep, R., Ramya, K., Siva, P., & Raghavan, R.. (2024). Improving crop production using an agro-deep learning framework in precision agriculture. *BMC Bioinformatics, 25*(1), 341. Advance online publication. DOI: 10.1186/s12859-024-05970-9 PMID: 39487390

Lu, J., Li, J., Fu, H., Tang, X., Liu, Z., Chen, H., Sun, Y., & Ning, X. (2024). Deep learning for multi-source data-driven crop yield prediction in northeast China. *Agriculture, 14*(6), 794. DOI: 10.3390/agriculture14060794

Lu, M., Li, Y., Yu, M., Zhang, Q., Zhang, Y., Liu, B., & Wang, M. (2023). Spatiotemporal prediction of radar echoes based on ConvLSTM and multisource data. *Remote Sensing (Basel), 15*(5), 1279. Advance online publication. DOI: 10.3390/rs15051279

Luo, L., Sun, S., Xue, J., Gao, Z., Zhao, J., Yin, Y., Gao, F., & Luan, X. (2023). Crop yield estimation based on assimilation of crop models and remote sensing data: A systematic evaluation. *Agricultural Systems, 210*, 103711. Advance online publication. DOI: 10.1016/j.agsy.2023.103711

Machen, R., & Pearce, W. (2025). Anticipating the Challenges of <scp>AI</scp> in Climate Governance: An Urgent Dilemma for Democracies. WIREs Climate Change, 16(2). Portico. DOI: 10.1002/wcc.70002

MacPherson, J., Voglhuber-Slavinsky, A., Olbrisch, M., Schöbel, P., Dönitz, E., Mouratiadou, I., & Helming, K. (2022). Future agricultural systems and the role of digitalization for achieving sustainability goals. A review. *Agronomy for Sustainable Development, 42*(4), 70. Advance online publication. DOI: 10.1007/s13593-022-00792-6 PMID: 35818482

Madani, M., Gray, S., Dyussenov, M., & Tadjine, F. (2025). Editorial: Climate change, natural resources, and human security in governance and society: vulnerabilities and adaptation strategies. *Frontiers in Political Science, 7*, 1621475. Advance online publication. DOI: 10.3389/fpos.2025.1621475

Maestrini, B., Mimić, G., Van Oort, P., Jindo, K., Brdar, S., Athanasiados, I., & Van Evert, F. (2022). Mixing process-based and data-driven approaches in yield prediction. *European Journal of Agronomy, 139*, 126569. Advance online publication. DOI: 10.1016/j.eja.2022.126569

Maheshwari, G. (2025). Governance of Nature-Based Solutions in Indian Climate Policy Framework. Handbook of Nature-Based Solutions to Mitigation and Adaptation to Climate Change, 2353–2368. DOI: 10.1007/978-3-031-34967-6_73

Ma, L., Liu, Y., Zhang, X., Ye, Y., Yin, G., & Johnson, B. A. (2019). Deep learning in remote sensing applications: A meta-analysis and review. *ISPRS Journal of Photogrammetry and Remote Sensing, 152*, 166–177. DOI: 10.1016/j.isprsjprs.2019.04.015

Mălina , C., Matei, F., Pop, I., Sălăgean, T., & Mălina , A. (2025). *Individual and Synergistic Contributions of GIS, Remote Sensing, and AI in Advancing Climate-Resilient Agriculture*. AgriEngineering., DOI: 10.3390/agriengineering7070230

Mallick, T., Murphy, J., Bergerson, J. D., Verner, D. R., Hutchison, J. K., & Levy, L. A. (2024). Analyzing regional impacts of climate change using natural language processing techniques. arXiv preprint arXiv:2401.06817

Mallik, S., Chakraborty, A., Podder, K., Talukdar, S., Rahman, A., & Mishra, U. (2025). Enhancing soil moisture prediction with explainable AI: Integrating IoT and multi-sensor remote sensing data through soft computing. *Applied Soft Computing, 180*, 113406. DOI: 10.1016/j.asoc.2025.113406

Mandla, V. R., Zhu, X., Tiwari, A., & Shoab, M. (2025). Editorial: Emerging trends and advancements of geoinformatics applications in earth and environmental systems. *Frontiers in Earth Science (Lausanne), 13*, 1641510. Advance online publication. DOI: 10.3389/feart.2025.1641510

Mansfield, L. A., & Sheshadri, A. (2024). Uncertainty quantification of a machine learning subgrid-scale parameterization for atmospheric gravity waves. Journal of Advances in Modeling Earth Systems, 16(7), e2024MS004292. https://doi.org/DOI: 10.1029/2024MS004292

Mansoor, S., Iqbal, S., Popescu, S., Kim, S., Chung, Y., & Baek, J. (2025). Integration of smart sensors and IOT in precision agriculture: Trends, challenges, and future prospects. *Frontiers in Plant Science*, *16*, 1587869. Advance online publication. DOI: 10.3389/fpls.2025.1587869 PMID: 40438737

Maranhão, R. de A., & Marchi, C. M. D. F. (2025). Urban Solid Waste Management in Salvador and Climate Mitigation: Theoretical Foundations, Legal Framework, Actions and Discourses of Public Authorities. *Revista de Gestão Social e Ambiental*, *19*(7), e012975. DOI: 10.24857/rgsa.v19n7-101

Maričić, T., & Pantić, M. (2025). Just Green Transition of Coal-Mining Cities. The Palgrave Handbook of Just Green Transitions in the Western Balkans and Beyond, 183–202. DOI: 10.1007/978-3-031-95075-9_11

Marinello, F., Zou, X., Liu, Z., Zhu, X., Zhang, W., Qian, Y., Li, Y., Karunathilake, E., Le, A., Heo, S., Chung, Y., & Mansoor, S. (2023). The Path to Smart Farming: Innovations and Opportunities in Precision Agriculture. *Agriculture*, *13*(8), 1593. Advance online publication. DOI: 10.3390/agriculture13081593

Martin, A., Liu, Z., & Zhang, Y.. (2024). *XAI-powered smart agriculture framework for improving food production and sustainability*. IEEE., DOI: 10.1109/ACCESS.2024.3492973

Materia, S., García, L., Van Straaten, C. O. S., Mamalakis, A., Cavicchia, L., Coumou, D., De Luca, P., Kretschmer, M., & Donat, M. (2024). Artificial intelligence for climate prediction of extremes: State of the art, challenges, and future perspectives. *Wiley Interdisciplinary Reviews: Climate Change*, *15*(6), e914. Advance online publication. DOI: 10.1002/wcc.914

McCormack, L., Bendechache, M., Lewis, D., & Huyskes, D. (2025). Trust and transparency in AI: Industry voices on data, ethics, and compliance. AI &amp. *AI & Society*. Advance online publication. DOI: 10.1007/s00146-025-02654-7

Meadow, A. M., Ferguson, D. B., Guido, Z., Horangic, A., Owen, G., & Wall, T. (2015). Moving toward the Deliberate Coproduction of Climate Science Knowledge. *Weather, Climate, and Society*, *7*(2), 179–191. DOI: 10.1175/WCAS-D-14-00050.1

Meckling, J., & Allan, B. B. (2020). The evolution of ideas in global climate policy. *Nature Climate Change*, *10*(5), 434–438. DOI: 10.1038/s41558-020-0739-7

Mehryar, S., Yazdanpanah, V., & Tong, J. (2024). AI and climate resilience governance. *iScience, 27*(6), 109812. DOI: 10.1016/j.isci.2024.109812 PMID: 38784017

Meimand, H. M., Chen, J., Kneeshaw, D., Bakhtyari, M., & Peng, C. (2025). Burned Area Detection in the Eastern Canadian Boreal Forest Using a Multi-Layer Perceptron and MODIS-Derived Features. *Remote Sensing (Basel), 17*(13), 2162. DOI: 10.3390/rs17132162

Meng, X., & Karniadakis, G. E. (2020). Physics-informed neural networks for high-dimensional PDEs: Forward and inverse problems. *Journal of Computational Physics, 401*, 109020. DOI: 10.1016/j.jcp.2019.109020

Mergel, I., Edelmann, N., & Haug, N. (2019). Defining digital transformation: Results from expert interviews. *Government Information Quarterly, 36*(4), 101385. DOI: 10.1016/j.giq.2019.06.002

Metheniti, V., Parasyris, A., Pereira, R. S., & Kazanjian, G. (2025). Recent Advancements and Challenges in Artificial Intelligence for Digital Twins of the Ocean. *Climate (Basel), 14*(1), 3. DOI: 10.3390/cli14010003

Meyer, H., & Pebesma, E. (2021). Predicting into unknown space? Estimating the area of applicability of spatial prediction models. *Methods in Ecology and Evolution, 12*(9), 1620–1633. DOI: 10.1111/2041-210X.13650

Meyer, H., & Pebesma, E. (2022). Machine learning-based global maps of ecological variables and the challenge of assessing them. *Nature Communications, 13*(1), 2208. DOI: 10.1038/s41467-022-29838-9

Milà, C., Mateu, J., Pebesma, E., & Meyer, H. (2022). Nearest neighbour distance matching leave-one-out cross-validation for map validation. *Methods in Ecology and Evolution, 13*(6), 1304–1316. DOI: 10.1111/2041-210X.13851

Miralles, P., Scannapieco, A. F., Jagadam, N., Baranwal, P., Faldu, B., Abhang, R., & Batul, B. (2021). Machine learning in earth observation operations: A review. 72nd International Astronautical Congress (IAC), Orynbaikyzy, A., Gessner, U., Mack, B., & Conrad, C. (2020). Crop type classification using fusion of sentinel-1 and sentinel-2 data: Assessing the impact of feature selection, optical data availability, and parcel sizes on the accuracies. *Remote Sensing, 12*(17), 2779.

Mohammed, S., & Yusuf, H.. (2025). *The role of AI in transforming precision agriculture*. Precision Agriculture and Sustainability., DOI: 10.14719/pst.6175

Mohan, R. N. V. J., Rayanoothala, P. S., & Sree, R. P. (2025). Next-gen agriculture: Integrating AI and XAI for precision crop yield predictions. *Frontiers in Plant Science, 15*, 1451607. DOI: 10.3389/fpls.2024.1451607 PMID: 39845494

Mohr, S., & Kühl, R. (2021). Acceptance of artificial intelligence in German agriculture: An application of the technology acceptance model and the theory of planned behavior. *Precision Agriculture*, *22*(6), 1816–1844. DOI: 10.1007/s11119-021-09814-x

Mohyuddin, G., Khan, M., Haseeb, A., Mahpara, S., Waseem, M., & Saleh, A. (2024). Evaluation of Machine Learning Approaches for Precision Farming in Smart Agriculture Systems: A Comprehensive Review. *IEEE Access : Practical Innovations, Open Solutions*, *12*, 60155–60184. DOI: 10.1109/ACCESS.2024.3390581

Mökander, J., & Floridi, L. (2021). Ethics-Based Auditing to Develop Trustworthy AI. *Minds and Machines*, *31*(2), 323–327. DOI: 10.1007/s11023-021-09557-8

Morchid, A., Alami, R., Raezah, A., & Sabbar, Y. (2023). Applications of Internet of Things (IoT) and sensor technology to increase food security and agricultural Sustainability: Benefits and challenges. *Ain Shams Engineering Journal*. Advance online publication. DOI: 10.1016/j.asej.2023.102509

Morice, C. P., Kennedy, J. J., Rayner, N. A., Winn, J. P., Hogan, E., Killick, R. E., Dunn, R. J. H., Osborn, T. J., Jones, P. D., & Simpson, I. R. (2021). An Updated Assessment of Near-Surface Temperature Change From 1850: The HadCRUT5 Data Set. Journal of Geophysical Research: Atmospheres, 126(3). Portico. DOI: 10.1029/2019JD032361

Moskolaï, W. R., Abdou, W., Dipanda, A., & Kolyang, . (2021). Application of deep learning architectures for satellite image time series prediction: A review. *Remote Sensing (Basel)*, *13*(23), 4822. Advance online publication. DOI: 10.3390/rs13234822

Mugambiwa, S. (2025). Climate Governance from the Bottom: Decentralized Adaptation and the Politics of Rural Livelihoods in Southern Africa. Journal of Cultural Analysis and Social Change, 631-639.

Muhammad, A., Khan, Z., Khan, J., Mashori, A., Ali, A., Jabeen, N., Han, Z., & Li, F. (2025). A comprehensive review of crop stress detection: Destructive, non-destructive, and ML-based approaches. *Frontiers in Plant Science*, *16*, 1638675. Advance online publication. DOI: 10.3389/fpls.2025.1638675 PMID: 40978773

Mukherjee, S., & Hastak, M. (2017). A Novel Methodological Approach to Estimate the Impact of Natural Hazard-Induced Disasters on Country/Region-Level Economic Growth. *International Journal of Disaster Risk Science*, *9*(1), 74–85. DOI: 10.1007/s13753-017-0156-3

Nair, A., Smitha Chandran, S., & Varangalil, N. (2025). Modelling and Forecasting of Climate Change Effects Using Artificial Intelligence Techniques. Climate Resilience: Impact of Quantum Computing and Artificial Intelligence on Urban Planning, 127–143. DOI: 10.1007/978-3-032-06791-3_7

Nair, M. N., Mani, S. K., Polinga, I. M., & Baiju, K. R. (2026). Eco-Social Resilience and Nature-Based Solutions: Local Strategies for Disaster Risk Reduction Across the Globe. Eco-Social Work for Disaster Risk Reduction, 71–87. DOI: 10.1007/978-981-95-4885-9_4

Narassimhan, E., Gallagher, K. S., Koester, S., & Alejo, J. R. (2018). Carbon pricing in practice: A review of existing emissions trading systems. *Climate Policy*, *18*(8), 967–991. DOI: 10.1080/14693062.2018.1467827

Nasir, N., Ramzan, S., Ali, B., Jabbar, S., Raza, A., Kim, C., Syafrudin, M., & Fitriyani, N. L. (2025). Interpretable deep multimodal-based tomato disease diagnosis and severity estimation. *Scientific Reports*, *15*(1), 37919. Advance online publication. DOI: 10.1038/s41598-025-21611-4 PMID: 41162432

Neil Adger, W., Arnell, N. W., & Tompkins, E. L. (2005). Successful adaptation to climate change across scales. *Global Environmental Change*, *15*(2), 77–86. DOI: 10.1016/j.gloenvcha.2004.12.005

Newell, P., & Mulvaney, D. (2013). The political economy of the 'just transition.' The Geographical Journal, 179(2), 132–140. Portico. DOI: 10.1111/geoj.12008

Nik, V. M., & Javanroodi, K. (2025). Adaptive reinforcement learning for energy management – A progressive approach to boost climate resilience and energy flexibility. *Advances in Applied Energy*, *17*, 100213. DOI: 10.1016/j.adapen.2025.100213

Nunes, L. J. R. (2023). The Rising Threat of Atmospheric CO2: A Review on the Causes, Impacts, and Mitigation Strategies. *Environments (Basel, Switzerland)*, *10*(4), 66. DOI: 10.3390/environments10040066

Nu ă, F., Mustata Wilson, G., Boateng, G. O., & Nuta, A. C. (Eds.). (2025). *Environmental degradation, health, and socioeconomic impacts*. Frontiers Research Topics., DOI: 10.3389/978-2-8325-7166-8

Nwokolo, S. C., Singh, R., Khan, S., Kumar, A., & Luthra, S. (2023). Impacts of Climate Change in Africa. Africa's Path to Net-Zero, 237–262. https://doi.org/DOI: 10.1007/978-3-031-44514-9_7

Obi, J. N., & Ojo, E. (2025). Climate Justice, Renewable Energy, and Food Security in Sub-Saharan Africa: A Bibliometric Analysis of Emerging Research Trends. Energy Transition, Climate Action and Sustainable Agriculture, 303–328. https://doi.org/DOI: 10.1007/978-3-031-83165-2_16

Oddo, P. C., Bolten, J. D., Kumar, S. V., & Cleary, B. (2024). Deep convolutional LSTM for improved flash flood prediction. *Frontiers in Water*, *6*, 1346104. Advance online publication. DOI: 10.3389/frwa.2024.1346104

Ojo, M., & Zahid, A. (2022). Deep Learning in Controlled Environment Agriculture: A Review of Recent Advancements, Challenges, and Prospects. *Sensors (Basel)*, *22*(20), 7965. Advance online publication. DOI: 10.3390/s22207965 PMID: 36298316

Olawade, D. B., Wada, O. Z., David-Olawade, A. C., Fapohunda, O., Ige, A. O., & Ling, J. (2024). Artificial intelligence potential for net zero sustainability: Current evidence and prospects. *Next Sustainability*, *4*, 100041. DOI: 10.1016/j.nxsust.2024.100041

Oliveira, L., Moreira, A., & Silva, M. (2021). Advances in Agriculture Robotics: A State-of-the-Art Review and Challenges Ahead. *Robotics (Basel, Switzerland)*, *10*(2), 52. DOI: 10.3390/robotics10020052

Olivetti, L., & Messori, G. (2024). Advances and prospects of deep learning for medium-range extreme weather forecasting. *Geoscientific Model Development*, *17*, 2347–2358. DOI: 10.5194/gmd-17-2347-2024

Omotoso, A. B., & Omotayo, A. O. (2023). The interplay between agriculture, greenhouse gases, and climate change in Sub-Saharan Africa. *Regional Environmental Change*, *24*(1), 1. Advance online publication. DOI: 10.1007/s10113-023-02159-3

Oraby, G. A. E. M., & Orabi, A. A. E. M. (2025). Government Policies in Egypt and Indonesia to Mitigate Carbon Emissions: A Comparative Study. *Journal of Geographical Sciences and Education*, *3*(4), 343–356. DOI: 10.69606/geography.v3i4.371

Oreskes, N. (2004). The Scientific Consensus on Climate Change. *Science*, *306*(5702), 1686–1686. DOI: 10.1126/science.1103618 PMID: 15576594

Orkuma, M. (2025). The Geopolitics of Climate Change: Global Power Relations and the Pursuit of Climate Justice. .DOI: 10.14293/PR2199.002193.v1

Overpeck, J. T., Meehl, G. A., Bony, S., & Easterling, D. R. (2011). Climate Data Challenges in the 21st Century. *Science*, *331*(6018), 700–702. DOI: 10.1126/science.1197869 PMID: 21311006

Ozdogan, M., Yang, Y., Allez, G., & Cervantes, C. (2010). Remote sensing of irrigated agriculture: Opportunities and challenges. *Remote Sensing (Basel)*, *2*(9), 2274–2304. DOI: 10.3390/rs2092274

Padhiary, M., Saha, D., Kumar, R., Sethi, L., & Kumar, A. (2024). Enhancing Precision Agriculture: A Comprehensive Review of Machine Learning and AI Vision Applications in All-Terrain Vehicles for Farm Automation. *Smart Agricultural Technology*, *8*, 100483. Advance online publication. DOI: 10.1016/j.atech.2024.100483

Pandey, M., Mishra, V. N., Kumari, M., Janizadeh, S., & Costache, R. (2026). Editorial: Advanced geospatial data analytics for environmental sustainability: current practices and future prospects. *Frontiers in Remote Sensing*, *6*, 1761905. Advance online publication. DOI: 10.3389/frsen.2025.1761905

Patil, Y., Ramachandran, H., Sundararajan, S., & Srideviponmalar, P. (2025). Comparative analysis of machine learning models for crop yield prediction across multiple crop types. *SN Computer Science*, *6*(1), 64. DOI: 10.1007/s42979-024-03602-w

Patro, P., M, S., N, S., Das, S., V P, A. K., & R, D. (2025). Big Data-Based Remote-Sensed Intelligent Visual Analytics for Environmental and Earth Monitoring Systems. Journal of Applied Science and Technology Trends, 35–50. .DOI: 10.38094/jastt605589

Pavlova, O., Liashenko, O., Pavlov, K., Demianiuk, O., Vitkovskyi, Y., Jakóbik, K., Piwowarczyk, Z., & Karpinska, N. (2026). Utility-Based Evaluation of National Climate Policies: A Multi-Criteria Framework for Global Assessment. DOI: 10.20944/preprints202601.1360.v1

Perez, C., Jones, E. M., Kristjanson, P., Cramer, L., Thornton, P. K., Förch, W., & Barahona, C. (2015). How resilient are farming households and communities to a changing climate in Africa? A gender-based perspective. *Global Environmental Change*, *34*, 95–107. DOI: 10.1016/j.gloenvcha.2015.06.003

Petraki, D., Gazoulis, I., Kokkini, M., Danaskos, M., Kanatas, P., Rekkas, A., & Travlos, I. (2025). Digital Tools and Decision Support Systems in Agroecology: Benefits, Challenges, and Practical Implementations. *Agronomy (Basel)*, *15*(1), 236. Advance online publication. DOI: 10.3390/agronomy15010236

Petropoulos, G., & Mantas, P.. (2025). *Interpretable machine learning framework for pulse crop yield prediction*. MDPI Applications., DOI: 10.3390/app15137074

Piekutowska, M., & Niedbała, G. (2025). Review of Methods and Models for Potato Yield Prediction. *Agriculture*, *15*(4), 367. Advance online publication. DOI: 10.3390/agriculture15040367

Ponnurajan, M., & Iqbal, A.. (2025). *Federated Learning for crop yield prediction*. IEEE., DOI: 10.1109/ASSIC64892.2025.11158569

Prakash, A., Totin, E., Kemp, G., Kerr, R. B., & Roberts, D. (2025). Bridging the gap: Promoting gender equity in climate change adaptation in the global south. *PLOS Climate*, *4*(6), e0000556. DOI: 10.1371/journal.pclm.0000556

Price, I., Sanchez-Gonzalez, A., Alet, F., Andersson, T. R., El-Kadi, A., Masters, D., Ewalds, T., Stott, J., Mohamed, S., Battaglia, P., Lam, R., & Willson, M. (2025). Probabilistic weather forecasting with machine learning. *Nature*, *637*, 84–90. DOI: 10.1038/s41586-024-08252-9

Priya, N. K., Khatri, A., Karki, P., Samota, S. D., Vishwakarma, S. K., Sukdeve, E. K., Tripathi, S., & Pathak, A. K. (2025). The Important Role of Extension Services in Strengthening the Capacity of Farmers' Resilience to Climate Change in India. *Journal of Experimental Agriculture International*, *47*(3), 204–223. DOI: 10.9734/jeai/2025/v47i33329

Prodhan, F., Zhang, J., Hasan, S., Sharma, T., & Mohana, H. (2022). A review of machine learning methods for drought hazard monitoring and forecasting: Current research trends, challenges, and future research directions. *Environmental Modelling & Software*, *149*, 105327. DOI: 10.1016/j.envsoft.2022.105327

Qiao, M., He, X., Cheng, X., Li, P., Luo, H., Zhang, L., & Tian, Z. (2021). Crop yield prediction from multi-spectral, multi-temporal remotely sensed imagery using recurrent 3D convolutional neural networks. *International Journal of Applied Earth Observation and Geoinformation*, *102*, 102436. DOI: 10.1016/j.jag.2021.102436

Qin, F., & Li, Y.. (2025). Predicting the effect of mechanical soil compaction on soybean crop yields. *Agricultural Systems*. Advance online publication. DOI: 10.1016/j.inpa.2025.09.002

Rabbi, M. F., Popp, J., Máté, D., & Kovács, S. (2022). Energy Security and Energy Transition to Achieve Carbon Neutrality. *Energies*, *15*(21), 8126. DOI: 10.3390/en15218126

Raftowicz, M. (2021). The Climate Crisis as a Product of Globalization. SHS Web of Conferences, 92, 06029. DOI: 10.1051/shsconf/20219206029

Raissi, M., Perdikaris, P., & Karniadakis, G. E. (2019). Physics-informed neural networks: A deep learning framework for solving forward and inverse problems involving nonlinear partial differential equations. *Journal of Computational Physics*, *378*, 686–707. DOI: 10.1016/j.jcp.2018.10.045

Ramaiah, M., Settu, P., & Ravi, V. (2025). Artificial Intelligence Techniques Enabled Soil Moisture Estimation Frameworks Using Remote Sensing Satellite Images: Challenges and Future Directions- Review. *Wiley Interdisciplinary Reviews. Data Mining and Knowledge Discovery*, *15*(3), e70032. Advance online publication. DOI: 10.1002/widm.70032

Ramesh, V., & Kumaresan, P. (2025). Stacked ensemble model for accurate crop yield prediction using machine learning techniques. *Environmental Research Communications*, *7*(3), 035006. DOI: 10.1088/2515-7620/adb9c0

Randeniya, J. N., Haigh, R., & Amaratunga, D. (2025). Responsible intelligence: Ethical AI governance for climate prediction in the Australian context. *AI and Ethics*, *6*(1), 81. Advance online publication. DOI: 10.1007/s43681-025-00909-z

Rane, J., Kaya, Ö., Mallick, S. K., & Rane, N. L. (2024). Smart farming using artificial intelligence, machine learning, deep learning, and ChatGPT: Applications, opportunities, challenges, and future directions. https://doi.org/DOI: 10.70593/978-81-981271-7-4_6

Rashidi Mohammadi, A., Petrović, E. K., & Perkins, N. (2025). Mapping forestry practices against sustainable development goal targets. *Frontiers in Sustainability*, *6*, 1534449. Advance online publication. DOI: 10.3389/frsus.2025.1534449

Rasp, S., Hoyer, S., Merose, A., Langmore, I., Battaglia, P., Russell, T., Sanchez-Gonzalez, A., Yang, V., Carver, R., Agrawal, S., & Chantry, M. (2024). WeatherBench 2: A benchmark for the next generation of data-driven global weather models. Journal of Advances in Modeling Earth Systems, 16(6), e2023MS004019. https://doi.org/DOI: 10.1029/2023MS004019

Rasp, S., Pritchard, M. S., & Gentine, P. (2018). Deep learning to represent subgrid processes in climate models. *Proceedings of the National Academy of Sciences of the United States of America*, *115*(39), 9684–9689. DOI: 10.1073/pnas.1810286115 PMID: 30190437

Raza, A., Shahid, M. A., Zaman, M., Miao, Y., Huang, Y., Safdar, M., Maqbool, S., & Muhammad, N. E. (2025). Improving Wheat Yield Prediction with Multi-Source Remote Sensing Data and Machine Learning in Arid Regions. *Remote Sensing (Basel)*, *17*(5), 774. DOI: 10.3390/rs17050774

Razavi-Termeh, S. V., Sadeghi-Niaraki, A., Ali, F., Pradhan, B., & Choi, S.-M. (2025). Optimizing ensemble learning for satellite-based multi-hazard monitoring and susceptibility assessment of landslides, land subsidence, floods, and wildfires. *Scientific Reports*, *15*(1), 30968. Advance online publication. DOI: 10.1038/s41598-025-15381-2 PMID: 40846725

Reichstein, M. (2017). *How Can Artificial Intelligence Enhance Our Understanding of the Earth System?* Latest Thinking., DOI: 10.21036/LTPUB10819

Reichstein, M., Benson, V., Blunk, J., Camps-Valls, G., Creutzig, F., Fearnley, C. J., Han, B., Kornhuber, K., Rahaman, N., Schölkopf, B., Tárraga, J. M., Vinuesa, R., Dall, K., Denzler, J., Frank, D., Martini, G., Nganga, N., Maddix, D. C., & Weldemariam, K. (2025). Early warning of complex climate risk with integrated artificial intelligence. *Nature Communications*, *16*(1), 2564. Advance online publication. DOI: 10.1038/s41467-025-57640-w PMID: 40089483

Reichstein, M., Camps-Valls, G., Stevens, B., Jung, M., Denzler, J., Carvalhais, N., & Prabhat, . (2019). Deep learning and process understanding for data-driven Earth system science. *Nature*, *566*(7743), 195–204. DOI: 10.1038/s41586-019-0912-1 PMID: 30760912

Reuß, F., Greimeister-Pfeil, I., Vreugdenhil, M., & Wagner, W. (2021). Comparison of long short-term memory networks and random forest for Sentinel-1 time series based large scale crop classification. *Remote Sensing (Basel)*, *13*(24), 5000. Advance online publication. DOI: 10.3390/rs13245000

Rizwan, A., & Saleem, Y. (2025). Stakeholders Role to Combat Climate Change: Initiatives Adopted and Resource Integration in Response to Sustainable Development Goal 13 in Pakistan. Disasters, Climate Change and Sustainability in South Asia, 231–253. DOI: 10.1007/978-3-032-04750-2_11

Robinson, M., & Shine, T. (2018). Achieving a climate justice pathway to 1.5 °C. *Nature Climate Change*, *8*(7), 564–569. DOI: 10.1038/s41558-018-0189-7

Rockström, J., Williams, J., Daily, G., Noble, A., Matthews, N., Gordon, L., & Steduto, P. (2017). Sustainable intensification of agriculture for human prosperity and global sustainability. *Ambio*, *46*(1), 4–17. DOI: 10.1007/s13280-016-0793-6 PMID: 27405653

Rodge, R. R., & Rajan, R. (2026). Frontier technologies and remote sensing applications for climate change mitigation and adaptation. Geospatial Techniques in Climate Change Monitoring and Environmental Mapping, 1–16. .DOI: 10.1016/B978-0-443-36396-2.00015-9

Rohde, F., Gossen, M., Wagner, J., & Santarius, T. (2021). Sustainability challenges of Artificial Intelligence and Policy Implications. Ökologisches Wirtschaften - Fachzeitschrift, 36(O1), 36–40. DOI: 10.14512/OEWO360136

Rolnick, D., Donti, P. L., Kaack, L. H., Kochanski, K., Lacoste, A., Sankaran, K., Ross, A. S., Milojevic-Dupont, N., Jaques, N., Waldman-Brown, A., Luccioni, A. S., Maharaj, T., Sherwin, E. D., Mukkavilli, S. K., Kording, K. P., Gomes, C. P., Ng, A. Y., Hassabis, D., Platt, J. C., & Bengio, Y. (2022). Tackling Climate Change with Machine Learning. *ACM Computing Surveys*, *55*(2), 1–96. DOI: 10.1145/3485128

Ronneberger, O., Fischer, P., & Brox, T. (2015). U-net: Convolutional networks for biomedical image segmentation. *International Conference on Medical image computing and computer-assisted intervention*

Rudin, C. (2019). Stop explaining black box machine learning models for high stakes decisions and use interpretable models instead. *Nature Machine Intelligence*, *1*(5), 206–215. DOI: 10.1038/s42256-019-0048-x PMID: 35603010

Running, S. W., Nemani, R. R., Heinsch, F. A., Zhao, M., Reeves, M., & Hashimoto, H. (2004). A Continuous Satellite-Derived Measure of Global Terrestrial Primary Production. *Bioscience*, *54*(6), 547. DOI: 10.1641/0006-3568(2004)054[0547:ACSMOG]2.0.CO;2

Saha, S., Kucher, O., Utkina, A., & Rebouh, N. (2025). Precision agriculture for improving crop yield predictions: A literature review. *Frontiers in Agronomy*, *7*, 1566201. Advance online publication. DOI: 10.3389/fagro.2025.1566201

Sah, S., Haldar, D., Singh, R., Das, B., & Nain, A. S. (2024). Rice yield prediction through integration of biophysical parameters with SAR and optical remote sensing data using machine learning models. *Scientific Reports*, *14*(1), 21674.

Sakamoto, T., Yokozawa, M., Toritani, H., Shibayama, M., Ishitsuka, N., & Ohno, H. (2005). A crop phenology detection method using time-series MODIS data. *Remote Sensing of Environment*, *96*(3-4), 366–374. DOI: 10.1016/j.rse.2005.03.008

Saleem, M., Potgieter, J., & Arif, K. (2021). Automation in Agriculture by Machine and Deep Learning Techniques: A Review of Recent Developments. *Precision Agriculture*, *22*(6), 2053–2091. DOI: 10.1007/s11119-021-09806-x

Samek, W., Montavon, G., Lapuschkin, S., Anders, C. J., & Müller, K.-R. (2021). Explaining deep neural networks and beyond: A review of methods and applications. *Proceedings of the IEEE*, *109*(3), 247–278. DOI: 10.1109/JPROC.2021.3060483

Sarewitz, D., & Pielke, R. A.Jr. (2007). The neglected heart of science policy: Reconciling supply of and demand for science. Environmental Science &amp. *Environmental Science & Policy*, *10*(1), 5–16. DOI: 10.1016/j.envsci.2006.10.001

Sarku, R., Clemen, U., & Clemen, T. (2023). The Application of Artificial Intelligence Models for Food Security. *Agriculture*, *13*(10), 2037. Advance online publication. DOI: 10.3390/agriculture13102037

Schlosberg, D. (2007). 1 Defining Environmental Justice. Defining Environmental Justice, 3–10. https://doi.org/DOI: 10.1093/acprof:oso/9780199286294.003.0001

Schlosberg, D., & Collins, L. B. (2014). From environmental to climate justice: climate change and the discourse of environmental justice. WIREs Climate Change, 5(3), 359–374. Portico. .DOI: 10.1002/wcc.275

Schlosberg, D., Collins, L. B., & Niemeyer, S. (2017). Adaptation policy and community discourse: Risk, vulnerability, and just transformation. *Environmental Politics*, *26*(3), 413–437. DOI: 10.1080/09644016.2017.1287628

Schreck, J. S., Gagne, D. J., Becker, C., Chapman, W. E., Elmore, K., Fan, D., Gantos, G., Kim, E., Kimpara, D., Martin, T., Molina, M. J., Przybylo, V. M., Radford, J., Saavedra, B., Willson, J., & Wirz, C. (2024). Evidential deep learning: Enhancing predictive uncertainty estimation for Earth system science applications. *Artificial Intelligence for the Earth Systems*, *3*(4), e230093. DOI: 10.1175/AIES-D-23-0093.1

Schultz, M. G., Betancourt, C., Gong, B., Kleinert, F., Langguth, M., Leufen, L. H., Mozaffari, A., & Stadtler, S. (2021). Can deep learning beat numerical weather prediction? *Philosophical Transactions of the Royal Society A*, *379*(2194), 20200097. DOI: 10.1098/rsta.2020.0097

Scoville-Simonds, M., Jamali, H., & Hufty, M. (2020). The Hazards of Mainstreaming: Climate change adaptation politics in three dimensions. *World Development*, *125*, 104683. DOI: 10.1016/j.worlddev.2019.104683

Sebastianelli, A., Mauro, F., Delilbasic, A., Di Stasio, P., Fan, F., Meoni, G.,… Ullo, S. (2025). Quantum Machine Learning for Earth Observation: A Review and Future Prospects.

Šestak, M., & Copot, D. (2023). Towards Trusted Data Sharing and Exchange in Agro-Food Supply Chains: Design Principles for Agricultural Data Spaces. *Sustainability (Basel)*, *15*(18), 13746. Advance online publication. DOI: 10.3390/su151813746

Shams, S., Chowdhury, T., ... (2024). XAI-CROP: Explainable AI-based Crop Recommendation System for Precision Agriculture. *Springer*. https://link.springer.com/article/10.1007/s00521-023-09391-2

Shams, M. Y., Gamel, S. A., & Talaat, F. M. (2024). Enhancing crop recommendation systems with explainable artificial intelligence: A study on agricultural decision-making. *Neural Computing & Applications*, *36*(11), 5695–5714. DOI: 10.1007/s00521-023-09391-2

Sharifi, A. (2021). Co-benefits and synergies between urban climate change mitigation and adaptation measures: A literature review. *The Science of the Total Environment*, *750*, 141642. DOI: 10.1016/j.scitotenv.2020.141642 PMID: 32858298

Sharma, R., Bansal, R., Munna, A. S., & Mrabet, J. (2026). Reimagining Development: Environmental Sustainability and the Global Green Transition. Aligning ESG Goals with the UNs Sustainable Development Goals Planning, 249–263. DOI: 10.1007/978-3-032-01715-4_13

Sharman, N. (2022). Inter-State climate technology transfer under the UNFCCC: A benefit-sharing approach. Review of European, Comparative & International Environmental Law, 31(3), 435–446. Portico. DOI: 10.1111/reel.12454

Shastri, S., Kumar, S., Mansotra, V., & Salgotra, R. (2025). Advancing crop recommendation system with supervised machine learning and explainable artificial intelligence. *Scientific Reports*, *15*(1), 25498. DOI: 10.1038/s41598-025-07003-8 PMID: 40664775

Shawon, S. M., Ema, F. B., Mahi, A. K., Niha, F. L., & Zubair, H. T. (2025). Crop yield prediction using machine learning: An extensive and systematic literature review. *Smart Agricultural Technology*, *10*, 100718. DOI: 10.1016/j.atech.2024.100718

Sheehan, M. C. (2022). 2021 Climate and Health Review – Uncharted Territory: Extreme Weather Events and Morbidity. *International Journal of Health Services*, *52*(2), 189–200. DOI: 10.1177/00207314221082452 PMID: 35229682

Shen, C., Appling, A. P., Gentine, P., Bandai, T., Gupta, H., Tartakovsky, A., & Lawson, K. (2023). Differentiable modelling to unify machine learning and physical models for geosciences. *Nature Reviews. Earth & Environment*, *4*(7), 552–567. DOI: 10.1038/s43017-023-00450-9

Silva, J., Heerwaarden, J., Reidsma, P., Laborte, A., Tesfaye, K., & Ittersum, M. (2023). Big data, small explanatory and predictive power: Lessons from random forest modeling of on-farm yield variability and implications for data-driven agronomy. *Field Crops Research*, *302*, 109063. Advance online publication. DOI: 10.1016/j.fcr.2023.109063 PMID: 37840838

Singh, S. S., & Jeganathan, C. (2025). Examining the Role of Geospatial Technologies and AI in Disaster Risk Management and Response, Including Hazard Mapping, Vulnerability Assessment, and Real-Time Monitoring. Advanced Geospatial Intelligence and AI for Environmental Resilience and Sustainable Development, 335–366. DOI: 10.1007/978-3-032-03714-5_14

Singh, G., Moncrieff, G., Venter, Z., Cawse-Nicholson, K., Slingsby, J., & Robinson, T. B. (2024). Uncertainty quantification for probabilistic machine learning in earth observation using conformal prediction. *Scientific Reports*, *14*, 16166. DOI: 10.1038/s41598-024-65954-w

Singh, G., Singh, S., Sethi, G., & Sood, V. (2022). Deep learning in the mapping of agricultural land use using Sentinel-2 satellite data. *Geographies*, *2*(4), 691–700. DOI: 10.3390/geographies2040042

Singh, S., Jakhar, S. K., & Kulhar, K. S. (2024). Harnessing nanotechnology and artificial intelligence for precision agriculture in smart cities. In *E3S Web of Conferences* (Vol. 540, p. 13002). EDP Sciences., DOI: 10.1051/e3sconf/202454013002

Sishodia, R., Ray, R., & Singh, S. (2020). Applications of Remote Sensing in Precision Agriculture: A Review. *Remote Sensing (Basel)*, *12*(19), 3136. DOI: 10.3390/rs12193136

Skotnicky, P., Puccio, A., & Das, S. (2025). The AI Compass: Navigating Ethical Dilemmas in Tech-Driven Sustainability. Generative AI for a Net-Zero Economy, 111–128. DOI: 10.1007/978-981-96-8015-3_7

Slater, L. J., Arnal, L., Boucher, M.-A., Chang, A. Y.-Y., Moulds, S., Murphy, C., Nearing, G., Shalev, G., Shen, C., Speight, L., Villarini, G., Wilby, R. L., Wood, A., & Zappa, M. (2023). Hybrid forecasting: Blending climate predictions with AI models. *Hydrology and Earth System Sciences*, *27*(9), 1865–1889. DOI: 10.5194/hess-27-1865-2023

Song, C., Woodcock, C. E., Seto, K. C., Lenney, M. P., & Macomber, S. A. (2001). Classification and change detection using Landsat TM data: When and how to correct atmospheric effects? *Remote Sensing of Environment*, *75*(2), 230–244. DOI: 10.1016/S0034-4257(00)00169-3

Soomro, J. A., Suhrab, M., & Abbasi, B. A. (2025). Role of Emerging Technologies in Promoting Sustainable Environment and Public Health. Environment and Public Health, 485–507. .DOI: 10.1007/978-3-031-99770-9_230

Soussi, A., Zero, E., Sacile, R., Trinchero, D., & Fossa, M. (2024). Smart Sensors and Smart Data for Precision Agriculture: A Review. *Sensors (Basel)*, *24*(8), 2647. Advance online publication. DOI: 10.3390/s24082647 PMID: 38676264

Stainforth, D. A., Allen, M. R., Tredger, E. R., & Smith, L. A. (2007). Confidence, uncertainty and decision-support relevance in climate predictions. Philosophical Transactions of the Royal Society A: Mathematical, Physical and Engineering Sciences, 365(1857), 2145–2161. DOI: 10.1098/rsta.2007.2074

Steckel, J. C., Jakob, M., Flachsland, C., Kornek, U., Lessmann, K., & Edenhofer, O. (2016). From climate finance toward sustainable development finance. WIREs Climate Change, 8(1). Portico. DOI: 10.1002/wcc.437

Stede, M., & Patz, R. (2021). The Climate Change Debate and Natural Language Processing. *Proceedings of the 1st Workshop on NLP for Positive Impact*. DOI: 10.18653/v1/2021.nlp4posimpact-1.2

Stephens, G. L., Slingo, J. M., Rignot, E., Reager, J. T., Hakuba, M. Z., Durack, P. J., Worden, J., & Rocca, R. (2020). Earth's water reservoirs in a changing climate. *Proceedings - Royal Society. Mathematical, Physical and Engineering Sciences*, *476*(2236), 20190458. Advance online publication. DOI: 10.1098/rspa.2019.0458 PMID: 32398926

Stevens, B., & Bony, S. (2013). What Are Climate Models Missing? *Science*, *340*(6136), 1053–1054. DOI: 10.1126/science.1237554 PMID: 23723223

Subeesh, A., & Mehta, C. (2021). *Automation and digitization of agriculture using artificial intelligence and the Internet of Things*. Artificial Intelligence in Agriculture., DOI: 10.1016/j.aiia.2021.11.004

Sudharson, K., Alekhya, B., Abinaya, G., Rohini, C., Arthi, S., & Dhinakaran, D. (2023, February). Efficient soil condition monitoring with IoT enabled intelligent farming solution. In 2023 IEEE International Students' Conference on Electrical, Electronics and Computer Science (SCEECS) (pp. 1-6). IEEE. doi:DOI: 10.1109/SCEECS57921.2023.10063050

Sumbul, G., de Wall, A., Kreuziger, T., Marcelino, F., Costa, H., & Benevides, P. (2021). BigEarthNet-MM: A large-scale, multimodal, multilabel benchmark archive for remote sensing image classification and retrieval. *IEEE Geoscience and Remote Sensing Magazine*, *9*(3), 174–180. DOI: 10.1109/MGRS.2021.3089174

Syari, M., & Zhang, X.. (2025). *IoT-based smart farming for spinach cultivation*. IEEE Xplore., DOI: 10.1109/ICCIT65724.2025.11167551

Taheri Hosseinkhani, N. (2025). Artificial Intelligence and Large Language Models in Energy Systems and Climate Strategies: Economic Pathways to Cost-Effective Emissions Reduction and Sustainable Growth. SSRN Electronic Journal. DOI: 10.2139/ssrn.5385513

Tahir, H. A., Alayed, W., & Hassan, W. U. (2025). A Federated Explainable AI Framework for Smart Agriculture: Enhancing Transparency, Efficiency, and Sustainability. *IEEE Access : Practical Innovations, Open Solutions*, *13*, 97567–97584. DOI: 10.1109/ACCESS.2025.3571340

Takaloo, S., Xu, A. H., Zaidan, L., Irannejad, M., & Yavuz, M. (2024). Towards point-of-care single biomolecule detection using next generation portable nanoplasmonic biosensors: A review. *Biosensors (Basel)*, *14*(12), 593. DOI: 10.3390/bios14120593 PMID: 39727858

Talari, G., Cummins, E., McNamara, C., & O'Brien, J. (2021). State-of-the-art review of Big Data and web-based Decision Support Systems (DSS) for food safety risk assessment with respect to climate change. *Trends in Food Science & Technology*. Advance online publication. DOI: 10.1016/j.tifs.2021.08.032

Talukder, M., Islam, M. M., Uddin, M. A., Hasan, K. F., Sharmin, S., Alyami, S. A., & Moni, M. A. (2024). Machine learning-based network intrusion detection for big and imbalanced data using oversampling, stacking feature embedding and feature extraction. *Journal of Big Data*, *11*(1), 33. Advance online publication. DOI: 10.1186/s40537-024-00886-w

Tamasiga, P., Ouassou, E., Onyeaka, H., Bakwena, M., Happonen, A., & Molala, M. (2023). Forecasting disruptions in global food value chains to tackle food insecurity: The role of AI and big data analytics – A bibliometric and scientometric analysis. *Journal of Agriculture and Food Research*, *14*, 100819. Advance online publication. DOI: 10.1016/j.jafr.2023.100819

Tan, M., & Liu, Y.. (2025). Optimized Random Forest framework for rice yield prediction. *Agronomy Journal*. Advance online publication. DOI: 10.3390/agronomy15102273

Tantoh, H. B., Ebhuoma, E. E., & Leonard, L. (2022). Indigenous Women's Vulnerability to Climate Change and Adaptation Strategies in Central Africa: A Systematic Review. Indigenous Knowledge and Climate Governance, 53–66. https://doi.org/ DOI: 10.1007/978-3-030-99411-2_5

Tariq, M., Ali, U., Abbas, S., Hassan, S., Naqvi, R. A., Khan, M. A., & Jeong, D. (2024). Corn leaf disease: Insightful diagnosis using VGG16 empowered by explainable AI. *Frontiers in Plant Science, 15*, 1402835. Advance online publication. DOI: 10.3389/fpls.2024.1402835 PMID: 38988642

Tasha, S. (2025). A Review of Artificial Intelligence Applications in Climate Change Mitigation. *International Journal of Environment and Climate Change, 15*(5), 443–449. DOI: 10.9734/ijecc/2025/v15i54864

Thakur, N., & Yadav, A. N. (2025). Nanotechnology in agriculture: A review on precision farming and sustainable crop production. *BioNanoScience, 15*(2), 243. DOI: 10.1007/s12668-025-01861-2

Thornton, P. K., Kristjanson, P. M., Kruska, R. L., & Reid, R. S. (2004). Mapping livestock and poverty: A tool for targeting research and development. *BSAP Occasional Publication, 33*, 37–50. DOI: 10.1017/S1463981500041649

Torre, D., Gao, J., & Macinnis-Ng, C. (2021). Remote sensing-based estimation of rice yields using various models: A critical review. *Geo-Spatial Information Science, 24*(4), 580–603. DOI: 10.1080/10095020.2021.1936656

Trenberth, K. E., Fasullo, J. T., & Shepherd, T. G. (2015). Attribution of climate extreme events. *Nature Climate Change, 5*(8), 725–730. DOI: 10.1038/nclimate2657

Trentin, C., Ampatzidis, Y., Lacerda, C., & Shiratsuchi, L. (2024). Tree Crop Yield Estimation and Prediction Using Remote Sensing and Machine Learning: A Systematic Review. *Smart Agricultural Technology, 9*, 100556. Advance online publication. DOI: 10.1016/j.atech.2024.100556

Trivedi, A., & Juyal, S. A. (2026). Smart Cities and Urban Circularity: Advancing Sustainable Transformation through Digital Twins, AI, and Capacity Building. Green Horizons: Advancing Climate Solutions Through Green Finance, Circular Economy, and ESG, 339–356. DOI: 10.1007/978-3-032-03484-7_18

Tucker, J., Daoud, M., Oates, N., Few, R., Conway, D., Mtisi, S., & Matheson, S. (2014). Social vulnerability in three high-poverty climate change hot spots: What does the climate change literature tell us? *Regional Environmental Change, 15*(5), 783–800. DOI: 10.1007/s10113-014-0741-6

Tupas, M. E., Roth, F., Bauer-Marschallinger, B., & Wagner, W. (2024). Assessment of time-series-derived no-flood references for sar-based Bayesian flood mapping. *GIScience & Remote Sensing, 61*(1), 2427304. DOI: 10.1080/15481603.2024.2427304

Turgut, O., Kok, I., & Ozdemir, S. (2024). AgroXAI: Explainable AI-Driven Crop Recommendation System for Agriculture 4.0. *arXiv*. https://doi.org/DOI: 10.1109/BigData62323.2024.10825771

Turnhout, E., Dewulf, A., & Hulme, M. (2016). What does policy-relevant global environmental knowledge do? The cases of climate and biodiversity. *Current Opinion in Environmental Sustainability*, *18*, 65–72. DOI: 10.1016/j.cosust.2015.09.004

Ukoba, K., Onisuru, O. R., Jen, T.-C., Madyira, D. M., & Olatunji, K. O. (2025). Predictive modeling of climate change impacts using Artificial Intelligence: A review for equitable governance and sustainable outcome. *Environmental Science and Pollution Research International*, *32*(17), 10705–10724. DOI: 10.1007/s11356-025-36356-w PMID: 40183866

Ullah, I., Ali, E., Osama, S., Murtaza, A., & Ullah, F. (2025). <b>Harnessing AI for Climate Change Mitigation: Strategies Challenges and Opportunities</b>. *Journal of Political Stability Archive*, *3*(3), 963–977. . DOI: 10.63468/jpsa.3.3.65

Upadhyaya, A. N., Debnath, P., Rani, P., Selvakumar, P., Manjunath, T. C., & Bhattacharya, S. (2025). AI-Based Disaster Prediction and Early Warning Systems. AI and Emerging Technologies for Emergency Response and Smart Cities, 55–74. DOI: 10.4018/979-8-3693-9770-1.ch003

Upadhyay, A., Chandel, N., Singh, K., Chakraborty, S., Nandede, B., Kumar, M., Subeesh, A., Upendar, K., Salem, A., & Elbeltagi, A. (2025). Deep learning and computer vision in plant disease detection: A comprehensive review of techniques, models, and trends in precision agriculture. *Artificial Intelligence Review*, *58*(3), 92. Advance online publication. DOI: 10.1007/s10462-024-11100-x

URL-1. UN National Inventory Reports. https://unfccc.int/process-and-meetings/transparency-and-reporting/reporting-and-review/reporting-and-review-under-the-paris-agreement/national-inventory-reports last access date: 24.01.2026

Usigbe, M., Asem-Hiablie, S., Uyeh, D., Iyiola, O., Park, T., & Mallipeddi, R. (2023). Enhancing resilience in agricultural production systems with AI-based technologies. *Environment, Development and Sustainability*, *26*(9), 1–29. DOI: 10.1007/s10668-023-03588-0

Uyar, N., & Uyar, A. (2025). Assessing Climate Change Impacts on Cropland and Greenhouse Gas Emissions Using Remote Sensing and Machine Learning. *Atmosphere (Basel)*, *16*(4), 418. DOI: 10.3390/atmos16040418

Valle, D., Izbicki, R., & Leite, R. V. (2023). Quantifying uncertainty in land-use land-cover classification using conformal statistics. *Remote Sensing of Environment*, *295*, 113682. DOI: 10.1016/j.rse.2023.113682

van Eck, N. J., & Waltman, L. (2017). Citation-based clustering of publications using CitNetExplorer VOSviewer. https://doi.org/DOI: 10.1007/s11192-017-2300-7

van Eck, N. J., & Waltman, L. (2021). VOSviewer goes online! (Part 1). https://doi.org/DOI: 10.59350/mmsjw-79h07

van Eck, N. J., & Waltman, L. (2009). Software survey: VOSviewer, a computer program for bibliometric mapping. *Scientometrics*, *84*(2), 523–538. DOI: 10.1007/s11192-009-0146-3 PMID: 20585380

Van Klompenburg, T., Kassahun, A., & Catal, C. (2020). Crop yield prediction using machine learning: A systematic literature review. *Computers and Electronics in Agriculture*, *177*, 105709. DOI: 10.1016/j.compag.2020.105709

Van Wittenberghe, S., Sabater Medina, N., Cendrero Mateo, M. P., Tenjo Gil, N. C., Moncholi Estornell, A., Alonso Chordá, L., & Moreno Méndez, J. F. (2021). Towards the quantitative and physically-based interpretation of solar-induced vegetation fluorescence retrieved from global imaging.

Vasilakos, C., Tsekouras, G. E., & Kavroudakis, D. (2022). LSTM-based prediction of Mediterranean vegetation dynamics using NDVI time-series data. *Land (Basel)*, *11*(6), 923. Advance online publication. DOI: 10.3390/land11060923

Vázquez-Canteli, J. R., & Nagy, Z. (2019). Reinforcement learning for demand response: A review of algorithms and modeling techniques. *Applied Energy*, *235*, 1072–1089. DOI: 10.1016/j.apenergy.2018.11.002

Vercillo, S., Huggins, C., & Cochrane, L. (2021). How is gender investigated in African climate change research? A systematic review of the literature. *Ambio*, *51*(4), 1045–1062. DOI: 10.1007/s13280-021-01631-w PMID: 34669170

Verma, S., Srivastava, K., Tiwari, A., & Verma, S. (2023). Deep learning techniques in extreme weather events: A review. arXiv preprint arXiv:2308.10995.

Vidhya, S., Sushith, M., Angelo, K. M., & Geetha, R. (2026). Climate Change Prediction Model on Smart Forestry 5.0 Using Novel Enhanced Streaming Gradient Boosted Trees-Based Optimisation Framework. Smart Forestry 5.0 for Powered AI Sustainable Environment and Climate, 75–102. .DOI: 10.1007/978-3-031-99086-1_5

Villa-Henriksen, A., Edwards, G., Pesonen, L., Green, O., & Sørensen, C. (2020). Internet of Things in arable farming: Implementation, applications, challenges, and potential. *Biosystems Engineering, 191*, 60–84. DOI: 10.1016/j.biosystemseng.2019.12.013

Vinuesa, R., Azizpour, H., Leite, I., Balaam, M., Dignum, V., Domisch, S., Felländer, A., Langhans, S. D., Tegmark, M., & Fuso Nerini, F. (2020). The role of artificial intelligence in achieving the Sustainable Development Goals. *Nature Communications, 11*(1), 233. Advance online publication. DOI: 10.1038/s41467-019-14108-y PMID: 31932590

Walker, D. W., Vergopolan, N., Cavalcante, L., Smith, K. H., Agoungbome, S. M. D., Almagro, A., Apurv, T., Dahal, N. M., Hoffmann, D., Singh, V., & Xiang, Z. (2024). Flash Drought Typologies and Societal Impacts: A Worldwide Review of Occurrence, Nomenclature, and Experiences of Local Populations. *Weather, Climate, and Society, 16*(1), 3–28. DOI: 10.1175/WCAS-D-23-0015.1

Wang, C., Jiang, S., Zheng, Y., Han, F., Kumar, R., Rakovec, O., & Li, S. (2024). Distributed hydrological modelling with physics-encoded deep learning: A general framework and its application in the Amazon. *Water Resources Research, 60*, e2023WR036170. DOI: 10.1029/2023WR036170

Wang, D., Cao, W., Zhang, F., Li, Z., Xu, S., & Wu, X. (2022). A Review of Deep Learning in Multiscale Agricultural Sensing. *Remote Sensing (Basel), 14*(3), 559. DOI: 10.3390/rs14030559

Wang, L., Li, R., Zhang, C., Fang, S., Duan, C., Meng, X., & Atkinson, P. M. (2022). UNetFormer: A UNet-like transformer for efficient semantic segmentation of remote sensing urban scene imagery. *ISPRS Journal of Photogrammetry and Remote Sensing, 190*, 196–214. DOI: 10.1016/j.isprsjprs.2022.06.008

Wang, M., Fang, X., & Zhang, K. (2024). Exploring a pathway to optimise the carbon tax policy in terms of the economy, the environment and health: A scenario-based system dynamics approach. *Heliyon, 10*(10), e31093. DOI: 10.1016/j.heliyon.2024.e31093 PMID: 38803911

Wang, M., Yu, H., Yang, Y., Lin, X., Guo, H., Li, C., Zhou, Y., & Jing, R. (2021). Unlocking emerging impacts of carbon tax on integrated energy systems through supply and demand co-optimization. *Applied Energy, 302*, 117579. DOI: 10.1016/j.apenergy.2021.117579

Wang, X., Jiang, D., & Lang, X. (2018). Climate Change of 4°C Global Warming above Pre-industrial Levels. *Advances in Atmospheric Sciences, 35*(7), 757–770. DOI: 10.1007/s00376-018-7160-4

Wang, Y., Khodadadzadeh, M., & Zurita-Milla, R. (2023). Spatial+: A new cross-validation method to evaluate geospatial machine learning models. *International Journal of Applied Earth Observation and Geoinformation, 121*, 103364. DOI: 10.1016/j.jag.2023.103364

Wang, Z., Wang, J., Yang, K., Wang, L., Su, F., & Chen, X. (2022). Semantic segmentation of high-resolution remote sensing images based on a class feature attention mechanism fused with Deeplabv3+. *Computers & Geosciences, 158*, 104969. Advance online publication. DOI: 10.1016/j.cageo.2021.104969

Wang, Z., Zhao, Y., & Wang, B. (2018). A bibliometric analysis of climate change adaptation based on massive research literature data. *Journal of Cleaner Production, 199*, 1072–1082. DOI: 10.1016/j.jclepro.2018.06.183

Weber, R., & Cisneros, E. (2018). Assessing Monitoring and Early Warning Systems for Food Security Risk.

Weiss, M., Jacob, F., & Duveiller, G. (2020). Remote sensing for agricultural applications: A meta-review. *Remote Sensing of Environment, 236*, 111402. Advance online publication. DOI: 10.1016/j.rse.2019.111402

Werther, M., Odermatt, D., Simis, S. G. H., Gurlin, D., Lehmann, M. K., Kutser, T., Gupana, R., Varley, A., Hunter, P. D., Tyler, A. N., & Spyrakos, E. (2022). A Bayesian approach for remote sensing of chlorophyll-a and associated retrieval uncertainty in oligotrophic and mesotrophic lakes. *Remote Sensing of Environment, 283*, 113295. DOI: 10.1016/j.rse.2022.113295

Williams, J., Chin-Yee, S., Maslin, M., Barnsley, J., Costello, A., Lang, J., McGlade, J., Mulugetta, Y., Taylor, R., Winning, M., & Parikh, P. (2023). Africa and climate justice at COP27 and beyond: Impacts and solutions through an interdisciplinary lens. *UCL Open Environment, 5*. Advance online publication. DOI: 10.14324/111.444/ucloe.000062 PMID: 37671394

Wu, Z., Hou, B., & Jiao, L. (2021). Multiscale CNN with autoencoder regularization joint contextual attention network for SAR image classification. *IEEE Transactions on Geoscience and Remote Sensing, 59*(2), 1200–1213. DOI: 10.1109/TGRS.2020.3004911

Xia, S., Yang, Y., & Poon, J. P. H. (2025). Ensuring a carbon-neutral future for artificial intelligence. *The Innovation Energy, 2*(1), 100071. DOI: 10.59717/j.xinn-energy.2024.100071

Xie, L., Huang, T., Zheng, X., Liu, Y., Wang, M., Vittal, V., Kumar, P. R., Shakkottai, S., & Cui, Y. (2022). Energy system digitization in the era of AI: A three-layered approach toward carbon neutrality. *Patterns (New York, N.Y.)*, *3*(12), 100640. DOI: 10.1016/j.patter.2022.100640 PMID: 36569552

Xiong, X., Tian, Q., Rodriguez, L. F., & Lin, T. (2022). Data-Driven Modeling for Crop Mapping and Yield Estimation. *Sensing, Data Managing, and Control Technologies for Agricultural Systems*, 131-158.

Xiong, C., Ma, H., Liang, S., He, T., Zhang, Y., Zhang, G., & Xu, J. (2023). Improved global 250 m 8-day NDVI and EVI products from 2000–2021 using the LSTM model. *Scientific Data*, *10*(1), 800. Advance online publication. DOI: 10.1038/s41597-023-02695-x PMID: 37963885

Xu, L., Zhang, H., Wang, C., Zhang, B., & Liu, M. (2018). Crop classification based on temporal information using sentinel-1 SAR time-series data. *Remote Sensing (Basel)*, *11*(1), 53. DOI: 10.3390/rs11010053

Yadav, A., & Yadav, K. (2025). Harnessing Nanotechnology and Artificial Intelligence to Revolutionize Sustainable Smart Agriculture. In Yadav, S. K., & Kumari, A. (Eds.), *Nanobiotechnology in Agri-food Sector. Sustainability Sciences in Asia and Africa*. Springer., DOI: 10.1007/978-981-96-7756-6_7

Yan, C., Fan, X., Fan, J., & Wang, N. (2022). Improved U-Net remote sensing classification algorithm based on multi-feature fusion perception. *Remote Sensing (Basel)*, *14*(5), 1118. Advance online publication. DOI: 10.3390/rs14051118

Yang, X., Cao, W., Lu, Y., & Zhou, Y. (2022). Hyperspectral image transformer classification networks. *IEEE Transactions on Geoscience and Remote Sensing*, *60*, 1–15. DOI: 10.1109/TGRS.2022.3171551

Yang, X., Yang, X., Zhang, C., & Wang, J. (2023). SAR image classification using Markov random fields with deep learning. *Remote Sensing (Basel)*, *15*(3), 617. Advance online publication. DOI: 10.3390/rs15030617

Yang, Z., Luo, X., & Yu, Y. (2024). Bayesian physics-informed neural networks with uncertainty quantification for environmental modelling. *Environmental Modelling & Software*, *173*, 105331. DOI: 10.1016/j.envsoft.2023.105331

Yao, S., Yuan, X., Jiang, L., Xiong, T., & Zhang, J. (2020). Recent progress on fullerene-based materials: Synthesis, properties, modifications, and photocatalytic applications. *Materials (Basel)*, *13*(13), 2924. DOI: 10.3390/ma13132924 PMID: 32629789

Yao, T., Thompson, L., Yang, W., Yu, W., Gao, Y., Guo, X., Yang, X., Duan, K., Zhao, H., Xu, B., Pu, J., Lu, A., Xiang, Y., Kattel, D. B., & Joswiak, D. (2012). Different glacier status with atmospheric circulations in Tibetan Plateau and surroundings. *Nature Climate Change*, *2*(9), 663–667. DOI: 10.1038/nclimate1580

Yeh, C.-H., Chien, T.-W., & Chou, P.-H. (2022). Citation analysis of the 100 top-cited articles on discectomy via endoscopy since 2011 using alluvial diagrams: Bibliometric analysis. *European Journal of Medical Research*, *27*(1), 169. Advance online publication. DOI: 10.1186/s40001-022-00782-0 PMID: 36050803

Yılmaz, O., & Alkan, M. (2026). The Potential of Machine Learning and Multi-Criteria Decision-Making for Data-Driven Land Policies in Türkiye. The International Archives of the Photogrammetry, Remote Sensing and Spatial Information Sciences, XLVIII-4/W17-2025, 339–344. DOI: 10.5194/isprs-archives-XLVIII-4-W17-2025-339-2026

You, J., Li, X., Low, M., Lobell, D., & Ermon, S. (2017). Deep gaussian process for crop yield prediction based on remote sensing data. Proceedings of the AAAI conference on artificial intelligence, Zhu, Y., Wu, S., Qin, M., Fu, Z., Gao, Y., Wang, Y., & Du, Z. (2022). A deep learning crop model for adaptive yield estimation in large areas. *International Journal of Applied Earth Observation and Geoinformation*, *110*, 102828.

Yuan, J., Zhang, Y., Zheng, Z., Yao, W., Wang, W., & Guo, L. (2024). Grain Crop Yield Prediction Using Machine Learning Based on UAV Remote Sensing: A Systematic Literature Review. *Drones (Basel)*, *8*(10), 559. Advance online publication. DOI: 10.3390/drones8100559

Yuan, Y., Chen, X., Wang, J., & Lu, X. (2021). Vision transformer for remote sensing image classification. *IEEE Transactions on Geoscience and Remote Sensing*, *59*(11), 9623–9636. DOI: 10.1109/TGRS.2021.3058466

Yu, Y., Gao, S., Wang, J., Cheng, Q., Deng, H., & Shen, Y. (2026). Unraveling climate–land cover interactions: How SSP–RCP scenarios drive ecosystem service trade-offs in contrasting Yangtze and Yellow River Basins, China. *Landscape Ecology*, *41*(2), 34. Advance online publication. DOI: 10.1007/s10980-026-02295-1

Yu, Y., Li, Y., Zhang, Z., Gu, Z., Zhong, H., Zha, Q., Yang, L., Zhu, C., & Chen, E. (2020). A bibliometric analysis using VOSviewer of publications on COVID-19. *Annals of Translational Medicine*, *8*(13), 816–816. DOI: 10.21037/atm-20-4235 PMID: 32793661

Zain, M., Ma, H., Nuruzzaman, M., Chaudhary, S., Nadeem, M., Shakoor, N., & Ahamad, T. (2023). Nanotechnology based precision agriculture for alleviating biotic and abiotic stress in plants. *Plant Stress, 10*, 100239. DOI: 10.1016/j.stress.2023.100239

Zamil Adib, A. A., Chan, G., & Orji, R. (2025). Combining Data Visualization and Interactive Narrative: A Persuasive Approach to Raise Climate Change Awareness. Human Behavior and Emerging Technologies, 2025(1). Portico. .DOI: 10.1155/hbe2/7275480

Zhang, C., Song, J., Yin, X., Cai, J., Liang, Y., & Lu, J. (2025). Digital twin-based approaches for agricultural machinery damage prediction and maintenance: A review. *Journal of Computational Design and Engineering*, *12*(10), 87–117. DOI: 10.1093/jcde/qwaf097

Zhang, D., Leng, J., Li, X., He, W., & Chen, W. (2022). Three-stream and double attention-based DenseNet-BiLSTM for fine land cover classification of complex mining landscapes. *Sustainability (Basel)*, *14*(19), 12465. Advance online publication. DOI: 10.3390/su141912465

Zhang, H., Wang, S., & Ding, J. (2025). Driving mechanism and nonlinear threshold identification of vegetation in China: Based on causal inference and machine learning. *Journal of Arid Land*, *17*(10), 1341–1360. DOI: 10.1007/s40333-025-0110-y

Zhang, R., Zhu, H., Chang, Q., & Mao, Q. (2025). A Comprehensive Review of Digital Twins Technology in Agriculture. *Agriculture*, *15*(9), 903. Advance online publication. DOI: 10.3390/agriculture15090903

Zhang, S., Bauer, N., Luderer, G., & Kriegler, E. (2014). Role of technologies in energy-related CO2 mitigation in China within a climate-protection world: A scenarios analysis using REMIND. *Applied Energy*, *115*, 445–455. DOI: 10.1016/j.apenergy.2013.10.039

Zhang, Y., Ye, M., Zhu, G., Liu, Y., Guo, P., & Yan, J. (2024). FFCA-YOLO for small object detection in remote sensing images. *IEEE Transactions on Geoscience and Remote Sensing*, *62*, 1–15. DOI: 10.1109/TGRS.2024.3363057

Zhao, Q., Pervaiz, A., Makhmudov, S., & Fatima, A. (2025). Do environmental stringent policies enhance the impact of renewable energy on CO_2 reduction? Evidence from BRICS-T economies. *Journal of Environmental Management*, *345*, 126464. DOI: 10.1016/j.jenvman.2025.126464

Zhao, S., Qing, L., & Wang, C. (2026). Editorial: Moving towards sustainable development: exploring the impact of land-use policies on land green utilization efficiency. *Frontiers in Environmental Science*, *14*, 1767889. Advance online publication. DOI: 10.3389/fenvs.2026.1767889

Zhao, S., Zhang, D., & Karniadakis, G. E. (2023). Multi-fidelity physics-informed neural networks: Leveraging low- and high-fidelity data for predictive modelling. *Computer Methods in Applied Mechanics and Engineering*, *389*, 115599. DOI: 10.1016/j.cma.2021.115599

Zhong, L., Hu, L., & Zhou, H. (2019). Deep learning based multi-temporal crop classification. *Remote Sensing of Environment*, *221*, 430–443. DOI: 10.1016/j.rse.2018.11.032

Zhu, X. X., Tuia, D., Mou, L., Xia, G.-S., Zhang, L., Xu, F., & Fraundorfer, F. (2017). Deep Learning in Remote Sensing: A Comprehensive Review and List of Resources. *IEEE Geoscience and Remote Sensing Magazine*, *5*(4), 8–36. DOI: 10.1109/MGRS.2017.2762307

Ziervogel, G., Cowen, A., & Ziniades, J. (2016). Moving from Adaptive to Transformative Capacity: Building Foundations for Inclusive, Thriving, and Regenerative Urban Settlements. *Sustainability (Basel)*, *8*(9), 955. DOI: 10.3390/su8090955

Zolfaghari, A. (2026). Role of Technology and Innovation in Sustainable Entrepreneurship. Sustainable Entrepreneurship, 121–140. .DOI: 10.1007/978-3-032-04909-4_7

Zou, L., & Li, C.. (2025). *Ensemble-quantile regression for crop yield prediction.* IEEE., DOI: 10.1109/ISSSR65654.2025.00057

Zscheischler, J., Westra, S., van den Hurk, B. J. J. M., Seneviratne, S. I., Ward, P. J., Pitman, A., AghaKouchak, A., Bresch, D. N., Leonard, M., Wahl, T., & Zhang, X. (2018). Future climate risk from compound events. *Nature Climate Change*, *8*(6), 469–477. DOI: 10.1038/s41558-018-0156-3

About the Contributors

Mehmet Çelik is an Assoc. Prof. of Physical Geography. He focuses on the interaction between climate variables, nature and the environment. His multidisciplinary research interests include remote sensing and GIS studies involving climate, natural resource management, and disasters. Dr. Çelik uses multispectral and hyperspectral remote sensing and GIS to monitor changes to ecosystems and analyse the contribution of natural and anthropogenic factors to these changes.

***

Sharif Eime Akhter is pursuing a bachelor's degree in Computer Science and Engineering at American International University-Bangladesh (AIUB), Bangladesh. Her academic interests include machine learning, data science, explainable AI, medical imaging, agricultural analytics, and predictive modeling.

Anorgul Ashirova is working as an Associate Professor at the Department of General Professional Sciences, Mamun University, Khiva, Uzbekistan

Selvaraj Damodaran is working as a professor in the Department of Electronics and Communication Engineering at Panimalar Engineering College. He obtained his B.E Degree from Madras University, ME degree from Anna University and Ph.D. from Sathyabama University. He has a vast teaching experience of 21 years. He has published more than 75 papers at various refereed journals and conferences. His area of interest is in the field of signal processing, image processing, Biomedical Instrumentation, wireless communication, Artificial Intelligence, Machine learning, Data Mining and Antenna Design. He is a member of IEEE and life member of IETE.

G. Pradeepa is an Assistant Professor of Department of Information Technology in Vivekanandha College of Technology for Women, Tiruchengode and has 11 years of teaching experience. I have obtained her B.E in Computer Science and

Engineering from Anna University, Chennai in 2007, and M.E in Computer Science Engineering from Anna University, Chennai during 2013.I have published many papers in National and Interna

Dipta Justin Gomes is an Assistant Professor in the Department of Computer Science at American International University-Bangladesh (AIUB). He holds master's degrees in Computer Science from the University of Ulster, Northern Ireland, and AIUB, and a bachelor's degree in Computer Information Systems from AIUB. His research interests span machine learning, computer vision, algorithms, image processing, natural language processing, graph theory, and computational geometry. He has published extensively in journals, conferences, and edited volumes, with research contributions in deep learning, medical and agricultural AI, intelligent systems, and image analysis. He previously taught at Notre Dame College, Dhaka, and has been recognized for academic excellence with Summa Cum Laude honors and national-level innovation achievements.

Deepak Gupta is a distinguished academician with over 20 years of teaching experience and expertise in Computer Science and Engineering. He holds a Ph.D. in Computer Engineering and has an extensive research portfolio, including many international journal papers, multiple conference proceedings, and several books. His research interests span machine learning, artificial intelligence, and data mining. Currently, Dr. Gupta serves as an Associate Professor and Head of Department of CSE (Emerging Branches) at ITM Gwalior, MP, India. He has also contributed significantly to the field through patents and copyrights and actively engages as a reviewer, conference chair and editor.

Sk Muktadir Hossain is a Computer Science Engineer and researcher specializing in machine learning, deep learning, computer vision, bioinformatics, and explainable AI. He received his B.Sc. in Computer Science and Engineering from American International University-Bangladesh (AIUB), with a major in Information Systems (Data Science). He has worked as an Academic Instructor and Research Assistant at AIUB and as a Research Assistant at Elite Lab, USA. His research spans medical imaging, Bengali speech technologies, crop classification, disease prediction, time-series analysis, and vision-language models. He has contributed to several conference papers, journal manuscripts, and book chapters, with a strong interest in developing interpretable and impactful AI systems for real-world applications.

Thiyagarajan Manoharan, B.E., M.Tech., (Ph.D.), is currently working as an Assistant Professor in the Department of Computer Science and Engineering at R.M.K. Engineering College, Kavaraipettai, Chennai. He is pursuing his

Ph.D. at Anna University, Chennai, with a research focus on Precision Agriculture and Information Security.He completed his M.Tech. in Computer Science and Engineering from SRM University and his B.E. in Computer Science and Engineering from Anna University, Chennai. His research interests include Machine Learning, AIoT, Cyber Security, and Cloud Computing. He has 17 years of teaching and research experience across various engineering colleges and universities. He has published several research papers in international journals and has also presented his work at international conferences. He can be contacted at mtn.cse@rmkec.ac.in.

Kazi Samia Mostofa is pursuing a bachelor's degree in Computer Science and Engineering at American International University-Bangladesh (AIUB), Bangladesh. Her academic interests include machine learning, explainable AI, agricultural informatics, data mining, medical imaging, and predictive analytics.

Parimala Devi Muthusamy received B.E. (Electronics and Communication) from Bharathiyar University, Coimbatore, M.E. (VLSI Design) from Anna University, Chennai, M.B.A. (Human Resource Management) from Anna University and Ph.D (Information and Communication Engineering) from Anna University, Chennai. Currently, she is Associate Professor in the Velalar College of Engineering and Technology, Erode. She has published 15 papers in national and international journals (SCI, SCOPUS) and conferences. She has published 15 book chapters with IEEE, Springer, IET, Taylor & Francis. She has been the reviewer for many SCI, SCIE, SCOPUS indexed journals and received best paper awards at international conferences. She has two Grant Patents and 5 Copyrights for software (Government of India) She has published a book named "Advanced Microprocessors and Microcontrollers" with University Science Press, New Delhi. She is a lifetime member of IETE. Her research interests include VLSI &SoC, IoT, Block chain and Machine & Deep learning.

Brindha P. received B.E. (Electronics and Communication Engineering) from Anna University of Technology, Chennai, M.E. (Applied Electronics) from Anna University, Chennai and Ph.D (Information and Communication Engineering) from Anna University, Chennai. Currently, she is Associate Professor in the Erode Sengunthar Engineering College, Erode. She has published 10 papers in national and international journals and conferences. She has one Grant Patent and 1 Copyrights for software (Government of India). She has published a book named "Advanced Microprocessors and Microcontrollers" with University Science Press, New Delhi. He is a lifetime member of IETE. Her research interests include VLSI, IoT, AI, Machine Learning & Deep learning.

Mahfujur Rahman is an Assistant Professor in the Department of Computer Science at American International University-Bangladesh (AIUB). He earned his B.Sc. and M.Sc. in Computer Science from AIUB, specializing in Intelligent Systems. He previously served as a Teaching Assistant, Instructor, and Lecturer at AIUB before being promoted to Assistant Professor in 2023. His research interests include deep learning, artificial intelligence, neural networks, natural language processing, bioinformatics, data mining and business intelligence, image processing, medical imaging, biomedical applications, and machine-to-machine data analytics. In addition to his academic work, he has industry experience in web application development and has supervised undergraduate thesis projects for over five years. He is also involved in reviewing for local and international peer-reviewed journals and conferences.

Md Sadiqur Rahman is pursuing a bachelor's degree in Computer Science and Engineering at American International University-Bangladesh (AIUB), Bangladesh. His research interests include machine learning, deep learning, explainable artificial intelligence, agricultural prediction systems, and intelligent data analysis.

Khurramov Ruslan is affiliated with the Department of Information Technology and Exact Sciences, Termez University of Economics and Service, Termez, Uzbekistan. His academic interests span information technology, applied computing, and exact sciences, with a focus on developing analytical and technology-driven solutions for contemporary educational and socio-economic challenges. He is actively involved in teaching, research, and academic development initiatives within the university.

Edwin Raja S., Associate Professor, Department of Computer Science and Engineering in Vel Tech Rangarajan Dr.Sagunthala R&D Institute of Science and Technology, Chennai. He has completed his Ph.D. in Information and Communication Engineering from Anna University-chennai in the year 2021. He obtained his M.Tech. (Network Engineering) in the year 2010 from Kalasalingam University, Krishnan kovil and B.E. (Computer Science and Engineering) in the year 2006 from C.S.I Institute of Technology, Thovalai. He has more than thirteen plus years of teaching experience and his area of interests are Cyber Security, Big Data Analytics, Machine Learning & Computer Networks. He has published more than 25 research articles in various journals and conferences. Also, he has published 3 patents, acted as reviewer in reputed journals.

Brindha Samarasam received her Undergraduate degree in Biomedical Instrumentation Engineering and Postgraduate degree in Applied Electronics. She is currently working as an Assistant Professor at Nandha Engineering College,

Erode, with over 13 years of teaching experience, and is a Research Scholar at Avinashilingam Institute for Home Science and Higher Education for Women, School of Engineering. She has published 15 research papers in reputed national and international journals and conferences, one book chapter, and has served as a reviewer for six journals. She has published three patents and has actively participated in more than 20 national and international conferences. Her research interests include medical image analysis, machine learning, deep learning, Explainable AI (XAI), and computer-aided diagnosis, with a focus on uterine carcinoma.

Ayswarjo Sarkar is pursuing a bachelor's degree in Computer Science and Engineering at American International University-Bangladesh (AIUB), Bangladesh. Her research interests include machine learning, interpretable artificial intelligence, agricultural data analysis, classification systems, and intelligent decision-support methods.

Turobova Sevara is a faculty member in the Department of Social Sciences Education at Termez University of Economics and Service, Termez, Uzbekistan. She is engaged in teaching and academic activities related to social sciences education, with a focus on strengthening students' conceptual understanding, critical thinking, and civic awareness. Her academic interests include social science pedagogy, educational development, and the application of innovative teaching methodologies to improve learning outcomes in higher education. She actively contributes to curriculum implementation and scholarly initiatives that support quality education and interdisciplinary learning.

Maheswari Subramanian completed Under Graduation in Biomedical Engineering at Sri Ramakrishna Engineering College, Coimbatore. She has completed her Post Graduation in Applied Electronics at Anna University Regional Centre, Coimbatore and Pursuing Ph.D in the gynecological disorders using imaging modality. She is having 14 years of Teaching Experience. At present she is working as an Assistant Professor in the department of Biomedical Engineering at Velalar College of Engineering and Technology, Thindal, Erode, Tamilnadu. She published her research works in 8 International Journals and presented papers in many National and International IEEE Conferences. She has been the reviewer for many indexed journals and reviewed papers related to gynecological disorders. She organized FDP's, National Conferences and STTP's funded by CSIR, INSA and AICTE respectively. She is an active member of BMESI, ISTE, IEEE – EMB Society and IEEE – WIE. Also the Branch coordinator of IEEE-EMBS Student branch & WIE affinity group of VCET. She had conducted many events for the benefit of IEEE students firm. She received Fund for the project development under

Niral Thiruvizha 2.0 and exhibited in the zonal presentation. She received three patents and three copyrights for lab manuals approved by Government of India. Her area of interests are Biomaterials, Biomechanics, Diagnostic and Therapeutic Equipments, Rehabilitation Engineering, Radiological Equipments, medical image processing and analysis using Machine Learning and Deep Learning Techniques. She gave lectures on artificial intelligence, machine Learning and Big Data Analytics in AICTE sponsored STTPs.

Kazi Tanvir is a Lecturer in the Department of Computer Science at American International University-Bangladesh (AIUB), Bangladesh. He received his M.Sc. in Data Science from Vellore Institute of Technology (VIT), India, and his B.Sc. in Computer Science and Engineering from AIUB, with a major in Computational Theory. His research interests include big data, deep learning, artificial intelligence, bioinformatics, computer vision, medical imaging, image processing, and biostatistics. He has contributed to conference publications and other scholarly outputs, including papers, datasets, book chapters, and patents, and is particularly interested in research at the intersection of machine learning, explainable AI, and applied data science.

Index

A

B

C

D

E

F

G

H

L

M

N

O

P

R

S

T

U

www.ingramcontent.com/pod-product-compliance
Lightning Source LLC
LaVergne TN
LVHW060006031026
849161LV00025B/1507
9798260007785